Many-Particle Dynamics and Kinetic Equations

Mathematics and Its Applications

Volume 420

Many-Particle Dynamics and Kinetic Equations

by

C. Cercignani

Department of Theoretical Mechanics,
Polytecnico di Milano,
Milano, Italy

V. I. Gerasimenko

and

D. Ya. Petrina

Department of Mathematical Physics,
Institute of Mathematics,
Ukrainian Academy of Sciences,
Kiev, Ukraine

SPRINGER-SCIENCE+BUSINESS MEDIA, B.V.

A C.I.P. Catalogue record for this book is available from the Library of Congress.

ISBN 978-0-7923-4696-8 ISBN 978-94-011-5558-8 (eBook)
DOI 10.1007/978-94-011-5558-8

Printed on acid-free paper

The manuscript was translated from Russian by
K. Petrina and V. Gredzhuk

Contents

Introduction

As our title suggests, there are two aspects in the subject of this book. The first is the mathematical investigation of the dynamics of infinite systems of interacting particles and the description of the time evolution of their states. The second is the rigorous derivation of kinetic equations starting from the results of the aforementioned investigation.

As is well known, statistical mechanics started in the last century with some papers written by Maxwell and Boltzmann. Although some of their statements seemed statistically obvious, we must prove that they do not contradict what mechanics predicts. In some cases, in particular for equilibrium states, it turns out that mechanics easily provides the required justification. However things are not so easy, if we take a step forward and consider a gas is not in equilibrium, as is, *e.g.*, the case for air around a flying vehicle.

Questions of this kind have been asked since the dawn of the kinetic theory of gases, especially when certain results appeared to lead to paradoxical conclusions. Today this matter is rather well understood and a rigorous kinetic theory is emerging. The importance of these developments stems not only from the need of providing a careful foundation of such a basic physical theory, but also to exhibit a prototype of a mathematical construct central to the theory of non-equilibrium phenomena of macroscopic size.

The advancement of science has offered something more than just new tools and concepts. The use of electronic computers allowed numerical experiments on systems with a large number of degrees of freedom. Among these one finds unusual experiments, such as the calculation of a certain problem of molecular dynamics and the study of the effect of the time inversion on the evolution quantity. Boltzmann had far-reaching ideas, which were not understood by his contemporaries and are clearly confirmed by these computer experiments.

Boltzmann had underlined that his kinetic equation for the distribution function depended, not only upon the laws of mechanics, but also on the initial conditions and the fact that we consider the limit of infinitely many degrees of freedom. In view of the fact that we claim validity for kinetic equations with these restrictions, we do not have to worry about some famous paradoxes.

The states of infinite systems of particles are described by infinite sequences of reduced distribution functions which satisfy an infinite system of integro-differential equations known as the Bogolubov-Born-Green-Kirkwood-Yvon (BBGKY) hierarchy. The first intent of this book is to construct solutions of this hierarchy in some function spaces. The second goal is to derive from the solutions thus obtained the kinetic, or Boltzmann, equation in a suitable limit and to link together the two most important branches of contemporary nonequilibrium statistical mechanics: the BBGKY hierarchy and the Boltzmann equation. This book is the result of a cooperative effort of people working in these two branches of statistical mechanics.

Usually the kinetic equations are derived at a phenomenological level; for example, the Boltzmann equation is obtained by calculating the number of pairwise collisions between molecules. The problem of a derivation of the Boltzmann equation on the basis of exact dynamics in the framework of classical mechanics was for long time a challenge for many generations of scientists.

The first successful attempt was made by Bogolubov. who derived the Boltzmann equation from the BBGKY hierarchy. He used a special solution in which all distribution functions depend on time through the one-particle distribution function and are represented by power series in the density. The Boltzmann equation was obtained as the first order term in this density expansion.

Cohen and Green proposed an approach in which they also used the special solution constructed by Bogolubov. In order to construct this solution, they expressed the one-particle distribution function at time $t = 0$ through the same function at time $t \neq 0$ in a way analogous to the method used for equilibrium states, where the activity is expressed in terms of the density from the expression of the equilibrium one-particle distribution function.

This was a great success of statistical mechanics. But, during about thirty years, rigorous results concerning the problem of proving the convergence of the functional series representing the distribution functions through the initial distribution function an density were lacking.

Some particular results concerning the solutions of the BBGKY hierarchy were obtained by Gallavotti, Lanford and Lebowitz, later by Sinai and Suchov, for one-dimensional systems of particles interacting via a smooth short-range potential in the case of initial data which are suitable perturbations of equilibrium states. In their papers the solution was not represented by a series in the density and the sequence of initial distribution functions.

At the very beginning of the seventies Petrina proposed to consider the BBGKY hierarchy as an abstract evolution equation in suitable function spaces and reduced the problem of its solution to constructing a group (or a semigroup) of evolution operators. It was first proved in the quantum case that there exists a group of evolution operators and a solution of the BBGKY hierarchy in the space of sequences of trace-class operators. For the classical case the group of evolution operators related to the BBGKY hierarchy was constructed and, on this basis, the existence and uniqueness of its solutions were proved in the space of sequences of summable functions.

The sequences of distribution functions in this space describe states with finite average numbers of particles, whereas to construct states with infinitely many particles it is necessary to perform the thermodynamic limit, *i.e.* to let the average number of particles and the volume of the system go to infinity with constant number density. From a mathematical standpoint the thermodynamic limit means that the group of evolution operators must be extended from the space of sequences of summable functions to the space of sequences of functions bounded with respect to positions and exponentially decreasing with respect to the squared momenta. It was proved for one-dimensional systems of hard spheres interacting through a short-range potential that there exist solutions of the BBGKY hierarchy on a finite time interval for arbitrary initial data in this function space. For certain initial data which are local perturbations of equilibria, these solutions can be continued on an arbitrary time interval.

In order to justify the procedure of taking the thermodynamic limit, a new conception in the theory of Hamilton equations was used, *i.e.* the conception of domain of interaction, which reflects the fact that since the particles interact through a short range potential, they do not act on each other at distances larger than the potential range. At a mathematical level this means that the integrands in the expressions for the distribution functions are in fact different from zero just in a bounded domain in the configuration space.

In spite of this substantial achievement, the problem of rigorous justification of the derivation of the Boltzmann equation remained unsolved. It became clear that for this goal it is necessary to simplify the model describing the particle interactions. The best candidate appeared to be a model of hard spheres interacting only when the particles touch each other and then undergo elastic scattering. In virtue of the singular character of this interaction, the problem of deriving the BBGKY hierarchy had a special character, different from the case of a space-distributed potential. It was formally solved by Grad and Cercignani, who also pointed out that, in the Boltzmann-Grad limit, one can obtain, from the BBGKY hierarchy, the so called Boltzmann hierarchy from which the Boltzmann equation for hard spheres follows if the initial data satisfy a chaos assumption.

For the BBGKY hierarchy of hard spheres Petrina and Gerasimenko, using the technique of groups of evolution operators, obtained solutions global in time in the space of summable functions. Lanford was the first to prove that an iteration series converges on a finite time interval provided that the initial data belong to the space of sequences of functions which are bounded with respect to positions and exponentially decreasing with respect to the squared momenta. By combining the representation of the solution in terms of evolution operators on one hand and iterative series on the other hand, Petrina and Gerasimenko continued the solutions on arbitrarily large time intervals for special initial data which are local perturbations of equilibria.

Thus everything was prepared for a rigorous justification of the procedure of taking the Boltzmann-Grad limit. This problem was first discussed by Cercignani and some aspects of the proof of the pointwise convergence of the distribution functions in the Boltzmann-Grad limit were discussed by Lanford. A complete proof of the existence of the Boltzmann-Grad limit was established in a series of papers by Petrina and Gerasimenko.

For a very long time, researches on the Boltzmann equation and the BBGKY hierarchy were independently performed by different people with their own methods and almost without intersection. Results established for the Boltzmann equation were not used for the BBGKY hierarchy and *vice versa.* The important works of Bogolubov, Cohen and Green are excluded from this statement, but they were performed on a purely formal level. The first substantial link was the justification of the Boltzmann-Grad limit which allowed a derivation of the Boltzmann hierarchy and equation from the BBGKY hierarchy.

Recently, Petrina and Gerasimenko obtained some results concerning a proof of convergence of the series, used by Bogolubov, Cohen and Green, and representing special solutions for summable functions, in the case of factorized initial data.

After these general remarks, let us outline the contents of the present book.

In Chapter I we present the Hamiltonian dynamics of finite systems of hard spheres interacting through a smooth, short-range potential, define the evolution operator, derive the Liouville equation and the BBGKY hierarchy. Some information about equilibrium states is also given.

Due to the singular character of the potential of hard spheres, the presentation of this material is rather complicated and we have given a completely rigorous proof of the existence of global in time solutions of the Hamilton equations, the group property and isometricity of the evolution operators, the existence of their infinitesimal generators and of the solution of the corresponding Liouville equation. Some very important and delicate details concerning the domain of definition of the infinitesimal operator of the group of evolution operators are given in two appendices.

In Chapter II the BBGKY hierarchy for systems of hard spheres interacting *via* a smooth pair potential is derived. The family of evolution operators for the same hierarchy (parameterized by time t) is also obtained. These operators are shown to be bounded and strongly continuous, and to form a group in the space L^1 of summable functions. The infinitesimal generator of this group is defined on a suitable everywhere dense subset $L_0 \in L^1$ and coincides with the operator appearing in the right hand side of the BBGKY hierarchy. As a consequence, the hierarchy itself can be considered as an abstract evolution equation in L^1 with the infinitesimal generator in the right hand side. It follows that the BBGKY hierarchy possesses a unique solution in L^1. We show that the widely used iteration series can be reduced to solutions expressed *via* the evolution operator.

These solutions of the hierarchy in L^1 describe states of finite systems, with an arbitrarily fixed average number of particles. In order to describe the states of infinite systems, it is necessary to consider solutions of the hierarchy in a larger space than the space L^1 of sequences of summable functions. This means that the evolution operator must be extended to a space containing the states of infinite systems.

In Chapter III we give a meaning to the evolution operator in the space E_ξ of sequences of functions bounded with respect to positions and decreasing exponentially in the squared momenta by taking the thermodynamic limit. For systems of hard spheres in several dimensions, we prove the existence of solutions in the space E_ξ for a finite time interval. For a subset of initial data, corresponding to locally perturbed equilibrium states, these solutions are continued to an arbitrarily large time interval.

For one dimensional systems of hard spheres interacting *via* a smooth short-range potential, a similar result is obtained, by using the new idea of domain of interaction.

In both cases, the solutions we obtain are the thermodynamic limit of solutions of finite systems.

In Chapter IV we present a completely rigorous justification of the existence of the Boltzmann-Grad limit in a finite time interval for the case of three-dimensional systems of hard spheres and initial data from the space E_ξ.

To this aim, we have used the representation of the solutions of the BBGKY hierarchy and the Boltzmann hierarchy by an iteration series. The exact mathematical meaning of the corresponding integrals is also discussed. We define the domain of initial data in the phase space of finitely many particles such that the particles can interact. We show that the volumes of the corresponding cones in momentum space, for fixed positions of the particles, are proportional to the square of the ratio between the diameter and the distances between particles.

The fact that the volumes of the cones depend on the distances between the particles requires a careful consideration when taking the Boltzmann-Grad limit, term by term, in the iteration series.

It is necessary to estimate the contribution of these "dangerous" sets of initial data in phase space to the various terms of the iteration series, in order to show that it tends to zero in the Boltzmann-Grad limit. In some previous justifications of the Boltzmann-Grad limit, these circumstances were not taken into account and, consequently these papers will not be discussed in detail in this book.

The justification of the existence of the Boltzmann-Grad limit of equilibrium states in the framework of the canonical as well as the grand canonical ensemble are given.

In the last chapter, Chapter V, we present Bogolubov's method of derivation of the Boltzmann equation directly from the BBGKY hierarchy.

The formal solutions, in which every distribution function depends on time through the one-particle distribution function, are represented by a power series in the density.

Bogolubov's principle of weakening of correlations is discussed and it is shown that the Boltzmann equation can be obtained if one uses the terms of zero-th order of the two-particle distribution function. Cohen's approach is also presented.

For initial data with chaos property we show that the Cauchy problem for the BBGKY hierarchy can be reduced to the corresponding initial value problem for a generalized kinetic equation.

CHAPTER I

The BBGKY Hierarchy

1.1. Introduction

The aim of this book is to give a mathematical exposition of the foundations of classical statistical mechanics with particular concern for the description of nonequilibrium states for a gas in the Boltzmann-Grad limit. This area has been reaching a certain degree of maturity in recent years, although many problems are still open.

Within the scope of this book, the molecules of a gas will be assumed to be particles that move and interact according to Newton's laws. A basic point in the definition of the system is the choice of the interaction potential. Although many potentials have been proposed (among them those named after Lennard-Jones and Morse) the most reasonable description is to assume that the particles are perfectly elastic spheres rather than mass points. An additional smooth potential may be added to simulate repulsion and the weaker part of the attraction potential. The use of a hard core potential has several advantages, including the fact that we may hide behind the infinitely high potential barrier our willingness to neglect quantum effects, and the availability of an intuitive picture in the form of a three-dimensional billiard game. Thus, e. g., if no external forces, such as gravity, are assumed to act on the molecules and the smooth part of the potential is ignored, each molecule will move in a straight line unless it happens to strike another sphere or a solid wall.

Although the rules generating the dynamics of these systems are easy to prescribe, their mathematical formulation requires a certain amount of care and the phenomena associated with the dynamics are not so simple. They are actually rather difficult to understand, especially if one is interested in the asymptotic behavior of the system for long times (ergodic properties) or in the case when the number of spheres is very large (kinetic and hydrodynamical limits). Both aspects of the dynamics of hard core interaction systems are relevant when dealing with a gas, but we shall concentrate upon the problem of discussing the behavior of this system when the number of the particles is very large. This because we know from physics that there are about $2.7 \cdot 10^{19}$ molecules in a cubic centimeter of a gas at atmospheric pressure and a temperature of 0°C.

In statistical mechanics all possible states of a system of particles are described by the solutions of the BBGKY hierarchy. As is well-known [Bo][BG1][BG2][Ki][Yv], the BBGKY hierarchy is a chain of an infinite number of integrodifferential equations for the infinite sequence of distribution functions [PGM2].

For systems of a fixed finite number of particles the BBGKY hierarchy consists of an finite chain of equations and is equivalent to the Liouville equation. For systems of an infinite number of particles the BBGKY hierarchy consists of an infinite chain of equations and is obtained as the thermodynamic limit of the BBGKY hierarchy of finite systems.

The BBGKY hierarchy allows to describe both the equilibrium and non-equilibrium states from a common point of view. Non-equilibrium states are characterized by the solutions of the initial value problem for this hierarchy and, correspondingly, equilibrium states are characterized by the solutions of a steady BBGKY hierarchy.

We also point out that in various asymptotic limits the solutions of the initial value problem for the BBGKY hierarchy tend to certain limit distribution functions, which are governed by nonlinear kinetic equations [Bo][Ce1][Co1][Gra1]. Therefore, phenomenological kinetic equations (for example, such as Boltzmann equations or hydrodynamic equations for a continuous medium) can be derived rigorously from a BBGKY hierarchy. Hence the BBGKY hierarchy forms the basis of the statistical theory of real systems of particles [Bo][Ce1][PGM2].

From a mathematical point of view the investigations of the initial value problem for the BBGKY hierarchy for systems of infinitely many particles and also of the asymptotic behaviour of its solutions are of considerable interest and very difficult. It will be performed in many steps.

In this chapter the necessary preliminary information about BBGKY hierarchy and about an evolution of systems of finitely many particles will be introduced. The information concerning evolution of systems, consisting of a finite number of particles, is necessary for the construction of solutions of the BBGKY hierarchy.

The plan of Chapter 1 is as follows.

In Sections 2 and 3 we investigate the Hamilton dynamics of finite classical systems of hard-core particles interacting via a smooth potential.

Owing to the singular nature of the interaction, the exposition of this material is not traditional [PG1][PG2][PG3][PGM1]. Also in Section 3 we consider the Liouville equation for such systems and prove an existence theorem of solutions with initial data from the space of summable functions. In the following the existence of solutions of the inital value problem for BBGKY hierarchy will be treated for this class of potentials.

In Section 4 we introduce the BBGKY hierarchy. As is well-known, this hierarchy was first derived by Bogolyubov [Bo], Born, Green [BG1], Kirkwood [Ki], and Yvon [Yv] for systems with a smooth potential. For systems with hard-core potentials the BBGKY hierarchy was derived by Grad [Gra1] and Cercignani [Ce1][Ce2]. Recently, results about the mathematical justification of this hierarchy have been obtained [PGM1][PGM2][PG1][PG2][PG3]. In the next chapter another method of mathematical derivation of the BBGKY hierarchy will also be proposed.

In Section 5 we discuss the problem of the construction of steady solutions of the BBGKY hierarchy, with main concern for the equilibrium distribution functions. With the help of the thermodynamic limit procedure we define the Gibbs distribution functions for systems of an infinite number of particles. It is natural to choose a suitably perturbed equilibrium distribution function as initial data for the BBGKY hierarchy. Hence in that section we shall give the definition of a class of states close to equilibrium, the so-called locally perturbed equilibrium states.

1.2 The Hamilton Dynamics of a System of Particles with a Hard Core

1.2.1 The Hamilton equations Consider a system of N identical particles of unit mass which move in a bounded domain $\Lambda \subset \mathbb{R}^\nu$ of ν–dimensional Euclidean space $\mathbb{R}^\nu, \nu \geqslant 1$ and interact via a pair potential Φ. Every particle is characterized by its position $q \equiv (q^1, \ldots, q^\nu) \in \Lambda \subset \mathbb{R}^\nu$ and momentum $p \equiv (p^1, \ldots, p^\nu) \in \mathbb{R}^\nu$. It is convenient to combine q and p into a single point $x \equiv (q,p)$ of the phase space $\Lambda \times \mathbb{R}^\nu$. The set of all possible values of positions and momenta of particles forms the phase space of the system $\Lambda^N \times \mathbb{R}^{\nu N}$.

The Hamiltonian of the system of N particles has the form

$$H_N^\Lambda = \sum_{i=1}^N \frac{p_i^2}{2} + \sum_{i<j=1}^N \Phi(q_i - q_j) + \sum_{i=1}^N u^\Lambda(q_i), \tag{1.2.1}$$

where $u^{\Lambda}(q)$ is the potential of the external field constraining the particles inside Λ such that

$$u^{\Lambda}(q) = \begin{cases} +\infty, & \text{if } q \notin \Lambda \\ 0, & \text{if } q \in \Lambda_{\delta}. \end{cases}$$

$\Lambda_{\delta} \subset \Lambda$ contains all points of Λ with dist $(q, \partial\Lambda) \geqslant \delta > 0$ for some fixed $\delta > 0$, and we suppose that $u^{\Lambda}(q)$ is a smooth non-negative function in Λ that converges to $+\infty$ as q tends to the boundary $\partial\Lambda$. We shall also consider systems of particles moving in an unbounded domain $\Lambda = \mathbb{R}^{\nu}$. In this case the function $u^{\Lambda}(q)$ is identically zero and the Hamiltonian of such systems will be denoted by H_N. Let us denote the position and momentum of the i–th particle at time t by $\big(Q_i(t), P_i(t)\big) \equiv X_i(t) \in \Lambda \times \mathbb{R}^{\nu}$, $i \in (1, \ldots, N)$. These functions are defined as a solution of the initial value problem for the Hamilton equations

$$\frac{dP_i(t)}{dt} = -\frac{\partial H_N^{\Lambda}(X(t))}{\partial Q_i(t)} \quad , \quad \frac{dQ_i(t)}{dt} = \frac{\partial H_N^{\Lambda}(X(t))}{dP_i(t)}, \tag{1.2.2}$$

$$P_i(0) = p_i \, , Q_i(0) = q_i \, , i \in (1, \ldots, N),$$

where $\{q_i, p_i\}_{i=1}^{N}$ are the initial values of the positions and momenta of N particles.To indicate the dependence of the solutions $X(t) \equiv \big(X_1(t), \ldots, X_N(t)\big)$ on the initial values $x \equiv (x_1, \ldots, x_N)$ in explicit fashion, we shall write $X_i(t) = X_i(t, x_1, \ldots, x_N) = X_i(t, x)$ or $X(t) = X(t, x)$. The solution $X(t, x)$ for $t \in \mathbb{R}^1$ and for fixed x defines a trajectory in phase space. It should be noted that Eqs. (1.2.2) are written for the sake of convenience in vector form. Each equation in this system contains, of course, ν equations.

As is well-known, if the interaction potential Φ satisfies some smoothness conditions, then the Cauchy problem (1.2.2) for the Hamilton equations has a unique solution, global in time [Arn2]. However we shall consider systems of particles, which interact via a singular potential as hard spheres. Let us prove the existence theorem for the phase trajectories $X(t, x)$ of such systems of particles.

The interaction potential, i.e. the function Φ in (1.2.2), is assumed to possess the following properties. There exists a finite number $a > 0$ such that

$$\Phi(q) = +\infty \ , \text{ if } |q| < a. \tag{1.2.3}$$

Such potentials are called hard-core potentials. Obviously, for configurations of a system with a hard-core potential the following inequalities must be satisfied (q now denotes the position of particle's centre).

$$|q_i - q_j| \geqslant a \ , \ i \neq j \in (1, \dots, N).$$

We denote by $\mathbb{W}_N = \{(q_1, \dots q_n) \in \Lambda^N \Big| |q_i - q_j| < a$ for at least one pair (i, j) : $i \neq j \in (1, \dots, N)\}$ the set of forbidden configurations. Then for systems with hard–core potentials the phase space is the set $(\Lambda^N \setminus \mathbb{W}_N) \times \Re^{\nu N}$.

We suppose that the following condition holds

$$\Phi'(a+0) = 0 \quad . \tag{1.2.4}$$

This condition means that at the instants of collision t, i.e. $\big(Q_1(t), \dots, Q_N(t)\big) \in \partial \mathbb{W}_N \equiv \{\big(Q_1(t), \dots, Q_N(t)\big) \in \Lambda^N \Big| |Q_i(t) - Q_j(t)| = a$ for at least one pair (i, j) : $i \neq j \in (1, \dots, N)\}$, only the hard core interaction is effective. In this case completely elastic scattering occurs: the phase states of particles change instantaneously according to

$$\big(Q_i(t). P_i(t)\big) \to \big(Q_i(t), P_i^*(t)\big)$$
$$\big(Q_j(t), P_j(t)\big) \to \big(Q_j(t), P_j^*(t)\big).$$

The functions $P_i^*(t), P_j^*(t)$ are interpreted as the values of the momenta of the i–th and j–th particles after their collision and are given by the formulae

$$P_i^*(t) = P_i(t) - \eta \langle \eta, (P_i(t) - P_j(t)) \rangle,$$
$$P_j^*(t) = P_j(t) + \eta \langle \eta, (P_i(t) - P_j(t)) \rangle, \tag{1.2.5}$$

where η is the unit vector directed along the line connecting the centres of particles at the time of the collisions: $\eta = \big(Q_i(t) - Q_j(t)\big) \Big/ |Q_i(t) - Q_j(t)|$. Here $\langle \eta, (P_i - P_j) \rangle = \sum_{\alpha=1}^{\nu} \eta^\alpha (P_i^\alpha - P_j^\alpha)$ is the scalar product.

The relations (1.2.5) follow from energy and momentum conservation

$$P_i^{*\,2} + P_j^{*\,2} = P_i^2 + P_j^2,$$
$$P_i^* + P_j^* = P_i + P_j.$$

For a grazing collision $\langle (P_i - P_j), \eta \rangle = 0$ we set $P_i^* = P_i$ and $P_j^* = P_j$. In the one–dimensional case for the momenta after a collision we have

$$P_i^*(t) = P_j(t) \ , \ P_j^*(t) = P_i(t) \tag{1.2.5'}$$

i.e. the particles exchange their momenta.

For hard-core potentials (1.2.3,4), the particles can be considered as hard spheres of diameter a.

We also assume that the interaction at distance between these hard-spheres is given by a spherically symmetric finite potential $\Phi(q)$, defined for $|q| \geqslant a$, $\Phi(q) = \Phi(|q|)$ and

$$\Phi(q) = 0 \ , \ \text{if } |q| > R \tag{1.2.6}$$

where R is the range of the interaction, and the potential Φ is a twice continuously differentiable function

$$\Phi \in C^2 \left[a, R\right] . \tag{1.2.7}$$

The potential of the external field $u^{\Lambda}(q)$ is assumed to be a twice continuously differentiable function on any subset of Λ.

Let us consider now the problem of constructing the function $X(t, x)$ for the system with potential (1.2.3-4,1.2.6-7) in some detail.

1.2.2 Definition of the phase trajectories

The initial total energy of the system is finite if the particles have finite momenta and do not collide with each other or with the boundary $\partial\Lambda$ at $t = 0$. Total energy is an integral of motion, so it is also finite at $t > 0$. Up to the instant of the first collision, the potential Φ is bounded, and therefore particles are located inside Λ, their distance from $\partial\Lambda$ remains greater than a certain positive number. In fact, the total energy must tend to infinity when the particles approach to the boundary $\partial\Lambda$, but this contradicts the energy-conservation law. We denote the domain in which particles are located by $\Lambda_0 \subset \Lambda$. Clearly, for the same reason momenta remain bounded under evolution and are situated in a some ball S_{R_0} of finite radius R_0. In other words, the initial conditions define a domain $(\Lambda_0^N \backslash \mathbb{W}_N) \times S_{R_0}^N$ in the phase space in which the particles are located at $0 \leqslant t < t_1$ (t_1 is the time of the first collision).

By virtue of the fact that dist $(\Lambda_0, \partial\Lambda) > 0$, the second derivatives of Φ and u^{Λ} are bounded in Λ_0 . Let us define

$$L = \sup_{q\in\Lambda_0, |q|>a} \left\{ \left| \frac{\partial^2\Phi(q)}{\partial q^2} \right|, \left| \frac{\partial^2 u^{\Lambda}(q)}{\partial q^2} \right| \right\}.$$

Then for every convex subset of Λ_0, the first derivatives of the potentials $\Phi(q)$ and $u^{\Lambda}(q)$ satisfy the Lipschitz conditions

$$\left| \frac{\partial}{\partial q_1} \begin{pmatrix} \Phi(q_1) \\ u^{\Lambda}(q_1) \end{pmatrix} - \frac{\partial}{\partial q_2} \begin{pmatrix} \Phi(q_2) \\ u^{\Lambda}(q_2) \end{pmatrix} \right| \leqslant L\,|q_1 - q_2|\,, \tag{1.2.8}$$

It follows from this that the right–hand sides of the Hamilton Eqs. (1.2.2) in the domain $\Lambda_0 \times S_{R_0}$ satisfy the Lipschitz condition with a constant equal to $\max\{1, L\}$. Then by virtue of well–known results from the theory of ordinary differential equations [Arn2], it follows that the Cauchy problem (1.2.2) for the Hamilton equations with the abovementioned initial data has a unique solution $X(t,x)$ (for a certain interval of time) that is continuosly differentiable with respect to t and to the initial data x.

We now define the time interval of existence of the solution $X(t,x)$. Particles are located in Λ_0 at an arbitrary time, so the Lipschitz condition (1.2.8) holds for the right–hand sides of (1.2.2) with the same constant $\max\{1, L\}$ at an arbitrary time (excluding collisions). Thus the solution $X(t,x)$ can be continued in time up to the instant of a collision, i.e. $X(t,x) \in \partial(\Lambda_0^N \backslash \mathbb{W}_N) \times S_{R_0}^N$.

Let us denote by $t_{i_1j_1} < t_{i_2j_2} < \dots$ the times of the first, second, ... pair collisions of particles in system. Then the function $X(t,x)$ for $t \in [t_{i_1j_1}, t_{i_2j_2}]$ is defined in the following way. We consider the phase point $X^*(t_{i_1j_1}, x) \equiv \big(Q(t_{i_1j_1}, x),$ $P^*(t_{i_1j_1}, x)\big)$, (where $P^* \equiv (P_1, \dots, P_{i_1}^*, \dots, P_{j_1}^*, \dots, P_N)$ and $(P_{i_1}^*, P_{j_1}^*)$ are defined by equalities (1.2.5.)), when the particles have already exchanged their momenta, as initial data at $t = t_{i_1j_1}$. When $t \in \,]t_{i_1j_1}, t_{i_2j_2}[$ the system of particles satisfies the same conditions as for $t \in [0, t_{i_1j_1}[$, so the evolution is described by the Hamilton equations (1.2.2) and there exists a solution $X\big(t - t_{i_1j_1}, X^*(t_{i_1j_1}, x)\big)$ for $t \in [t_{i_1j_1}, t_{i_2j_2}[$ which possesses the same properties of differentiability with respect to time and new initial data $X^*(t_{i_1j_1}, x)$.

This procedure can be repeated over and over again if the number of collisions is finite on an arbitrary bounded time interval and if only pair collisions occur in the system.

Let us denote by $\mathcal{M}_1^0$ the set of initial data $x \in (\mathbb{R}^{\nu N} \times (\Lambda^N \backslash \mathbb{W}_N))$ for which a multiple (i.e., triple and of higher order) collision occur in the system and by $\mathcal{M}_2^0 \subset (\mathbb{R}^{\nu N} \times (\Lambda^N \backslash \mathbb{W}_N))$ the set of initial data for which an infinite number of collisions may occur on a bounded time interval. We show that these sets are the union of some hypersurfaces in the phase space of dimension less than that of the phase space $2N\nu$. The following lemma holds:

Lemma 1.2.1. *The set $\mathcal{M}_1^0 \subset (\mathbb{R}^{\nu N} \times (\Lambda^N \backslash \mathbb{W}_N))$ is of zero Lebesgue measure.*

Proof. . For the sake of simplicity let us consider the case of triple collisions. If three particles i, j, k collide at some $t > 0$, then at least the following equations hold

$$\begin{aligned} |Q_i(t,x) - Q_j(t,x)| &= a, \\ |Q_j(t,x) - Q_k(t,x)| &= a. \end{aligned} \tag{1.2.9}$$

At fixed t these equations define hypersurfaces in the phase space that are non–trivial. Indeed, all equations (1.2.9) cannot be satisfied identically for x belonging to some set of positive measure, because this would contradict the Liouville theorem (to be formulated at the end of this Section).

Clearly, triple collisions are possible only if the initial phase points x are situated on the intersection of these hypersurfaces, i.e. on a set of dimension less than that of the phase space of the system. The union of such hypersurfaces for all $t > 0$ is also of dimension less than that of the phase space and thus has Lebesgue measure zero. The investigation of the collisions of more than three particles can be carried out analogously. In what follows the set $\mathcal{M}_1^0$ will be excluded from our considerations.

Lemma 1.2.2. *The set $\mathcal{M}_2^0 \subset (\mathbb{R}^{\nu N} \times (\Lambda^N \backslash \mathbb{W}_N))$ is of zero Lebesgue measure.*

Proof. . For the interaction potential (1.2.3,4), (1.2.6,7) the forces acting upon a given particle by the others at any instant of time (except the collision instants) are bounded uniformly with respect to N. Consequently, inside finite time intervals between collisions the momenta of particles may have only finite increments, which approach zero as the lengths of the corresponding time intervals approach zero (and only at pair collision the momenta have jumps).

Since the energy of a system of N particles is bounded, the momentum of any particle is also bounded uniformly with respect to t and, consequently, changes of the particle positions are finite and approach zero if the lengths of the corresponding time intervals approach zero. We now assume that an infinite number of collisions between two particles, for example, between the first and the second ones during a finite time interval, is possible for some initial data x. We will denote the collision instants by $t_1, \ldots, t_n, \ldots$, where $\lim_{n \to +\infty} t_n = t_0 < \infty$ (for at least some subsequence).

The sequence of momenta for these particles obviously converges (for at least some subsequence):

$$\lim_{n \to \infty} P_i(t_n, x) = P_i(t_0, x),\ i = 1, 2$$

The increments of these momenta tend to zero when $n \to \infty$, because the lengths of the time interval between two subsequent collisions tends to zero and the forces acting upon the particles are bounded.

Therefore we have for the difference of momenta

$$\lim_{n \to \infty} |P_1(t_n, x) - P_2(t_n, x)| = 0, \tag{1.2.10}$$

i.e.

$$P_1(t_0, x) = P_2(t_0, x) \tag{1.2.10$'$}$$

Moreover, the sequences $Q_1(t_n, x), Q_2(t_n, x)$ are Cauchy and, consequently,

$$|Q_1(t_0, x) - Q_2(t_0, x)| = a \tag{1.2.10$''$}$$

The set of initial data $x \in \mathbb{R}^{\nu N} \times (\Lambda^N \backslash \mathbb{W}_N) \backslash \mathcal{M}_1^0$ for which conditions (1.2.10) are satisfied form a hypersurface of zero Lebesgue measure in phase space.

Now we assume that there exists a group of $s > 2$ particles such that for some initial datum x each of them has an infinite number of collisions with other particles of this group on finite time interval, and let $t_1, \ldots, t_n$ be the collision instants of the group, for which there holds $\lim_{n \to \infty} t_n = t_0$. As above, the sequence $Q_i(t_n, x)$ is Cauchy

$$\lim_{n \to \infty} Q_i(t_n, x) = Q_i(t_0, x),\ i \in (1, \ldots, s),$$

for these initial data $x \in \mathbb{R}^{\nu N} \times (\Lambda^N \backslash \mathbb{W}_N) \backslash \mathcal{M}_1^0$ and for any particle belonging to this group and hence for any pair of particles having an infinite number of collisions, we have the equation

$$|Q_i(t_0, x) - Q_j(t_0, x)| = a,\ (i, j) \in (1, \ldots, s).$$

Consequently, all s particles are in touch with each other at the instant of time t_0, i.e. multiple collision occurs, but this is impossible since the set $\mathcal{M}_1^0$ is eliminated from consideration. (It is also easy to show that momentum of these s particles coincide at t_0). We remark that in papers [Ga][I][MC] the estimates for the number of collisions were obtained for a system of finitely many hard spheres moving in all of $\mathbb{R}^3$. Thus in this case the set $\mathcal{M}_2^0$ is empty. In what follows this set will be excluded from our considerations.

Let us define the phase trajectory $X(t,x)$ for initial data $x \in \partial(\Lambda^N \backslash \mathbb{W}_N) \times \mathbb{R}^{\nu N})$. We exclude from our considerations the initial phase points in which three or more particles are in touch at the initial instant, because the set of their trajectories is a hypersurface of zero Lebesgue measure in phase space.Consider a phase point of the system such that initially two particles (e.g. the i–th and j–th) collide.

Since the collision occurs instantaneously, the phase points x and $x^* \equiv (x_1, \ldots, q_i, p_i^*, \ldots, q_j, p_j^*, \ldots, x_N)$ (where p_i^*, p_j^* are defined by (1.2.5) and η is the unit vector $a^{-1}(q_i - q_j)$) will be considered as identical in the sense that they determine the same initial phase state for the Hamilton system. As the initial data consist of two phase points x and x^*, the point should be chosen in such a way as to correspond to the state of two departing particles, i.e. $|Q_i(t) - Q_j(t)| > a$ when $t \in \mathbb{R}^1$. It follows from this requirement that for $\langle (q_i - q_j), (p_i - p_j) \rangle < 0$ and $t \geqslant 0$ we must take the phase point x^* as the initial data for the Hamilton Eqs. (1.2.2) (x remains the initial data for the evolution at $t \leqslant 0$). For the case of $\langle (q_i - q_j), (p_i - p_j) \rangle > 0$ the initial phase point for the Hamilton dynamics at $t \geqslant 0$ is x (while at $t \leqslant 0$ it is x^*). Thus for the function $X(t,x)$ for $x \in \partial(\Lambda^N \backslash \mathbb{W}_N) \times \mathbb{R}^{\nu N}$ the relation

$$X(t,x) = X(t,x^*) \tag{1.2.11}$$

is valid. The points x or x^* are chosen as above from the requirement that the state describes two departing particles.

Finally we define the phase trajectory of the system for $t \in \,]-\infty, 0]$. Between the collision times the function $X(t,x)$ is defined as a solution of the Hamilton equations (1.2.2), which are invariant with respect to time reversal, i.e. if $X(t,x)$ is a solution of the Cauchy problem (1.2.2) then $\overline{X}(-t,\overline{x}) \equiv \big(Q(-t,\overline{x}), -P(-t,\overline{x})\big)$ is a solution (1.2.2) with the initial data $\overline{x} \equiv (q, -p)$; moreover $\overline{X}(-t,\overline{x}) = X(t,x)$. This condition must also be satisfied at the collision instants.

1.2.3 Existence and some properties of the phase trajectories

The function $t_{ij}(x)$ is set equal to infinity $t_{ij}(x) = \infty$ if the i–th and j–th particles do not collide and $t_{ij}(x) < \infty$ if they do collide. We will investigate the properties of the function $t_{ij}(x)$ for initial data $x \in \mathbb{K}$ for which $t_{ij}(x) < \infty$.

For fixed positions (q_i, q_j), there is a cone in the momentum space described by (p_i, p_j), a subset of $\mathbb{K}$, which gives the possible collisions of the i-th and j-th particles. The domain $\mathbb{K}$ is the union of these cones for the admissible values of (q_i, q_j).

Let us define the collision times as functions of initial data. Also let $t_{ij}(x)$ be the instant of collision of i–th and j–th particles and let us define the set $\mathbb{K}$ by $\mathbb{K} = \{x \in \mathbb{R}^{\nu N} \times (\Lambda^N \backslash \mathbb{W}_N) \big| t_{ij}(x) < +\infty\}$. Each function $t_{ij}(x)$ is defined, for each pair (i, j) as the solution of the equation

$$|Q_i(t, x) - Q_j(t, x)| = a, \tag{1.2.12}$$

and the following lemma holds:

Lemma 1.2.3. *The functions $t_{ij}(x)$ are continuously differentiable functions with respect to initial data x almost everywhere in $\mathbb{R}^{\nu N} \times (\Lambda^N \backslash \mathbb{W}_N) \cap \mathbb{K}$.*

Proof. From Hamilton Eqs. (1.2.2) it follows that

$$\frac{\partial}{\partial t}\,|Q_i(t, x) - Q_j(t, x)|$$

$$= \big(|Q_i(t, x) - Q_j(t, x)|\big)^{-1} \langle \big(Q_i(t, x) - Q_j(t, x)\big), \big(P_i(t, x) - P_j(t, x)\big)\rangle$$

and we find that $\frac{\partial}{\partial t}\,|Q_i(t, x) - Q_j(t, x)| \neq 0$ if

$$\langle \big(Q_i(t, x) - Q_j(t, x)\big), \big(P_i(t, x) - P_j(t, x)\big)\rangle \neq 0.$$

In this case the Eq. (1.2.12) can be solved with respect to t, and, by the implicit function theorem, its solution $t_{ij}(x)$ is continuously differentiable with respect to x. If

$$\langle \big(Q_i(t, x) - Q_j(t, x)\big), \big(P_i(t, x) - P_j(t, x)\big)\rangle = 0 \tag{1.2.13}$$

then Eqs. (1.2.12-13) define hypersurfaces of lower dimension than that of phase space, and consequently the set $\mathcal{M}_3^0 \subset \mathbb{R}^{\nu N} \times (\Lambda^N \backslash \mathbb{W}_N)$ of initial phase points x

belonging to the intersection of the surfaces (1.2.12) and (1.2.13) is of Lebesgue measure zero.

It is obvious that the hypersurface $\mathcal{M}_3^0$ (see 1.2.12-13) coincides with the boundary $\partial\mathbb{K}$ of the domain $\mathbb{K}$ and consists of the initial points x for which particles touch at the collision instants.

The function $X(t,x)$ is continuously differentiable with respect to its initial data on every interval $]t_{i_{n-1}j_{n-1}}(x), t_{i_n j_n}(x)[$, $n \in (1,2,\dots)$ between the collisions. At the collision instants the function $X(t,x)$ is not continuous.Therefore we must consider the dependence of $X(t,x)$ on x in detail.

Lemma 1.2.4. *The phase trajectory $X(t,x)$ is a continuously differentiable function almost everywhere with respect to its initial data x on every time interval between collisions.*

Proof. According to Lemma 1.2.3 the function $X(t_{i_1j_1}(x),x)$ at the instant of the first collision $t_{i_1j_1}(x)$ is continuously differentiable with respect to the initial data x. The unit vector

$$\eta_{i_1j_1}(x) = \big(Q_{i_1}(t_{i_1j_1}(x),x) - Q_{j_1}(t_{i_1j_1}(x),x)\big) / \left|Q_{i_1}(t_{i_1j_1},x) - Q_{j_1}(t_{i_1j_1},x)\right|$$

is also continuously differentiable with respect to x. It follows that the momenta $P_i^*(t_{i_1j_1}(x),x)$ and $P_j^*(t_{i_1j_1}(x),x)$ (1.2.5) of departing particles will be continuously differentiable with respect to x. As the phase point $X^*(t_{i_1j_1}(x),x)$ is initial for the Hamilton evolution at $t > t_{i_1j_1}(x)$, the solution of the Hamilton equations $X\big(t - t_{i_1j_1}(x), X^*(t_{i_1j_1}(x),x)\big)$ is also continuously differentiable with respect to x. In the same way we can prove this for an arbitrary interval

$$\big]t_{i_{n-1}j_{n-1}}(x), t_{i_nj_n}(x)\big[\ , \ n \geqslant 2.$$

Let us remark that the trajectory $X(t,x)$ is continuous with respect to x on the boundary $\partial\mathbb{K}$, defined by (1.2.12-13), (or on $\mathcal{M}_3^0$), where $\langle \eta_{ij} \cdot (P_i(t_{ij},x) - P_j(t_{ij},x))\rangle = 0$ and thus $P_i^*(t_{ij},x) = P_i(t_{ij},x)$, $P_j^*(t_{ij},x) = P_j(t_{ij},x)$ according to (1.2.5). Outside $\mathbb{K}$ the i-th and j-th particles do not collide at $t = t_{ij}$.

Let us denote by $\mathcal{M}_N^0$ the set $\mathcal{M}_1^0 \cup \mathcal{M}_2^0 \cup \mathcal{M}_3^0 \subset \mathbb{R}^{\nu N} \times (\Lambda^N \backslash \mathbb{W}_N)$ which is of Lebesgue measure zero. We exclude this set from our considerations. Thus the phase trajectory $X(t,x)$ of the system of particles with a potential Φ satisfying the conditions (1.2.3-4), (1.2.6-7) is defined for $x \in \mathbb{R}^{\nu N} \times (\Lambda^N \backslash \mathbb{W}_N) \backslash \mathcal{M}_N^0$.

All the results obtained above are summarized by the following theorem.

Theorem 1.2.1. *If the interaction potential Φ satisfies the conditions (1.2.3-4),(1.2.6-7) then for $t \in \mathbb{R}^1$ and for almost all initial data x a unique function $X(t.x)$ exists such that in the time intervals between collisions it is continuously differentiable with respect to t and x and satisfies the Hamilton Eqs. (1.2.2).*

The theorem is also true in case $\Lambda = \mathbb{R}^\nu$.

Remark 1. The trajectory $X(t,x)$ exists for such initial data, when two particles touch at $t = 0$, $|q_i - q_j| = a$, $i \neq j = (1,\dots,N)$ and if there are no multiple collisions and so on for $t \gtrless 0$. Indeed these particles will depart, i. e. $|Q_i(t) - Q_j(t)| > a$ and, up to the instant of the next pair collision, the system will be in a state where the distances between the particle centers will be larger than a. If one takes this state as initial state, then the trajectory of the sytem can be constructed according to Theorem 1.2.1 for $t \in \mathbb{R}^1$. This means that the surfaces $|q_i - q_j| = a$, $i,j = 1,\dots,N$ do not coincide with the hyperplanes $\mathcal{M}_N^0$. We shall consider a more general situation in some detail in Chapter II (see Theorem 2.5.1). It will be shown, in fact, that even if all positions $q_1,\dots,q_N$ are fixed, the hyperplanes $\mathcal{M}_N^0$ are of lower dimension than the momentum space described by $p_1,\dots,p_N$ and consequently are of zero Lebesgue measure in this space. Let us note some properties of the phase trajectory $X(t,x)$.

1. The function $X(t,x)$ has the group property

$$X(t_1 + t_2, x) = X(t_1, X(t_2, x)) = X(t_2, X(t_1, x))$$

for all t_1 and t_2.

2. The function $X(t,x)$ is permutation–invariant, i.e.

$$X(t, x_{i_1},\dots,x_{i_N}) = \big(X_{i_1}(t,x),\dots,X_{i_N}(t,x)\big),$$

where $(i_1,\dots,i_N)$ is an arbitrary permutation of $(1,\dots,N)$.

3. The function $X(t,x)$ defines a mapping (a flow) of the phase space onto itself

$$T_t x = X(t,x)\ ,\ t \in \mathbb{R}^1\ ,\ x \in \mathbb{R}^{\nu N} \times (\Lambda^N \backslash \mathbb{W}_N) \backslash \mathcal{M}_n^0.$$

For each $t \in \mathbb{R}^1$ the flow T_t preserves phase volume (Liouville's Theorem) [Arn1] [PGM2]. **4**. Let us write down the Hamilton Eqs. (1.2.2) in the abstract form

$$\frac{d}{dt} X(t) = \{X(t), H_N^\Lambda\}, \tag{1.2.14}$$

where $\{., H_N^\Lambda\}$ is the Poisson bracket with the Hamiltonian H_N^Λ

$$\{., H_N^\Lambda\} =$$

$$\left(\sum_{i=1}^{N}\langle p_i, \frac{\partial}{\partial q_i}\rangle - \sum_{i\neq j=1}^{N}\langle\frac{\partial}{\partial q_i}\Phi(q_i - q_j), \frac{\partial}{\partial p_i}\rangle - \sum_{i=1}^{N}\langle\frac{\partial}{\partial q_i}u^\Lambda(q_i), \frac{\partial}{\partial p_i}\rangle\right) . \quad (1.2.15)$$

It follows from Theorem 1.2.1 that for small enough $t \in \mathbb{R}^1$ and for admissible initial data x on an arbitrary compact set $\mathcal{K}$ outside of $\mathbb{W}_N$ the function $X(t,x)$ can be written down in the form

$$X(t,x) = x + \{x, H_N^\Lambda\}t + o(t^2)(x), \quad (1.2.16)$$

where $o(t^2)(x)/t$ tends to zero uniformly in $x \in \mathcal{K}$ as $t \to 0$. From the expression (1.2.16) it follows that $X(t,x)$ converges to x uniformly in $x \in \mathcal{K}$ as $t \to 0$.

Remark 2. The set $\mathcal{M}_N^0$ consists of the union of the hypersurfaces (1.2.9-10), (1.2.12-13). For a fixed time t, in the phase space described by $X(t,x) \equiv (Q_1(t,x), \ldots, Q_N(t,x), P_1(t,x), \ldots, P_N(t,x))$, these hypersurfaces represent a set of lower dimension than the dimesnion of the phase space itself and the closure of $\mathcal{M}_N^0$ can be covered by a finite number of balls having a total volume less than ϵ for any given $\epsilon > 0$. The union of these sets corresponding to a finite time interval $|t| < t_0$ has also a finite volume, less than $2t_0\epsilon$. The mapping $T_t x = X(t,x)$ preserves the volume in phase space according to Liouville's theorem. and, therefore, the set $\mathcal{M}_N^0$ can be covered by balls of total volume less than $2t_0\epsilon$ as well in the phase space described by x. This means that for an arbitrary point $x \notin \overline{\mathcal{M}_N^0}$ there exists some neighborhood of nonzero measure which does not intersect $\overline{\mathcal{M}_N^0}$. Thus for arbitrary points $x \notin \overline{\mathcal{M}_N^0}$ we can assert that $X(t,x)$ is differentiable with respect to x.

Remark 3. The Hamilton dynamics of infinitely many particles system has been developed in papers [DF][Lan1][Lan2][MPP][PPT][Si1][Si2]. This problem has been satisfactorily solved only for one– and two–dimensional systems.

1.3 The Evolution Operator for a System with Finite Number of Particles

1.3.1 Definition of the evolution operator

In this section we will continue the investigation of Hamilton dynamics for a system of particles with the Hamiltonian (1.2.1,1.2.3-4,6-7).

We begin with the case $\Lambda = \mathbb{R}^\nu$. Let us introduce a set of measurable functions $f_N(x) \equiv f_N(x_1, \ldots, x_N)$ defined on the phase space $\mathbb{R}^{\nu N} \times (\mathbb{R}^{\nu N} \backslash \mathbb{W}_N)$ of an N–particle system, which are invariant under permutations of the arguments $x_1, \ldots, x_N$ and equal to zero on the set of forbidden configurations $\mathbb{W}_N$. Let us define on this set of functions the evolution operator $S^N(t)$, $t \in \mathbb{R}^1$ by the formula

$$(S^N(t) f_N)(x) = \begin{cases} f_N(X(t, x)), & \text{if } x \in (\mathbb{R}^{\nu N} \times (\mathbb{R}^{\nu N} \backslash \mathbb{W}_N)) \backslash \mathcal{M}_N^0, \\ 0, & \text{if } x \in \mathbb{R}^{\nu N} \times \mathbb{W}_N, \end{cases} \tag{1.3.1}$$

where $X(t, x)$ is the phase trajectory for the particles with the Hamiltonian H_N (Theorem 1.2.1). This operator is linear according to its definition and $S^N(0) = I$, where I is the unit operator.

In order to define the evolution operator (1.3.1) in a rigorous way, we must restrict ourselves to the functions of a special functional space.

We denote by $L^1(\mathbb{R}^{\nu N} \times (\mathbb{R}^{\nu N} \backslash \mathbb{W}_N))$ the linear Banach space of summable functions $f_N(x)$ symmetrical with respect to permutations of the arguments, equal to zero on the set $\mathbb{W}_N$, with a norm

$$\|f_N\| = \int dx_1 \ldots dx_N \, |f_N(x_1, \ldots, x_N)| = \int dx \, |f_N(x)| \qquad (dx = dx_1 \ldots dx_N) .$$

By $L_0^1(\mathbb{R}^{\nu N} \times (\mathbb{R}^{\nu N} \backslash \mathbb{W}_N))$ we denote the subset of the Banach space $L^1(\mathbb{R}^{\nu N} \times (\mathbb{R}^{\nu N} \backslash \mathbb{W}_N))$ consisting of continuously differentiable functions with compact support and equal to zero in some ε–neighbourhood of the forbidden configurations $\mathbb{W}_N$. Here $\varepsilon > 0$ is a number depending on the functions : $\varepsilon = \varepsilon(f_N)$. Clearly, the set $L_0^1(\mathbb{R}^{\nu N} \times (\mathbb{R}^{\nu N} \backslash \mathbb{W}_N))$ is everywhere dense in $L^1(\mathbb{R}^{\nu N} \times (\mathbb{R}^{\nu N} \backslash \mathbb{W}_N))$ and the space L^1 is the closure of the set L_0^1 with respect to the L^1 norm. If $x \in \partial(\mathbb{R}^{\nu N} \backslash \mathbb{W}_N) \times \mathbb{R}^{\nu N}$, then the following symmetry properties hold for the functions from $L_0^1(\mathbb{R}^{\nu N} \times (\mathbb{R}^{\nu N} \backslash \mathbb{W}_N))$

$$f_N(x_1, \ldots, q_i, p_i, \ldots, q_j, p_j, \ldots, x_N) = f_N(x_1, \ldots, q_i, p_i^*, \ldots, q_j, p_j^*, \ldots, x_N),$$
$$i \neq j \in (1, \ldots, N), \tag{1.3.2}$$

where the momenta p_i^*, p_j^* are related to p_i, p_j by (1.2.5).

We define the operator $S^N(t)$ (1.3.1) first on the functions from the set $L_0^1(\mathbb{R}^{\nu N} \times (\mathbb{R}^{\nu N} \backslash \mathbb{W}_N))$. Let us show that for every t the functions $\big(S^N(t) f_N(x)\big)$ are defined on the set $\mathbb{R}^{\nu N} \times (\mathbb{R}^{\nu N} \backslash \mathbb{W}_N) \backslash \mathcal{M}_N^0$. The trajectory $X(t,x)$ of the N–particle system is a continuous function of time and initial data inside time intervals between collisions and has jumps only at collision times when particles touch (Theorem 1.2.1). At the same time, functions $f_N \in L_0^1(\mathbb{R}^{\nu N} \times (\mathbb{R}^{\nu N} \backslash \mathbb{W}_N))$ are identically zero in some ε–neighbourhood of the set of forbidden configurations $\mathbb{W}_N$, and consequently the function $\big(S^N(t) f_N\big)(x)$ is continuous with respect to x just like $f_N(x)$ and with respect to t. It follows from the definition (1.3.1) that the function $\big(S^N(t) f_N\big)(x)$ has the same symmetry properties as the function $f_N(x)$. Hence the evolution operator (1.3.1) is defined correctly on functions $f_N \in L_0^1(\mathbb{R}^{\nu N} \times (\mathbb{R}^{\nu N} \backslash \mathbb{W}_N))$.

We now show that the function $\big(S^N(t) f_N\big)(x)$ is summable for arbitrary $f_N \in L_0^1(\mathbb{R}^{\nu N} \times (\mathbb{R}^{\nu N} \backslash \mathbb{W}_N))$. Using the Liouville Theorem we get

$$\begin{aligned} \|S^N(t) f_N\| &= \int dx \, |f_N(X(t,x))| = \int dx \, \det \left| \frac{\partial X(t,x)}{\partial x} \right| |f_N(X(t,x))| \\ &= \int dX(t) \, |f_N(X(t))| = \|f_N\|. \end{aligned}$$

It follows from this relation that the operator $S^N(t)$ is defined on $L_0^1(\mathbb{R}^{\nu N} \times (\mathbb{R}^{\nu N} \backslash \mathbb{W}_N))$ and is isometric : $\|S^N(t)\| = 1$.

The operator $S^N(t)$ can now be extended by continuity to the entire space $L^1(\mathbb{R}^{\nu N} \times (\mathbb{R}^{\nu N} \backslash \mathbb{W}_N))$ as an isometric operator.

If particles are contained in the bounded domain $\Lambda \subset \mathbb{R}^\nu$, the evolution operator $S_\Lambda^N(t)$ is defined on functions $f_{\Lambda,N} \in L^1(\mathbb{R}^{\nu N} \times (\Lambda^N \backslash \mathbb{W}_N))$, where we denote by $L^1(\mathbb{R}^{\nu N} \times (\Lambda^N \backslash \mathbb{W}_N))$ a subspace in $L^1(\mathbb{R}^{\nu N} \times (\mathbb{R}^{\nu N} \backslash \mathbb{W}_N))$ which consists of functions $f_{\Lambda,N}$ defined on the domain Λ of configuration space and equal to zero outside Λ.

Remark 4. Let the domain $\Lambda \subset \mathbb{R}^\nu$ be a ball of finite radius and let us assume that the particles are elastically reflected from the boundary $\partial\Lambda$ of this domain Λ and then $u^\Lambda(q) \equiv 0$. We denote by $L_0^1(\mathbb{R}^{\nu N} \times (\Lambda^N \backslash \mathbb{W}_N))$ the subspace of continuously differentiable functions with compact support and equal to zero in some ε–neighbourhood of $\partial(\Lambda^N \backslash \mathbb{W}_N)$. In this case for $q_i \in \partial\Lambda$, $i \in (1, \ldots, N)$ the following symmetry property holds for the functions from $L^1_0(\mathbb{R}^{\nu N} \times (\Lambda^N \backslash \mathbb{W}_N))$

$$f_N(x_1, \ldots, q_i, , p_i, \ldots, x_N) = f_N(x_1, \ldots, q_i, p_i^*, \ldots, x_N), \tag{1.3.2'}$$

where p_i^* is the momentum of the i–th particle after reflection from $\partial\Lambda$: $p_i^* = p_i - 2\eta\langle\eta, p_i\rangle$, $|\eta| = 1$ and η is normal to $\partial\Lambda$.

We need now some definitions from the theory of groups of operators [Ka][Kr] [RS][Yo]. We recall that a group of bounded operators $S^N(t)$, $t \in \mathbb{R}^1$ defined on $L^1(\mathbb{R}^{\nu N} \times (\mathbb{R}^{\nu N} \backslash \mathbb{W}_N))$ is called strongly continuous if

$$\lim_{\Delta t \to 0} \left\| S^N(t + \Delta t) f_N - S^N(t) f_N \right\| = 0$$

for every $f_N \in L^1(\mathbb{R}^{\nu N} \times (\mathbb{R}^{\nu N} \backslash \mathbb{W}_N))$. An operator $\mathcal{H}_N$ is called the infinitesimal generator of a strongly continuous group $S^N(t)$ if

$$\lim_{\Delta t \to 0} \left\| \frac{1}{\Delta t}(S^N(\Delta t) f_N - f_N) - \mathcal{H}_N f_N \right| = 0$$

for any $f_N \in \mathcal{D}(\mathcal{H}_N)$, where $\mathcal{D}(\mathcal{H}_N)$ is the domain of definition of $\mathcal{H}_N$, which is everywhere dense in L^1. The operator $\mathcal{H}_N$ is closed and the following equality holds for every $f_N \in \mathcal{D}(\mathcal{H}_N) \subset L^1$

$$\lim_{\Delta t \to 0} \frac{1}{\Delta t}(S^N(t + \Delta t) f_N - S^N(t) f_N) = S^N(t)\mathcal{H}_N f_N = \mathcal{H}_N S^N(t) f_N.$$

1.3.2 Properties of the evolution operator

The properties of the operator $S^N(t)$ in space $L^1(\mathbb{R}^{\nu N} \times (\mathbb{R}^{\nu N} \backslash \mathbb{W}_N))$ is described in the following theorem.

Theorem 1.3.1. *The family of operators $S^N(t)$, $t \in \mathbb{R}^1$ (1.3.1) is a strongly continuous one–parameter group of isometric operators in $L^1(\mathbb{R}^{\nu N} \times (\mathbb{R}^{\nu N} \backslash \mathbb{W}_N))$. On the set $L^1_0(\mathbb{R}^{\nu N} \times (\mathbb{R}^{\nu N} \backslash \mathbb{W}_N))$ the infinitesimal generator $\mathcal{H}_N$ of the group $S^N(t)$ is given by the Poisson bracket with boundary conditions on $\partial(\mathbb{R}^{\nu N} \backslash \mathbb{W}_N)$.*

Proof. The group property of the family of operators (1.3.1) $S^N(t_1 + t_2) = S^N(t_1)$ $S^N(t_2) = S^N(t_2) S^N(t_1)$ for any $t_1, t_2 \in \mathbb{R}^1$ follows from the definition (1.3.1) and the group property of the phase trajectory $X(t, x)$ (see sec. 1.2).

Now we prove that

$$\lim_{\Delta t\to 0} \left\|S^N(t+\Delta t)f_N - S^N(t)f_N\right\| = 0 \tag{1.3.3}$$

for every $f_N \in L_0^1(\mathbb{R}^{\nu N} \times (\mathbb{R}^{\nu N}\backslash \mathbb{W}_N))$. In fact, according to the group property and isometricity of $S^N(t)$, we have

$$\left\|S^N(t+\Delta t)f_N - S^N(t)f_N\right\| = \left\|S^N(\Delta t)f_N - f_N\right\|$$

$$= \int dx \left|f_N\left(X(\Delta t, x)\right) - f_N(x)\right|$$

As was already mentioned above (see (1.2.16)), the functions $X(\Delta t, x) \to x$ as $\Delta t \to 0$ uniformly in x from any compact set outside $\mathbb{W}_N \cup \mathcal{M}_N^0$ and therefore $\left|f_N\big(X(\Delta t, x)\big) - f_N(x)\right| \to 0$ uniformly with respect to x on compact sets as $\Delta t \to 0$. Thus we can pass to the limit $\Delta t \to 0$ under the integral sign and consequently the validity of (1.3.3) is proved.

According to the fact that $L_0^1(\mathbb{R}^{\nu N} \times (\mathbb{R}^{\nu N}\backslash \mathbb{W}_N))$ is everywhere dense in $L^1(\mathbb{R}^{\nu N} \times (\mathbb{R}^{\nu N}\backslash \mathbb{W}_N))$, the strong continuity of the group $S^N(t)$ in $L^1(\mathbb{R}^{\nu N} \times (\mathbb{R}^{\nu N}\backslash \mathbb{W}_N))$ follows from (1.3.3) and the boundednes of the group $S^N(t)$.

It follows from the general properties of isometric groups of operators [RS,Yo] that there exists a closed infinitesimal generator $\mathcal{H}_N$ of the group $S^N(t)$ (1.3.1). Let us prove that this operator $\mathcal{H}_N$ coincides with the Poisson bracket on $L_0^1(\mathbb{R}^{\nu N} \times (\mathbb{R}^{\nu N}\backslash \mathbb{W}_N))$. To do this we first differentiate the functions $\big(S^N(t)f_N\big)(x)$ with respect to time using point–by–point convergence. Let $f_N \in L_0^1(\mathbb{R}^{\nu N} \times (\mathbb{R}^{\nu N}\backslash \mathbb{W}_N))$; then for each fixed point $x \in (\mathbb{R}^{\nu N} \times (\mathbb{R}^{\nu N}\backslash \mathbb{W}_N))\backslash \mathcal{M}_N^0$ we obtain

$$\left(\frac{d}{dt}S^N(t)f_N\right)(x) = \lim_{\Delta t\to 0}\left(S^N(t)\frac{1}{\Delta t}\big(S^N(\Delta t) - I\big)f_N\right)(x)$$
$$= \big(S^N(t)\{f_N, H_N\}\big)(x), \tag{1.3.4.a}$$

where $\{., H_N\}$ is the Poisson bracket (1.2.15) with Hamiltonian H_N. Here we take into account that this expression is non–zero only in the time intervals between collisions, where $X(t, x)$ is completely determined by the Hamilton Eqs. (1.2.2) and that the operator $S^N(t)$ preserves the boundedness of the functions f_N.

Let us calculate this derivative in another way. The function $\big(S^N(t)f_N\big)(x)$ for $t \in \mathbb{R}^1$ may be different from zero in a neighborhood of forbidden configurations where $f_N(x)$ is zero. If x is an internal point of the set $\mathbb{R}^{\nu N} \times (\mathbb{R}^{\nu N} \backslash \mathbb{W}_N) \backslash \mathcal{M}_N^0$, then

$$\left(\frac{d}{dt} S^N(t) f_N\right)(x) = \lim_{\Delta t \to 0} \left(\frac{1}{\Delta t}(S^N(\Delta t) - I) S^N(t) f_N\right)(x)$$
$$= \left\{ \left(S^N(t) f_N\right), H_N \right\}(x). \tag{1.3.4.b}$$

Here we have used the fact that the function $\left(S^N(t) f_N\right)(x)$ is differentiable with respect to x almost everywhere on the set $\mathbb{R}^{\nu N} \times (\mathbb{R}^{\nu N} \backslash \mathbb{W}_N) \backslash \mathcal{M}_N^0$. In fact, we have

$$\begin{pmatrix} \frac{\partial}{\partial q_i} \\ \frac{\partial}{\partial p_i} \end{pmatrix} (S^N(t) f_N)(x) = \sum_{k=1}^N \left(\left\langle \frac{\partial (S^N(t) f_N)(x)}{\partial Q_k(t,x)}, \begin{pmatrix} \frac{\partial}{\partial q_i} \\ \frac{\partial}{\partial p_i} \end{pmatrix} Q_k(t,x) \right\rangle \right.$$
$$\left. + \left\langle \frac{\partial (S^N(t) f_N)(x)}{\partial P_k(t,x)}, \begin{pmatrix} \frac{\partial}{\partial q_i} \\ \frac{\partial}{\partial p_i} \end{pmatrix} P_k(t.x) \right\rangle \right), \quad i \in (1, \dots, N)$$

The function $\left(S^N(t) f_N\right)(x)$ is differentiable with respect to $Q_k(t,x)$, $P_k(t,x)$ since

a) $f_N(x) \in L_0^1(\mathbb{R}^{\nu N} \times (\mathbb{R}^{\nu N} \backslash \mathbb{W}_N))$ is differentiable.

b) the expressions $\{\partial(S^N(t) f_N)(x)/\partial Q_k(t,x); \partial(S^N(t) f_N)(x)/\partial P_k(t,x)\}_{k=1}^N$ are equal to zero in the neighborhood of the collision instants,

c) the functions $\{Q_k(t,x), P_k(t,x)\}_{k=1}^N$ are differentiable with respect to x for any time except collision instants.

If $x \in \partial(\mathbb{R}^{\nu N} \times \mathbb{R}^{\nu N} \backslash \mathbb{W}_N)$, for example, $|q_i - q_j| = a$ for some pair $(i,j) : i \neq j \in (1, \dots, N)$ then in expression (1.3.4.b) at $\Delta t > 0$ and $\langle (q_i - q_j), (p_i - p_j) \rangle < 0$ we must replace the momenta p_i, p_j by p_i^*, p_j^* according to the relations (1.2.5). At $\Delta t < 0$ this replacement is made in $\{f_N, H_N\}$ if $\langle (q_i - q_j, (p_i - p_j)) \rangle > 0$. These conditions will be called the boundary conditions for the Poisson bracket. The functions $\left(S^N(t) f_N\right)(x)$, where $f_N \in L^1(\mathbb{R}^{\nu N} \times \mathbb{R}^{\nu N} \backslash \mathbb{W}_N)$, are well defined also on surfaces

$$|q_i - q_j| = a, (i,j) : i \neq j \in (1, \dots, N) ,$$

because the trajectory $X(t,x)$ exists for such initial data (see Remark 1).

The two expressions (1.3.4 a) and (1.3.4 b) obtained for the derivative of the function $\left(S^N(t) f_N\right)(x)$ are obviously identical in virtue of the group property of $S^N(t)$ and the expression (1.3.4.a) is continuous in $x \in (\mathbb{R}^{\nu N} \times (\mathbb{R}^{\nu N} \backslash \mathbb{W}_N)) \backslash \mathcal{M}_N^0$ and $t \in \mathbb{R}^1$, and thus the same holds for (1.3.4.b).

Let us prove now that the function $(S^N(t)f_N)(x)$ is differentiable in the norm of the space $L^1(\mathbb{R}^{\nu N} \times (\mathbb{R}^{\nu N}\backslash\mathbb{W}_N))$. For this purpose we show that an operator $\{H_N, .\}$ given by (1.2.15) is defined on $L_0^1(\mathbb{R}^{\nu N} \times (\mathbb{R}^{\nu N}\backslash\mathbb{W}_N))$. Every function $f_N \in L_0^1(\mathbb{R}^{\nu N} \times (\mathbb{R}^{\nu N}\backslash\mathbb{W}_N))$ is continuously differentiable and has compact support and, moreover, $\Phi \in C^2([a, R])$; therefore $\{H_N, f_N\}$ is a continuous function with compact support and is zero in an ε–neighbourhood of the set $\mathbb{W}_N$. Hence $\{H_N, f_N\}$ is summable. In other words, $L_0^1(\mathbb{R}^{\nu N} \times (\mathbb{R}^{\nu N}\backslash\mathbb{W}_N)) \subset \mathcal{D}(\mathcal{H}_N)$, where $\mathcal{D}(\mathcal{H}_N)$ is the domain of definition of $\mathcal{H}_N$.

For $f_N \in L_0^1(\mathbb{R}^{\nu N} \times (\mathbb{R}^{\nu N}\backslash\mathbb{W}_N))$ we then have

$$\left\| \frac{1}{\Delta t}\big(S^N(t+\Delta t)f_N - S^N(t)f_N\big) - S^N(t)\{f_N, H_N\} \right\|$$
$$= \left\| \frac{1}{\Delta t}\big(S^N(\Delta t)f_N - f_N\big) - \{f_N, H_N\} \right\|$$
$$= \int dx \left| \frac{1}{\Delta t}\big(f_N(X(\Delta t, x)) - f_N(x)\big) - \big(\{f_N, H_N\}\big)(x) \right| .$$

According to (1.2.16), the integrand in this expression converges to zero as $\Delta t \to 0$ uniformly in x on an arbitrary compact outside the forbidden configurations. Therefore, in this expression we can pass to the limit as $\Delta t \to 0$ under the integral sign and this means that the function $(S^N(t)f_N)(x)$ is differentiable in the norm $L^1(\mathbb{R}^{\nu N} \times (\mathbb{R}^{\nu N}\backslash\mathbb{W}_N))$. Hence, the infinitesimal generator $\mathcal{H}_N$ of the group $S^N(t)$ coincides on $\mathcal{D}(\mathcal{H}_N)$ with the Poisson bracket with boundary conditions on $\partial(\mathbb{R}^{\nu N}\backslash\mathbb{W}_N)$, and the following equalities hold

$$\lim_{\Delta t \to 0} \frac{1}{\Delta t}\big(S^N(t+\Delta t)f_N - S^N(t)f_N\big) = S^N(t)\mathcal{H}_N f_N = \mathcal{H}_N S^N(t)f_N.$$

Note too, that functions $(S^N(t)f_N)(x)$ for $f_N \in L_0^1(\mathbb{R}^{\nu N} \times (\mathbb{R}^{\nu N}\backslash\mathbb{W}_N))$ are continuously differentiable with respect to t and $x \in (\mathbb{R}^{\nu N} \times (\mathbb{R}^{\nu N}\backslash\mathbb{W}_N))\backslash\mathcal{M}_N^0$, and satisfy the following condition

$$f_N(X(t,x)) = f_N(X(t,x^*)) \tag{1.3.5}$$

if $x \in \partial(\mathbb{R}^{\nu N}\backslash\mathbb{W}_N) \times \mathbb{R}^{\nu N}$, according to (1.2.11).

Remark 5. It was shown above in Theorem 1.3.1 that the infinitesimal operator $\mathcal{H}_N$ of group $S_N(t)$ has different boundary conditions for $t > 0$ and $t < 0$.

This means that group $S_N(t)$, $-\infty < t < \infty$ in fact consists of two semigroups **1**)$S_N(t) = S_N^+(t), 0 \leqslant t < \infty$ and **2**)$S_N(t) = S_N^-(t), -\infty < t \leqslant 0$ with different infinitesimal operators. This means that the infinitesimal generator of $S_N^+(t)$ must be calculated only for $t > 0$, $\Delta t > 0$ and that of $S_N^-(t)$ only for $t < 0$, $\Delta t < 0$.

Theorem 1.3.1 remains valid for systems of particles contained in the bounded domain $\Lambda \subset \mathbb{R}^\nu$. In this case we must use the Poisson bracket with the Hamiltonian H_N^Λ. In fact, the discussion performed above was true for functions defined on compact sets in phase space. And now we add to the infinitesimal operator the terms connected with the potential $u^\Lambda(q)$ of the external field and consider only the bounded domain Λ_δ of configuration space. In the case when the particles can elastically reflect from $\partial\Lambda$, we must use the corresponding boundary conditions on $\partial\Lambda$ (1.3.2') for the Poisson bracket.

Remark 6. The operator $\mathcal{H}_N$, as infinitesimal generator of the group $S^N(t)$ of isometric operators, is closed on $\mathcal{D}(\mathcal{H}_N)$. It acts differently at $t > 0$ and $t < 0$ and we have already proved that on the set $L_0^1(\mathbb{R}^{\nu N} \times (\mathbb{R}^{\nu N} \backslash \mathbb{W}_N))$ it is given by the Poisson bracket with boundary conditions on $\partial(\mathbb{R}^{\nu N} \backslash \mathbb{W}_N)$. The question how to define the domain $\mathcal{D}(\mathcal{H}_M) \subset L^1(\mathbb{R}^{\nu N} \times (\mathbb{R}^{\nu N} \backslash \mathbb{W}_N))$ was discussed in the important paper [Kot1] (see Appendix I). Formally, the operator $\mathcal{H}_N$ at $t > 0$ acts as the pseudo–Liouville operator [BB][DE][EDHL]

$$(\mathcal{H}_N f_N)(x) = (\{f_N, H_N\})(x) + a^2 \sum_{1=i<j}^{N} \int_{\mathbb{R}^3 \times \mathbb{S}_+^2} d\eta \, \langle \eta, (p_i - p_j) \rangle \, \delta(q_j - q_i - a\eta)$$

$$\big[f_N(x_1, \ldots, q_i, p_i - \eta \langle \eta, (p_i - p_j) \rangle, \ldots, q_j, p_j + \eta \langle \eta, (p_i - p_j) \rangle, \ldots, x_N) \\ - f_N(x_1, \ldots x_N)\big],$$

where

$$\mathbb{S}_+^2 = \{\eta \in \mathbb{R}^3 \big| |\eta| = 1, \langle \eta, (p_i - p_j) \rangle > 0\}.$$

However, in all our further discussions we shall use $\mathcal{H}_N$ only on the set $L_0^1(\mathbb{R}^{\nu N} \times (\mathbb{R}^{\nu N} \backslash \mathbb{W}_N)) \subset \mathcal{D}(\mathcal{H}_N)$, where it is completely defined and coincides with the Poisson bracket with boundary conditions on $\partial(\mathbb{R}^{\nu N} \backslash \mathbb{W}_N)$. (see Appendix II for details).

1.3.3 Initial value problem for the Liouville equation The evolution of all possible states of classical systems of a finite number of particles is described by the Liouville equation [Ba][Bo][PGM2]. For the system under consideration the Liouville equation is the following evolution equation

$$\frac{d}{dt}f_N(t) = -\mathcal{H}_N f_N(t), \tag{1.3.6}$$

where $(-\mathcal{H}_N)$ is the infinitesimal generator of the group $S^N(-t)$ (1.3.1) (defined by Theorem 1.3.1). Let us stress that we shall use the following boundary conditions for the operator $(-\mathcal{H}_N)$: if $\langle (q_i - q_j), (p_i - p_j)\rangle > 0$, $t > 0$, then p_i, p_j are replaced by p_i^*, p_j^* and for $t < 0$ this replacement should be made if $\langle (q_i - q_j), (p_i - p_j)\rangle < 0$. In what follows these replacement rules will be called the boundary conditions for the Liouville equation.

The solutions of Eq (1.3.6) have the following meaning. The expression

$$\Big(\int dx_1 \dots dx_N f_N(t, x_1, \dots, x_N)\Big)^{-1} f_N(t, x_1, \dots, x_N) dx_1 \dots dx_N,$$

where $\big(\int dx f_N(t,x)\big)^{-1}$ is a normalization factor, is the probability of finding positions and momenta of the 1–st,..., N–th particles at the time t in the phase volumes $dx_1, \dots, dx_N$ with center in the phase points $x_1, \dots, x_N$, respectively.

Let us consider the initial value problem

$$f_N(0) = f_N^0 \in L^1(\mathbb{R}^{\nu N} \times (\mathbb{R}^{\nu N} \backslash \mathbb{W}_N)) \tag{1.3.6'}$$

for the Liouville equation in the Banach space $L^1(\mathbb{R}^{\nu N} \times (\mathbb{R}^{\nu N} \backslash \mathbb{W}_N))$.

Theorem 1.3.2. *The Cauchy problem (1.3.6) for the Liouville equation has a unique solution in the space of summable functions* $L^1(\mathbb{R}^{\nu N} \times (\mathbb{R}^{\nu N} \backslash \mathbb{W}_N))$ *and it is given by the formula*

$$f_N(t) = S^N(-t) f_N^0, \tag{1.3.7}$$

where $S^N(-t)$ *is the evolution operator (1.3.1). This solution is strong for initial data* $f_N^0 \in L_0^1(\mathbb{R}^{\nu N} \times (\mathbb{R}^{\nu N} \backslash \mathbb{W}_N)) \subset L^1(\mathbb{R}^{\nu N} \times (\mathbb{R}^{\nu N} \backslash \mathbb{W}_N))$ *and weak (generalized) for arbitrary initial data from* $L^1(\mathbb{R}^{\nu N} \times (\mathbb{R}^{\nu N} \backslash \mathbb{W}_N))$.

Proof. According to well–known theorems of functional analysis [Ka][Kr][RS][Yo], the Cauchy problem (1.3.6) has a unique solution in $L^1(\mathbb{R}^{\nu N} \times (\mathbb{R}^{\nu N} \backslash \mathbb{W}_N))$ given by formula (1.3.7). We have already proved in theorem 1.3 that for $f_N^0 \in L_0^1(\mathbb{R}^{\nu N} \times (\mathbb{R}^{\nu N} \backslash \mathbb{W}_N))$ this solution is strong.

Let us show that (1.3.7) represents, in general, a weak solution of the Cauchy problem (1.3.6). For this purpose we introduce a functional

$$(\varphi, f_N(t)) = \int dx \varphi(x) f_N(t, x),$$

where $\varphi(x)$ is continuously differentiable functions with compact support, equal to zero in a certain ε–neighbourhood of the set $\mathbb{W}_N$. As φ is bounded and $f_N(t)$ is summable, this functional exists.

Using the formula (1.3.7) and the isometry property of the operator $S^N(t)$, we can transform the functional $\big(\varphi, f_N(t)\big)$ as follows

$$\big(\varphi, f_N(t)\big) = \big(\varphi, S^N(-t) f_N^0\big) = \big(S^N(t)\varphi, f_N^0\big).$$

In Theorem 1.3.1 it was shown that the function $\big(S^N(t)\varphi\big)(x)$ is differentiable with respect to t, and, moreover, the expression

$$\left| \frac{1}{\Delta t}\big(S^N(t+\Delta t)\varphi - S^N(t)\varphi\big) - \{S^N(t)\varphi, H_N\} \right| \to 0$$

as $\Delta t \to 0$ uniformly in x on arbitrary compact sets outside $\mathbb{W}_N \cup \mathcal{M}_N^0$. Since the function $\big(\{S^N(t)\varphi, H_N\}\big)(x)$ is bounded the function $\big(\{S^N(t)\varphi, H_N\} f_N^0\big)(x)$ is summable and the functional $\big(\{S^N(t)\varphi, H_N\}, f_N^0\big)$ exists.

Thus we can differentiate the functional $\big(S^N(t)\varphi, f_N^0\big)$ with respect to t. In consequence we obtain

$$\begin{aligned} \frac{d}{dt}\big(\varphi, f_N(t)\big) &= \big(\{S^N(t)\varphi, H_N\}, f_N^0\big) = \big(S^N(t)\{\varphi, H_N\}, f_N^0\big) \\ &= \big(\{\varphi, H_N\}, S^N(-t) f_N^0\big) = \big(\{\varphi, H_N\}, f_N(t,x)\big). \end{aligned}$$

It follows from these equalities that the functions $f_N(t)$ defined by the formula (1.3.7) is a weak solution of the Cauchy problem (1.3.6) for the Liouville equation.

This theorem is also valid for functions $f_{\Lambda,N} \in L^1(\mathbb{R}^{\nu N} \times (\Lambda^N \backslash \mathbb{W}_N))$, provided that $S^N(t)$ and H_N are replaced by $S_\Lambda^N(t)$ and H_N^Λ respectively.

Remark 7. The equations of the characteristic curves for the Liouville equation (1.3.6) in the time intervals between collisions coincide with the Hamilton Eqs. (1.2.2). Hence, according to well–known theorems for partial differential equations, a solution of the Cauchy problem (1.3.6) exists in the class continuously differentiable functions that are zero in some neighbourhood of the set $\mathbb{W}_N$. This solution is unique and in the time intervals between collisions has the form

$$f_N(t,x) = f_N^0(X(-t,x)),$$

where $X(-t,x)$ is a solution of the equations of for the characteristics (1.2.2). Taking into consideration the boundary conditions on $\partial(\mathbb{R}^{\nu N}\backslash\mathbb{W}_N)$, we arrive at formula (1.3.7).

Remark 8. Let us consider the connection between the Hamilton dynamics and the Liouville equation. As is well-known [BB][PGM2], the class of a pure states of the Hamilton systems is described by the distribution functions

$$D_N(t,x_1,\ldots,x_N) = \frac{1}{N!}\sum_{\{i_1,\ldots,i_N\}}\prod_{k=1}^{N}\delta(x_{i_k} - X_k(t,x^0)),$$

where $X(t,x^0)$ is the phase point at which the system of N particles is situated at time t if it was at the phase point $x^0 \in (\mathbb{R}^{\nu N}\times(\mathbb{R}^{\nu N}\backslash\mathbb{W}_N))\backslash\mathcal{M}_N^0$ at $t=0$, and the summation is carried out over arbitrary permutations $\{i_1,\ldots,i_n\}$ of the set $\{1,\ldots,N\}$. All possible states are described by a superposition of pure states

$$D_N(t,x_1,\ldots,x_N) = \int dx^0 D_N^0(x^0)\frac{1}{N!}\sum_{\{i_1,\ldots,i_N\}}\prod_{k=1}^{N}\delta(x_{i_k} - X_k(t,x^0)), \tag{1.3.8}$$

where $D_N^0(x^0)$ is determined by the initial state

$$D_N(0,x_1,\ldots,x_N) = \int dx^0 D_N^0(x^0)\frac{1}{N!}\sum_{\{i_1,\ldots,i_N\}}\prod_{k=1}^{N}\delta(x_{i_k} - x_k^0).$$

Performing the integration in this expression, we obtain

$$D_N(0,x_1,\ldots,x_N) = D_N^0(x_1,\ldots,x_N),$$

i.e. the initial state is given by the probability distribution function $D_N^0(x)$.

Changing the variables in the integrand of (1.3.8) by $x^0 \to X(-t,x)$ and taking into account Liouville's theorem, we obtain

$$D_N(t,x_1,\ldots,x_N) = D_N(0,X_1(-t,x),\ldots,X_N(-t,x)). \tag{1.3.9}$$

Let us stress that the following relation should be valid at arbitrary time

$$D_N(t,x) = D_N(t,x^*), \qquad \text{if } x \in \partial(\mathbb{R}^{\nu N} \backslash \mathbb{W}_N) \times \mathbb{R}^{\nu N} \tag{1.3.10}$$

and $D_N(t,x) = 0$, if $x \in \mathbb{R}^{\nu N} \times \mathbb{W}_N$; $D_N(t,x) \geqslant 0$, otherwise. Clearly, the condition that the initial distribution function $D_N(0,x)$ be in $L^1_0(\mathbb{R}^{\nu N} \times (\mathbb{R}^{\nu N} \backslash \mathbb{W}_N))$ and non–negative provides the essential requirements on the distribution function.

To obtain the equation that gives the evolution law for a distribution function $D_N(t,x)$ we differentiate both sides of (1.3.9) with respect to time. For the functions $D_N(0) \in L^1_0(\mathbb{R}^{\nu N} \times (\mathbb{R}^{\nu N} \backslash \mathbb{W}_N))$ from Theorem 1.3.1 we have

$$\frac{\partial}{\partial t} D_N(t,x) = \{H_N(x), D_N(t,x)\} \tag{1.3.11}$$

with the corresponding boundary conditions on $\partial(\mathbb{R}^{\nu N} \backslash \mathbb{W}_N)$. This relation is obtained as a consequence of (1.3.9) but we can consider it as an equation for an unknown function $D_N(t,x)$. It is called the Liouville equation.

1.3.4 Statistical ensembles Let us now consider a system with a nonfixed (random) finite number of particles. Its state is described by an infinite sequence $D(t) = \{D_0, D_1(t,x_1), \ldots, D_n(t,x_1,\ldots,x_n), \ldots\}$ of distribution functions defined on the phase spaces of $0, 1, \ldots, n, \ldots$ particles respectively. The average number $\langle N \rangle$ of particles in the system is determined by

$$\langle N \rangle = \left(\sum_{n=0}^{\infty} \frac{1}{n!} \int d(x)_n D_n(t,(x)_n) \right)^{-1} \sum_{n=1}^{\infty} \frac{1}{(n-1)!} \int d(x)_n D_n(t,(x)_n)$$

where

$$(x)_n = (x_1 \ldots, x_n)\ ; \qquad d(x)_n = dx_1 \ldots dx_n$$

and is assumed to be finite. Here, conventionally, $D_0 = 1$ and the n-tuple integral reduces to the integrand for $n = 0$

The evolution of the state of a system with a random number of particles is determined as a solution of the Cauchy problem for an infinite sequence of independent Liouville equations

$$\frac{\partial}{\partial t} D_n(t) = -\mathcal{H}_n D_n(t),$$
$$D_n(0) = D_n^0 \ , \quad n = 1, 2 \ldots \tag{1.3.12}$$

The solution of this initial–value problem is given by the following formula:

$$D(t) = S(-t)D(0), \tag{1.3.13}$$

where $S(-t)$ is a direct sum of n–particle evolution operators (1.3.1), which is defined as follows.

Let us denote by $L^1 = \bigoplus_{n=0}^{\infty} L^1(\mathbb{R}^{\nu N} \times (\mathbb{R}^{\nu N} \backslash \mathbb{W}_N))$ the Banach space of the infinite sequences of functions $f = \{f_n(x_1, \dots, x_n)\}_{n \geqslant 0}$ $(f_0 \in C)$ which is a direct sum of spaces $L^1(\mathbb{R}^{\nu N} \times (\mathbb{R}^{\nu N} \backslash \mathbb{W}_N))$ with the norm

$$\|f\| = \sum_{n=0}^{\infty} \int d(x)_n \, |f_n((x)_n)| \quad .$$

Let us introduce, in this space, a direct sum of the evolution operators (1.3.1)

$$S(t) = \sum_{n=0}^{\infty} \oplus S^n(t), \quad S(0) = I.$$

$$\left(S(t)f\right)_n (x_1, \dots, x_n) = \left(S^n(t)f\right)(x_1, \dots, x_n) \, . \tag{1.3.14}$$

First we define an operator $S(t)$ in L^1 on the everywhere dense set $L_0^1 \subset L^1$ of finite sequences of functions $f_n \in L_0^1(\mathbb{R}^{\nu N} \times (\mathbb{R}^{\nu N} \backslash \mathbb{W}_N))$ ($f_n = 0$ for $n \geqslant n_0$, where n_0 is some finite number). On this set $\|S(t)\| = 1$ (Theorem 1.3.1). Then the operator $S(t)$ is extended from $L_0^1 \subset L^1$ to the whole space L^1 by continuity.

The operator (1.3.14) possesses the group property. Moreover, it is strongly continuous as a direct sum of strongly continuous isometric operators (1.3.1) and is strongly differentiable on the set $L_0^1 \subset L^1$. The infinitesimal generator $\mathcal{H}$ of the group $S(t)$ is defined as the direct sum of infinitesimal generators $\mathcal{H}_n$ (see Theorem 1.3.1)

$$(\mathcal{H}f)_n (x_1, \dots, x_n) = (\mathcal{H}_n f_n)(x_1, \dots, x_n) \, . \tag{1.3.15}$$

The sequence of Eqs. (1.3.12) can be considered as an abstract evolution equation in the space L^1, the operator in the right hand side being given by the infinitesimal generator $\mathcal{H}$ of the group $S(t)$. According to well-known theorems of functional analysis [Ka][Kr][RS][Yo] the Cauchy problem for that equation has a unique solution given by (1.3.13) in the space L^1 and $D(t) \in L^1$ if $D(0) \in L^1$. This solution is strong for $D(0) \in L_0^1$ and weak for arbitrary initial data in L^1.

It should be noted here that functions in L^1 describe the states of systems with a finite number of particles.

The above considerations concerning the operator $S(t)$ are also valid for the evolution operators $S_\Lambda(t)$ of a system of particles moving in a bounded domain $\Lambda \subset \mathbb{R}^\nu$. This operator acts in the space $L^1(\Lambda) = \bigoplus_{n=0}^{\infty} L^1(\mathbb{R}^{\nu N} \times (\Lambda^N \backslash \mathbb{W}_N))$.

In the following we shall consider only the more general case of systems of a random number of particles (non–equilibrium grand canonical ensembles in the terminology of statistical mechanics). All the constructions realized for this case can be easily extended to the case of the systems with a fixed number N of particles (non–equilibrium canonical ensembles), as their states are described by sequences from L^1 with only one non–zero component : $D^{(N)}(t) = \{0, \ldots, 0, D_N(t, x_1, \ldots, x_N), 0, \ldots\}$.

1.4 The Derivation of the BBGKY Hierarchy

1.4.1 The distribution functions

Let us consider now the connection between the Liouville equation and the BBGKY hierarchy. As was already mentioned, for systems with a finite number of particles these equations are equivalent and describe the evolution of all possible states of classical system of particles. It is important to note that the BBGKY hierarchy can also be used for the description of states of systems with infinitely many particles.

It was shown in the previous section that the state of a finite system of particles can be described by the solution $D(t)$ (1.3.13) of the Cauchy problem for the Liouville equations. There is an equivalent way of describing the state of such system by means of reduced distribution functions. In order to introduce these functions rigorously from a mathematical point of view let us give some definitions.

We define the following operator in the Banach space L^1 of sequences of summable functions (in a certain sense the analogue of the annihilation operator in quantum field theory):

$$\left(\int dx f\right)_n (x_1, \ldots, x_n) = \int dx_{n+1} f_{n+1}(x_1, \ldots, x_n, x_{n+1}). \tag{1.4.1}$$

This operator is defined and bounded in L^1, $\|\int dx\| = 1$. Consequently the operators $e^{\pm \int dx}$ are also defined in the space L^1 and can be represented by a series convergent in norm:

$$e^{\pm \int dx} = \sum_{n=0}^{\infty} \frac{1}{n!} \left(\pm \int dx \right)^n, \left\| e^{\pm \int dx} \right\| \leqslant e \, .$$

In particular, we have $e^{\int dx} e^{-\int dx} = I$, where I is unit operator.

We define now the sequence of reduced distribution functions $F(t) = \{F_n(t, x_1, \ldots, x_n)\}_{n \geqslant 0}$ by using the operator $\int dx$ in the following formula

$$F(t) = \left(e^{\int dx} D(t) \right)_0^{-1} e^{\int dx} D(t), \tag{1.4.2}$$

where $\left(e^{\int dx} D(t) \right)_0$ is the normalization factor (the grand partition function) and $D(t)$ is a solution of the Cauchy problem for the Liouville equation.

The sequence (1.4.2) exists and $F(t) \in L^1$ if $D(t) \in L^1$, $\left(e^{\int dx} D(t) \right)_0 \neq 0$, and the following estimate holds

$$\|F(t)\| \leqslant \left| \left(e^{\int dx} D(t) \right)_0^{-1} \right| \|D(t)\|.$$

Obviously the condition $\left(e^{\int dx} D(t) \right)_0 \neq 0$ is fulfilled by the distribution functions, because $D_s(t, x_1, \ldots, x_s) \geqslant 0, s \geqslant 1$, $D_0 = 1$. In what follows we shall suppose that $D(0) \in L^1$, and thus $D(t) \in L^1$.

From Liouville's theorem it follows that

$$\left(e^{\int dx} D(t) \right)_0 = \left(e^{\int dx} D(0) \right)_0 = \sum_{n=0}^{\infty} \frac{1}{n!} \int d(x)_n D_n(0, x_1, \ldots, x_n) \quad .$$

Then, taking this and the formula (1.3.13) into account, we obtain from (1.4.2)

$$F(t) = \left(e^{\int dx} D(0) \right)_0^{-1} e^{\int dx} S(-t) D(0) \tag{1.4.3}$$

or, componentwise,

$$F_s(t, x_1, \ldots, x_s) = \left(e^{\int dx} D(0) \right)_0^{-1} \sum_{n=0}^{\infty} \frac{1}{n!} \int dx_{s+1} \ldots dx_{s+n} \left(S^{s+n}(-t) D_{s+n}(0) \right) (x_1, \ldots, x_{s+n}) \, . \tag{1.4.3'}$$

The functions $D_n(0, x_1, \ldots. x_n)$ and $D_n(t, x_1, \ldots, x_n)$ are equal to zero on forbidden configurations. Obviously the functions $F_s(t, x_1, \ldots, x_s), s \geqslant 2$ (1.4.3') are also identically zero on these configurations. i.e. if $(x_1, \ldots, x_s) \in \mathbb{R}^{\nu s} \times \mathbb{W}_s$, and are defined almost everywhere on phase space together with $S_{s+n}(-t)D_{s+n}(0, x_1, \ldots, x_{s+n})$. From the boundary conditions (1.3.10) for $S(-t)D(0)$ it follows that the functions (1.4.3′) also satisfy the following boundary conditions

$$F_s(t, x_1, \ldots, x_s) = F_s(t, x_1, \ldots. x_i^*, \ldots, x_j^*, \ldots, x_s) \tag{1.4.4}$$

if $|q_i - q_j| = a$ for some pair $(i, j) : j \neq i \in (1, \ldots, s)$. It follows from the boundary conditions for the Liouville equation that for $t > 0$, $\langle (q_i - q_j), (p_i - p_j) \rangle > 0$, and for $t < 0$, $\langle (q_i - q_j), (p_i - p_j) \rangle < 0$, the function $F_s(t, x_1, \ldots, x_i^*, \ldots, x_j^*, \ldots, x_s)$ should be used.

For a system with a fixed N number of particles, the sequence $F(t)$ given by (1.4.2) is a finite one, i. e. $F(t) = \{F_1(t, x_1), \ldots, F_N(t, x_1, \ldots, x_N), 0, \ldots\}$, where it is clear that the last function (not vanishing identically) $F_N(t, x_1, \ldots x_N)$ coincides with $N! D_N(t, x_1, \ldots, x_N) \left(\int dx_1 \ldots dx_N D_N(t, x_1, \ldots, x_N)\right)^{-1}$.

The distribution functions (1.4.3′) are symmetrical with respect to arbitrary permutations of $x_1, \ldots, x_s$ and $F_s(t, x_1, \ldots, x_s) dx_1 \ldots dx_s$ is proportional to the probability of finding positions and momenta of the 1–st, ..., s–th particles at the time t in phase volumes $dx_1, \ldots dx_s$ with their centers at the phase points $x_1, \ldots x_s$, respectively, independently of the phase state of the other particles.

The formulae (1.4.3) express the sequence $F(t)$ in terms of the sequence $D(0)$. We want to express $F(t)$ in terms of the sequence $F(0)$. In order to do this we notice that the initial reduced distribution functions are defined as follows

$$F(0) = \left(e^{\int dx} D(0)\right)_0^{-1} e^{\int dx} D(0),$$

from which we can express $D(0)$ in terms of $F(0)$:

$$D(0) = \left(e^{\int dx} D(0)\right)_0 e^{-\int dx} F(0) \quad .$$

Substituting the last formula into (1.4.3), we obtain

$$F(t) = e^{\int dx} S(-t) e^{-\int dx} F(0) \quad , \tag{1.4.5}$$

or, componentwise:

$$F_s(t, x_1, \dots, x_s) = \sum_{n=0}^{\infty}\sum_{k=0}^{n} \frac{(-1)^k}{k!(n-k)!} \int dx_{s+1} \dots dx_{s+n} \times$$
$$\times\, S^{s+n-k}(-t, x_1, \dots, x_s, x_{s+1}, \dots, x_{s+n-k}) \times$$
$$\times\, F_{s+n}(0, x_1, \dots, x_s, x_{s+1}, \dots, x_{s+n-k}, \dots, x_{s+n}) \quad . \tag{1.4.5'}$$

This formula expresses the sequence $F(t)$ at the instant t in terms of the initial data $F(0)$. In what follows we shall prove rigorously that the functional series, determined by (1.4.5), converges and is the solution of the initial value problem for the BBGKY hierarchy if $F(0) \in L^1$.

1.4.2 The BBGKY hierarchy of equations

To derive the equations that give the evolution law for the sequence of distribution functions $F(t)$ (1.4.3) we differentiate both sides of (1.4.3) with respect to time. Let $D(0) \in L_0^1 \subset L^1$, then the functions $d/dt S^{s+n}(-t)D_{s+n}(0)$, $n = 1, 2, \dots$ depend continuously on $x_1, \dots x_s, x_{s+1}, \dots x_{s+n}$ almost everywhere and are absolutely integrable. Thus we can change in (1.4.3') the order of differentiation with respect to time and integration over $x_{s+1}, \dots, x_{s+n}$. For the same reasons we can take out of the integral sign the derivatives with respect to $x_1, \dots, x_s$. Moreover, all integrations in (1.4.3') are carried out at first over the domain, where $|q_i - q_j| > a + \varepsilon, i \neq j \in (1, \dots s+n)$, and in the final expression the parameter ε tends to zero.

Taking Theorem 1.3.1 into account we obtain

$$\begin{aligned}\frac{\partial}{\partial t}F(t) &= -\left(e^{\int dx} D(0)\right)_0^{-1} e^{\int dx}\mathcal{H}S(-t)D(0)\\ &= -\mathcal{H}F(t) - \left(e^{\int dx} D(0)\right)_0^{-1} \left[e^{\int dx}, \mathcal{H}\right] S(-t)D(0),\end{aligned} \tag{1.4.6}$$

where the bracket $[A, B]$ is the commutator of the operators A and B, $\mathcal{H}$ is the infinitesimal generator (1.3.15) of the group $S(t)$.

Using the following operator equality (to be proved in Chapter II)

$$\left[e^{\int dx}, \mathcal{H}\right] = \left[\int dx, \mathcal{H}\right] e^{\int dx},$$

we finally obtain [PGM2] from the identity (1.4.6), according to definition (1.4.3), the following formula

$$\frac{d}{dt}F(t) = -\mathcal{H}F(t) + \left[\mathcal{H}, \int dx\right] F(t). \tag{1.4.7}$$

or, componentwise:

$$\frac{\partial}{\partial t}F_s(t) = -\mathcal{H}_s F(t, x_1, \ldots, x_s) + \left([\mathcal{H}, \int dx]F(t)\right)_s (x_1, \ldots, x_s) \quad . \tag{1.4.7$'$}$$

We postpone the justification of this calculation to the next Chapter. This relation results in fact from an identity that connects the time derivative of the sequence $F(t)$ with respect of action of the operator $-\mathcal{H} + [\mathcal{H}, \int dx]$ upon the same sequence $F(t)$. This relation follows from the definition (1.4.3).

From another point of view, one can consider the identity (1.4.7) as the abstract equation for the sequence $F(t)$. Such equation (the chain of equations) is called the BBGKY hierarchy.

Let us determine the componentwise form the BBGKY hierarchy (1.4.7) in the case of a system of particles with hard–core potential (1.2.3,4). For this purpose we must calculate in L^1 the commutator of $\mathcal{H}$ and $\int dx$. In the next Chapter we shall prove that at $t > 0$ the following equality holds

$$\left(\left[\mathcal{H}, \int dx\right] F(t)\right)_s (x_1, \ldots, x_s) = a^{\nu-1} \sum_{i=1}^{s} \int_{\mathbb{R}^\nu \times \mathbb{S}_+^{\nu-1}} d\eta dp_{s+1} \langle \eta, (p_i - p_{s+1})\rangle$$

$$\times \Big(F_{s+1}(t; x_1, \ldots; q_i, p_i - \eta\langle\eta, (p_i - p_{s+1})\rangle; \ldots . x_s; q_i - a\eta, p_{s+1} + \eta\langle\eta, (p_i - p_{s+1})\rangle)$$

$$-F_{s+1}(t, x_1, \ldots, x_s; q_i + a\eta, p_{s+1})\Big) +$$

$$\sum_{i=1}^{s} \int dx_{s+1} \{\Phi(q_i - q_{s+1}), F_{s+1}(t, x_1, \ldots . x_{s+1})\}$$

$$\equiv I_1 + I_2. \tag{1.4.8}$$

where $\{.,.\}$ is the Poisson bracket (1.2.15) and $\mathbb{S}_+^{\nu-1} = \{\eta \in \mathbb{R}^\nu \,|\, |\eta| = 1, \langle\eta, (p_i - p_{s+1})\rangle > 0\}$. For the case $t < 0$ this equality should be modified correspondingly

$$\left(\left[\mathcal{H}, \int dx\right] F(t)\right)_s (x_1, \ldots, x_s) = a^{\nu-1} \sum_{i=1}^{s} \int_{\mathbb{R}^\nu \times \mathbb{S}_+^{\nu-1}} d\eta dp_{s+1} \langle \eta, (p_i - p_{s+1})\rangle$$

$$\times\Big(F_{s+1}(t,x_1,\dots,x_s;q_i-a\eta,p_{s+1})-F_{s+1}(t;x_1,\dots;q_i,p_i-\eta\langle\eta,(p_i-p_{s+1})\rangle;\dots,x_s;$$

$$q_i+a\eta,p_{s+1}+\eta\langle\eta,(p_i-p_{s+1})\rangle)\Big)\Big)+\sum_{i=1}^{s}\int dx_{s+1}\{\Phi(q_i-q_{s+1}),F_{s+1}(t,x_1,\dots,x_{s+1})\}. \tag{1.4.9}$$

Thus, the formula (1.4.7) and (1.4.8,9) give the explicit form of the BBGKY hierarchy. In particular, for systems of hard sphere particles the BBGKY hierarchy is determined by the expression I_1 and for interaction potentials without hard core by I_2.

To be precise, let us note that it is not clear whether the lower–dimensional integration in expression I_1 (1.4.8,9) has a meaning, because the functions $F_{s+n}(t,x_1,\dots,x_{s+n})$ are defined almost everywhere in the phase space. In Chapter 2 this problem will be considered in a more general situation. It will be proved that the integrals in (1.4.8), (1.4.9) are correctly defined.

It should be noted here that BBGKY hierarchy was defined above for finite systems of particles, i.e. systems for which the average number of particles

$$\langle N\rangle=\left(e^{\int dx}NF(t)\right)_0,$$

$$\langle N\rangle=\int dxF_1(t,x) \tag{1.4.10}$$

is finite. Actually, for $F_1(t)\in L^1$ we have $\langle N\rangle<\infty$.

In the case of a system of a fixed number N of particles, the BBGKY hierarchy consists of a system of a finite number of equations where the last equation for the function $F_N(t)$ coincides with the Liouville Eq. (1.3.6).

For a system of particles in a bounded domain $\Lambda\subset\mathbb{R}^\nu$ the sequence of reduced distribution functions $F_\Lambda(t)$ is defined similarly to (1.4.2) by the corresponding operator $\int\limits_\Lambda dx$ (acting in $L^1(\Lambda)$). The system of BBGKY equations in this case has the form

$$\frac{d}{dt}F_\Lambda(t)=-\mathcal{H}_\Lambda F_\Lambda(t)+\left[\mathcal{H}_\Lambda,\int\limits_\Lambda dx\right]F_\Lambda(t). \tag{1.4.11}$$

where $\mathcal{H}_\Lambda$ is the infinitesimal generator of the group $S_\Lambda(t)$.

As mentioned above, the BBGKY hierarchy can be used also for the description of systems of infinitely many particles. We notice, that the aim of statistical mechanics is just to construct the states of infinite systems. Formally the BBGKY hierarchy for these systems has the same form as for finite systems (1.4.7), (1.4.8-9) and it can be obtained rigorously as a result of a certain procedure called the thermodynamic limit, according to which $\Lambda \nearrow \mathbb{R}^{\nu}$, $\langle N\rangle = \int dx F_{\Lambda,1}(t,x) \to \infty$, $\langle N\rangle / |\Lambda| = \text{const} < \infty$. In other words, the domain $\Lambda =$ tends to all $\mathbb{R}^{\nu}$ and the volume of the system tends to infinity whereas the number of particles tends to infinity with a fixed density. For the sake of simplicity we shall suppose in what follows that the domain Λ is a ball of radius r, and $\Lambda \nearrow \mathbb{R}^{\nu}$ means that $r \to \infty$. This limiting procedure will be performed in Chapter III.

1.4.3 Non–symmetrical systems of particles: the BBGKY hierarchy

Let us consider in detail a one–dimensional system of particles interacting with their nearest neighbours via the hard–core pair potential (1.2.3,4). The simplest example is a hard rod particles system.

It is obvious that for the configurations of such system the following inequalities must be satisfied: $a + q_i \leqslant q_{i+1}$ (for arbitrary i), where a is the length of a particle. Thus, in this case, the Hamiltonian is not a function invariant with respect to permutations of the arguments $x \equiv (q_i, p_i) \in \mathbb{R}^1 \times \mathbb{R}^1$, and the natural way to number the particles is the numbering by means of the integers from the set $\mathbb{Z}^1 \backslash \{0\}$ (henceforth we shall call these systems non–symmetrical).

The Hamiltonian of the system of $N = N_1 + N_2$ particles has the form

$$H_N = \sum_{i \in \{-N_2,\ldots,-1,1,\ldots.N_1\}} p_i^2/2$$

$$+ \sum_{(i,i+1)\in\{(-N_2,-N_2+1),\ldots,(-1,1),\ldots,(N_1-1.N_1)\}} \Phi(q_i - q_{i+1}). \tag{1.4.12}$$

As the non–symmetrical system (1.4.12) is a special case of the systems considered in previous sections, then all results, obtained above, are also valid for these systems.

Let us denote by $L^1(\mathbb{R}^n \times (\mathbb{R}^n \backslash \mathbb{W}_n))$ the linear space of measurable functions $f_n(x_{-n_2}, \ldots, x_{n_1})$ $(n = n_1 + n_2)$, which are equal to zero on the forbidden set $\mathbb{W}_n \equiv \{(q_{-n_2}, \ldots, q_{n_1}) \in \mathbb{R}^{n_1+n_2} | a + q_i > q_{i+1}$ for at least one pair $(i, i+1) \in ((-n_2, -n_2+1), \ldots, (-1,1), \ldots, (n_1 - 1, n_1))\}$. There are different functions $f_n(x_{-n_2}, \ldots x_{n_1})$ with the same number n but different numbers n_1, n_2.

The norm of f_n is given by

$$\|f_n\| = \int dx_{-n_2} \dots dx_{n_1} \left| f_n \left(x_{-n_2}, \dots, x_{n_1} \right) \right|.$$

By analogy with the multi-dimensional case, we introduce the everywhere dense set $L_0^1(\mathbb{R}^n \times (\mathbb{R}^n \backslash \mathbb{W}_n)) \subset L^1(\mathbb{R}^n \times (\mathbb{R}^n \backslash \mathbb{W}_n))$, consisting of differentiable functions with compact support equal to zero in ε–neighborhood of $\mathbb{W}_n$.

We denote by L_α^1 the Banach space of sequences of functions $f = \{f_n(x_{-n_2}, \dots, x_{n_1})\}$ where $n = n_1 + n_2 \geq 0, n_1, n_2 \geq 0)$ with the norm

$$\|f\| = \sum_{n=0}^{\infty} \alpha^n \sum_{\substack{n_1, n_2 \geq 0 \\ n_1 + n_2 = n}} \int dx_{-n_2} \dots dx_{n_1} \left| f_n \left(x_{-n_2}, \dots, x_{n_1} \right) \right|, \tag{1.4.13}$$

with $\alpha > 1$ (see below). Let $L_\alpha^0 \subset L_\alpha^1$ denotes the subspace of finite sequences of functions from $L_0^1(\mathbb{R}^n \times (\mathbb{R}^n \backslash \mathbb{W}_n))$.

In the space L_α^1 we define the evolution operator $S(t) = \bigoplus_{n=0}^{\infty} S^n(t)$. The infinitesimal generator $\mathcal{H}$ of the group $S(t)$ is the direct sum of the operators $\mathcal{H}_n$, which (see Theorem 1.3.1) coincide on the set $L_0^1(\mathbb{R}^n \times (\mathbb{R}^n \backslash \mathbb{W}_n))$ with the Poisson bracket with boundary conditions on $\partial(\mathbb{R}^n \backslash \mathbb{W}_n)$, defined as follows. If $(x_{-n_2}, \dots, x_{n_1}) \in \partial(\mathbb{R}^{n_1+n_2} \backslash \mathbb{W}_{n_1+n_2}) \times \mathbb{R}^{n_1+n_2}$, for example, $|q_i - q_{i+1}| = a$ for some pair $(i, i+1)$, then at $t > 0$ and $p_i - p_{i+1} > 0$ we must replace the momenta p_i, p_{i+1} by p_i^*, p_{i+1}^* according to the expression (1.2.5′). At $t < 0$ this replacement is made in the Poisson bracket if $p_i - p_{i+1} < 0$.

In the space L_α^1 we also define the following operators

$$\left(\int dx^+ f \right)_n \left(x_{-n_2}, \dots, x_{n_1} \right) = \int dx_{n_1+1} f_{n+1} \left(x_{-n_2}, \dots, x_{n_1+1} \right), \tag{1.4.14}$$

$$\left(\int dx^- f \right)_n \left(x_{-n_2}, \dots, x_{n_1} \right) = \int dx_{-(n_2+1)} f_{n+1} \left(x_{-(n_2+1)}, \dots, x_{n_1} \right).$$

These operators are defined in L_α^1 and bounded $\| \int dx^{\mp} \| = \alpha^{-1}$. The operators $\left(1 - \int dx^{\pm}\right)^{-1}$ are represented by the following series, convergent in norm

$$\left(1 - \int dx^{\pm} \right)^{-1} = \sum_{n=0}^{\infty} \left(\int dx^{\pm} \right)^n,$$

$$\left(1 - \int dx^+ \right)^{-1} \left(1 - \int dx^- \right)^{-1} = \sum_{n=0}^{\infty} \sum_{\substack{n_1, n_2 \geqslant 0 \\ n_1 + n_2 = n}} \left(\int dx^+ \right)^{n_1} \left(\int dx^- \right)^{n_2}.$$

and the following inequalities hold

$$\| \left(1 - \int dx^{\pm}\right)^{-1} \| \leqslant \alpha(\alpha-1)^{-1}.$$

Please note that the space L^1_α with $\alpha > 1$ was introduced in order to make the norms of the operators $\int dx^{\pm}$ less than $\alpha^{-1} \leq 1$ and to guarantee the convergence of the series $\sum\limits_{n=0}^{\infty} (\int dx^{\pm})^n = (1 - \int dx^{\pm})^{-1}$. Let us now define the sequence of reduced distribution functions $F(t) = \{F_n(t, x_{-n_2}, \ldots, x_{n_1})\}_{n=n_1+n_2} > 0$ by the formula

$$F(t) = \left(\left(1 - \int dx^{+}\right)^{-1} \left(1 - \int dx^{-}\right)^{-1} D(t)\right)_0^{-1} \times \left(1 - \int dx^{+}\right)^{-1} \left(1 - \int dx^{-}\right)^{-1} D(t), \tag{1.4.15}$$

where $\left(\left(1 - \int dx^{+}\right)^{-1} \left(1 - \int dx^{-}\right)^{-1} D(t)\right)_0$ is the normalization factor (the grand partition function), $D(t)$ is a solution of the Cauchy problem for the Liouville equation

$$D(t) = S(-t)D(0).$$

According to the Liouville theorem, we have

$$\left(\left(1 - \int dx^{+}\right)^{-1} \left(1 - \int dx^{-}\right)^{-1} D(t)\right)$$

$$= \left(\left(1 - \int dx^{+}\right)^{-1} \left(1 - \int dx^{-}\right)^{-1} D(0)\right),$$

and the formula (1.4.15) can be represented in following form

$$F(t) = \left(\left(1 - \int dx^{+}\right)^{-1} \left(1 - \int dx^{-}\right)^{-1} D(0)\right)_0^{-1} \times \left(1 - \int dx^{+}\right)^{-1} \left(1 - \int dx^{-}\right)^{-1} S(-t)D(0), \tag{1.4.16}$$

or componentwise,

$$F_s(t, x_{-s_2}, \ldots, x_{s_1}) = \left(\left(1 - \int dx^+\right)^{-1} \left(1 - \int dx^-\right)^{-1} D(0)\right)_0^{-1}$$

$$\times \sum_{n=0}^{\infty} \sum_{\substack{n_1, n_2 \geq 0 \\ n_1 + n_2 = n}} \int dx_{-(n_2+s_2)} \ldots dx_{-(s_2+1)} dx_{s_1+1} \ldots dx_{s_1+n_1} \tag{1.4.16$'$}$$

$$\times \left(S^{s+n}(-t) D_{s+n}(0)\right) \left(x_{-(n_2+s_2)}, \ldots, x_{s_1+n_1}\right), \; s = s_1 + s_2, \; n = n_1 + n_1.$$

Let us stress that in the definition of the reduced distribution functions (1.4.16) the factor $\frac{1}{n!}$, which was present in definition of symmetrical distribution functions (1.4.3), is now absent . This is connected with nonsymmetrical nature of the distribution functions F_n. To obtain the equations that give the evolution law for the functions $F(t)$ (1.4.16), we differentiate both sides of (1.4.16) with respect to time. For $D(0) \in L^0_\alpha \subset L^1_\alpha$ we have

$$\frac{d}{dt} F(t) = -\left(\left(1 - \int dx^+\right)^{-1} \left(1 - \int dx^-\right)^{-1} D(0)\right)_0^{-1} \left(1 - \int dx^+\right)^{-1}$$

$$\times \left(1 - \int dx^-\right)^{-1} \mathcal{H} S(-t) D(0) =$$

$$-\mathcal{H} F(t) - \left(\left(1 - \int dx^+\right)^{-1} \left(1 - \int dx^-\right)^{-1} D(0)\right)_0^{-1}$$

$$\times \left[\left(1 - \int dx^+\right)^{-1} \left(1 - \int dx^-\right)^{-1}, \mathcal{H}\right] S(-t) D(0), \tag{1.4.17}$$

where the bracket $[A, B]$ is the commutator of two operators A and B, $(-\mathcal{H})$ is the infinitesimal generator of the group $S(-t)$. Using the following operator equality (to be proved in Chapter II)

$$\left[\left(1 - \int dx^+\right)^{-1} \left(1 - \int dx^-\right)^{-1}, \mathcal{H}\right]$$

$$= \left[\left(\int dx^+ + \int dx^-\right)^{-1}, \mathcal{H}\right] \left(1 - \int dx^+\right)^{-1} \left(1 - \int dx^-\right)^{-1},$$

the equality (1.4.17) is transformed into the following identity [Ger1]

$$\frac{d}{dt}F(t) = -\mathcal{H}F(t) + \left[\mathcal{H}, \int dx^+\right] F(t) + \left[\mathcal{H}, \int dx^-\right] F(t). \tag{1.4.18}$$

Let us consider this identity as an abstract equation for the sequence $F(t)$ of distribution functions. This is the BBGKY hierarchy for one–dimensional systems.

To give the explicit componentwise form of BBGKY hierarchy for one–dimensional system of particles with hard–core potential we use the following result from Chapter II [Ger1]:

The following equality holds at $t > 0$

$$\left(\left[\mathcal{H}, \int dx^+\right] F(t)\right)_s (x_{-s_2}, \ldots, x_{s_1}) = \int\limits_0^\infty dP P\Big(F_{s+1}(t, x_{-s_2}, \ldots, x_{s_1-1};$$

$$q_{s_1}, p_{s_1} + P; q_{s_1} + a, p_{s_1}) - F_{s+1}(t, x_{-s_2}, \ldots, x_{s_1}; q_{s_1} + a, p_{s_1} - P)\Big)$$

$$+ \int dx_{s_1+1} \left\{\Phi(q_{s_1} - q_{s_1+1}), F_{s+1}(t, x_{-s_2}, \ldots, x_{s_1+1})\right\}, \tag{1.4.19}$$

$$\left(\left[\mathcal{H}, \int dx^-\right] F(t)\right)_s (x_{-s_2}, \ldots, x_{s_1}) = \int\limits_0^\infty dP P\Big(F_{s+1}(t, q_{-s_2} - a, p_{-s_2};$$

$$q_{-s_2}, p_{-s_2} - P; x_{-(s_2-1)}, \ldots, x_{s_1}) - F_{s+1}(t, q_{-s_2} - a, p_{-s_2} + P; x_{-s_2}, \ldots, x_{s_1})\Big)$$

$$+ \int dx_{-(s_2+1)} \left\{\Phi(q_{-s_2} - q_{-(s_2+1)}), F_{s+1}(t, x_{-(s_2+1)}, \ldots, x_{s_1})\right\}.$$

It should be noted that the BBGKY hierarchy (1.4.18), (1.4.19) is defined for finite systems of particles, i.e. systems for which the average number of particles

$$\langle N \rangle = \left(\left(1 - \int dx^+\right)^{-1} \left(1 - \int dx^-\right)^{-1} N F(t)\right)_0,$$

or (1.4.20)

$$\langle N \rangle = \sum_{n=1}^{\infty} \sum_{\substack{n_1, n_2 \geqslant 0 \\ n_1 + n_2 = n}} \int dx_{-n_2} \ldots dx_{n_1} F_n(t, x_{-n_2}, \ldots, x_{n_1})$$

is finite. For $F(t) \in L^1_\alpha$ we have $\langle N \rangle < \infty$. As was mentioned above, the BBGKY hierarchy (1.4.18) can be used also for the description of systems of infinitely many particles (see Chapter III).

1.5 The Steady BBGKY Hierarchy

1.5.1 On the solutions of the steady BBGKY hierarchy

As was already mentioned in the introduction to this chapter, all possible states of statistical systems can be classified as equilibrium or non–equilibrium. Equilibrium states are described by distribution functions that are special solutions of the steady BBGKY hierarchy.

The steady BBGKY hierarchy has the form

$$-\mathcal{H}F + \left[\mathcal{H}, \int dx\right] F = 0, \tag{1.5.1}$$

where the operators $\mathcal{H}$ and $\left[\mathcal{H}, \int dx\right]$ are determined according to formulae (1.3.15) and (1.4.8,9) (or (1.4.19)).

The problem of constructing the formal solution for hierarchy (1.5.1) was examined in a series of papers [GS]. But till now this mathematical problem is not completely investigated.

Therefore, in statistical mechanics one assumes that the equilibrium states of many-particle systems are described by the Gibbs distribution functions (Gibbs's hypothesis).

Note, that the mathematical description of the Gibbs equilibrium states of systems of infinitely many particles is a well–known problem [BPK][D1][D2][LR] [Mi1][Mi2][PGM2][R][Thi] and forms the principal part of modern statistical mechanics.

In this section we construct equilibrium states of infinite systems of particles with a hard core. We obtain their states from those of finite systems of particles situated in a bounded domain Λ by using the thermodynamic limit procedure. Also we will describe one class of distribution functions close to equilibrium distribution functions, which will be used as initial data for the Cauchy problem for the BBGKY hierarchy.

1.5.2 Equilibrium distribution functions

According to Gibbs's hypothesis for a system with a finite number of particles (1.2.1) situated in a bounded domain Λ and with the n–particles Hamiltonian (1.2.1), the sequences of the Gibbs distribution functions F_Λ is defined by the following formula

$$F_\Lambda = \left(e^{\int\limits_\Lambda dx} \Psi_\Lambda \right)_0^{-1} e^{\int\limits_\Lambda dx} \Psi_\Lambda, \tag{1.5.2}$$

where $\Psi_\Lambda = (1, \Psi_{\Lambda,1}(x_1), \ldots, \Psi_{\Lambda,n}(x_1, \ldots, x_n), \ldots)$ is the sequence of special solutions of the steady Liouville equation: $\mathcal{H}_n^\Lambda \Psi_{\Lambda,n} = 0$, having the form

$$\Psi_{\Lambda,n}(x_1, \ldots, x_n) = \exp\left\{ -\beta \sum_{i=1}^n \left(\frac{p_i^2}{2} - \mu \right) - \beta \sum_{1=i<j}^n \Phi(q_i - q_j) - \beta \sum_{i=1}^n u^\Lambda(q_i) \right\}$$

Here the parameter $\beta \geqslant 0$, characterizing the average energy of the system, has the meaning of inverse temperature. μ is the chemical potential, characterizing the average number of particles $\langle N \rangle$ and $\left(e^{\int\limits_\Lambda dx \Psi_\Lambda} \right)_0$ is the grand partition function. It is obvious that the sequence (1.5.2) is a particular solution of the steady BBGKY hierarchy

$$-\mathcal{H}_\Lambda F_\Lambda + \left[\mathcal{H}_\Lambda, \int\limits_\Lambda dx \right] F_\Lambda = 0. \tag{1.5.3}$$

Notice, that the functions $\Psi_{\Lambda,n}$ can depend also on other additive integrals of motion according to Gibbs's hypothesis [Ba][Bo][Zu]. A more delicate questions arises when we consider systems which are perturbations of integrable systems. Then the theorem of Kolmogorov [Kol], Arnol'd [Arn3], Moser [Mo] (KAM) applies and the distribution is certainly not of the Gibbs type, at least for a finite number of degrees of freedom. The thermodynamic limit of these systems is open to conjectures, since no general results are known.

We do not consider the problem of ergodicity and refer the readers to the survey [Sz].

Let us now remark, that twice continuously differentiable finite interaction potentials with a hard core (1.2.3-4.6-7) satisfy the stability condition

$$\sum_{1=i<j}^n \Phi(q_i - q_j) \geqslant -Bn \ , \ B \geqslant 0 \tag{1.5.4}$$

and the regularity condition

$$\int dq \left| e^{-\beta\Phi(q)} - 1 \right| \equiv c(\beta) < \infty \tag{1.5.5}$$

For these potentials the distribution functions (1.5.2) exist and satisfy the following inequality [BPK][PGM2][R]:

$$\sum_{n=0}^{\infty} \int_{(\Lambda\times\mathbb{R}^{\nu})^n} d(x)_n \left| F_{\Lambda,n}(x_1,\ldots,x_n) \right|$$

$$\leqslant \left(e^{\int\limits_{\Lambda} dx} \Psi_{\Lambda} \right)_0^{-1} \exp\left\{ zVe^{\beta B} \right\} \sum_{n=0}^{\infty} z^n V^n e^{\beta B n},$$

where z is the activity $\left(z = e^{\beta\mu(2\pi/\beta)^{\nu/2}} \right)$, $V \equiv |\Lambda|$ is the volume of a domain $\Lambda \subset \mathbb{R}^{\nu}$. Hence the sequence (1.5.2) belongs for $0 < z < V^{-1}e^{-B\beta}$ to the space of sequences of summable functions $L^1(\Lambda)$ and describes the state of a system with finite a number of particles.

$$\langle N \rangle = \left(e^{\int\limits_{\Lambda} dx} NF_{\Lambda} \right)_0 < \infty .$$

To describe the equilibrium states of a system with infinitely many particles and inverse temperature β, we define a suitable Banach space, containing the distribution functions of such states.

We consider a linear space of sequences $f = \{f_0, f_1(x_1), \ldots, f_n(x_1,\ldots,x_n), \ldots\}$ of functions $f_n(x_1,\ldots,x_n)$, which are invariant with respect to permutations of $x_1,\ldots,x_n$ and are equal to zero on the set $\mathbb{W}_n$ of forbidden configurations and are continuous outside $\mathbb{W}_n$. The norm in this space is defined by

$$\|f\| = \sup_{n\geqslant 1} \sup_{x_1,\ldots,x_n} \xi^{-n} \left| f_n(x_1,\ldots,x_n) \right| \exp\left\{ \beta \sum_{i=1}^{n} \frac{p_i^2}{2} \right\}, \tag{1.5.6}$$

where $\xi > 0$ is some constant which will be defined later. This Banach space of sequences we denote by $L^{\infty}_{\xi,\beta}$ or for the sake of simplicity by L^{∞}.

The equilibrium state of a system with infinitely many particles is defined as the thermodynamic limit ($\Lambda \nearrow \mathbb{R}^{\nu}, \langle N \rangle \to \infty, \langle N \rangle/|\Lambda| = \frac{1}{v} < +\infty, \langle N \rangle = \int\limits_{\Lambda} dx F_{\Lambda.1}(x)$) of the sequence F_{Λ} of functions (1.5.2).

The sequence of Gibbs distribution functions $F = \{F_s(x_1, \ldots, x_s)\}_{s>0}$ for an infinite system is defined by the following formulae

$$F_s(x_1, \ldots, x_s) = \left(\frac{2\pi}{\beta}\right)^{-\frac{\nu}{2}s} \exp\left\{-\beta \sum_{i=1}^{s} \frac{p_i^2}{2}\right\} \rho_s(q_1, \ldots, q_s), \tag{1.5.7}$$

where $\rho = \{\rho_s(q_1, \ldots, q_s)\}_{s \geqslant 1}$ is a solution of the Kirkwood–Salsburg equation

$$\rho = z\alpha + zK\rho,$$
$$\alpha = \{1, 0, \ldots\}, \; K = K_1 + K_2, \tag{1.5.8}$$

The operators K_1 and K_2 are defined by the formulae

$$(K_1\rho)_s(q_1, \ldots, q_s) = e^{-\beta \sum\limits_{i=2}^{s} \Phi(q_1 - q_i)} (1 - \delta_{1,s})\rho(q_2, \ldots, q_s),$$

$$(K_2\rho)_s(q_1, \ldots, q_s) = e^{-\beta \sum\limits_{i=2}^{s} \Phi(q_1 - q_i)} \sum_{n=1}^{\infty} \frac{1}{n!} \int\limits_{\mathbb{R}^{\nu n}} dy_1 \ldots dy_n$$

$$\times \prod_{i=1}^{n} \varphi_{q_1}(y_i)\rho_{s+n-1}(q_2, \ldots, q_s, y_1, \ldots, y_n).$$

Here $\delta_{1,s}$ is a Kronecker symbol and $\varphi_{q_1}(y_i) = \exp\{-\beta\Phi(q_1 - y_i)\} - 1$. Let us introduce the space L_ξ^∞ of sequences $f = \{f_n(q_1, \ldots, q_n)\}_{n \geqslant 0}$ symmetrical functions equal to zero on the set $\mathbb{W}_n$ and continuous outside of them with the norm

$$\|f\| = \sup_{n \geqslant 0} \xi^{-n} \sup_{q_1 \ldots q_n} |f_n(q_1, \ldots, q_n)|$$

The solution (1.5.8) is represented by the series

$$\rho = \sum_{k=0}^{\infty} (zK)^k z\alpha \tag{1.5.9}$$

which is convergent in norm in L^∞.

The sequence F (1.5.7) exists for

$$|z| < \xi e^{-2\beta B} \exp\{-\xi C(\beta)\} \tag{1.5.10}$$

and belongs to the space L^∞ [PGM2][R]. (The parameter z is positive but we can also consider (1.5.8) for complex values of z). The results are true also for states F_Λ

of finite systems. The functions $F_{\Lambda.n}$, $F_\Lambda \in L^1(\Lambda)\cap L^\infty$ (1.5.2) converge uniformly in each compact subset $\mathbb{R}^{\nu n}\times(\mathbb{R}^{\nu n}\backslash \mathbb{W}_N)$ as $\Lambda \nearrow \mathbb{R}^\nu$, $\langle N\rangle = \int\limits_{\Lambda\times\mathbb{R}^\nu} dx F_{\Lambda,1}(x) \to \infty$, $\langle N\rangle/\Lambda = \frac{1}{v}$ to the limit functions F_n (1.5.7), for z, satisfying inequality (1.5.10). For arbitrary $z>0$ the limit sequence F exists and belongs to L^∞.

The property of differentiability of $F_{\Lambda.n}$ (1.5.2) and $F_n(x_1,\dots,x_n)$ in $p_1,\dots p_n$ is obvious. The functions $F_{\Lambda.n}(x_1,\dots,x_n)$ are also differentiable in $(q_1,\dots,q_n)\in(\mathbb{R}^{\nu n}\backslash \mathbb{W}_n)$, because the expression (1.5.2) for $F_{\Lambda.n}$ can be differentiated under the integral sign, and

$$\frac{\partial}{\partial q_i}F_{\Lambda.n}(x_1,\dots,x_n) = -\beta\frac{\partial u^\Lambda(q_i)}{\partial q_i}F_{\Lambda.n}(x_1,\dots,x_n) - \beta\frac{\partial H_n}{\partial q_i}F_{\Lambda.n}(x_1,\dots,x_n)$$
$$-\beta\int dx_{n+1}\sum_{i=1}^{n}\frac{\partial}{\partial q_i}\Phi(q_i-q_{n+1})F_{\Lambda,n+1}(x_1,\dots,x_{n+1}),\ i=1,\dots,s\,. \quad (1.5.11)$$

Since the expression on the right–hand side of (1.5.11) converges uniformly on every compact subset in $\mathbb{R}^{\nu n}\times(\mathbb{R}^{\nu n}\backslash \mathbb{W}_n)$ to the same expression with the limiting distribution function (1.5.7) as $\Lambda\nearrow\mathbb{R}^\nu$, $\langle N\rangle\to\infty$, $\langle N\rangle/|\Lambda| = \frac{1}{v}<\infty$, the limiting distribution functions F_n are also differentiable in $(q_1,\dots,q_n)\in\mathbb{R}^{\nu n}\backslash\mathbb{W}_n$ and

$$\frac{\partial}{\partial q_i}F_n(x_1,\dots,x_n) = -\beta\frac{\partial H_n}{\partial q_i}F_n(x_1,\dots,x_n)$$
$$-\beta\int dx_{n+1}\sum_{i=1}^{n}\frac{\partial}{\partial q_i}\Phi(q_i-q_{n+1})F_{n+1}(x_1,\dots,x_{n+1})\,.$$

1.5.3 Equilibrium states of non–symmetrical systems

Let us describe the equilibrium states of a one–dimensional system of particles interacting with their nearest neighbours via a hard–core potential (1.2.3,4) (a non–symmetrical system). For this system with a finite number of particles situated in bounded domain Λ. are sequence of Gibbs distribution functions F_Λ is defined by the following formula

$$F_\Lambda =$$
$$\left(\left(1-\int\limits_\Lambda dx^+\right)^{-1}\left(1-\int\limits_\Lambda dx^-\right)^{-1}\Psi_\Lambda\right)_0^{-1}\left(1-\int\limits_\Lambda dx^+\right)^{-1}\left(1-\int\limits_\Lambda dx^-\right)^{-1}\Psi_\Lambda \quad (1.5.12)$$

where $\Psi_\Lambda = \{\Psi_{\Lambda.n}(x_{-n_2},\ldots,x_{n_1})\}_{n=n_1+n_2\geq 0}$ is the sequence of special partial solutions of the steady Liouville equation : $\mathcal{H}_\Lambda F_\Lambda = 0$, which has the form

$$\Psi_{\Lambda,n}(x_{-n_2},\ldots,x_{n_1}) = \exp\left\{-\beta\sum_{i=-n_2}^{n_1}\left(\frac{p_i^2}{2}-\mu\right) - \beta\sum_{i=-n_2}^{n_1-1}\Phi(q_i-q_{i+1}) - \beta\sum_{i=-n_2}^{n_1} u^\Lambda(q_i)\right\}, \qquad \Psi_{\Lambda,0}=1$$

Here the parameter $\beta \geqslant 0$ is, as above, an inverse temperature, μ is the chemical potential and $\left(\left(1-\int_\Lambda dx^+\right)^{-1}\left(1-\int_\Lambda dx^-\right)^{-1}\Psi_\Lambda\right)_0$ is the grand partition function. Clearly, the sequence (1.5.12) is the solution of the steady BBGKY hierarchy (1.5.3), (1.4.19). For $0 < z < (\alpha V)^{-1}e^{-\beta B}$ it belongs to the space of sequences of summable functions $L^1_\alpha(\Lambda)$ and describes the state of a system with a finite number of particles (1.4.20).

The sequence of Gibbs distribution functions $F = \{F_s(x_{-s_2},\ldots,x_{s_1})\}_{s_1+s_2>0}$ for this system with infinite number of particles is defined by formulae [Ger1][PGM2]

$$F_s(x_{-s_2},\ldots,x_{s_1}) = \Psi(x_{-s_2},\ldots,x_{s_1})\left(1+z^2\int dxdy \times \left(e^{-\beta\Phi(x-y)}-1\right)\rho(x)\rho(y)\right)^{-1}\rho(q_{-s_2})\rho(q_{s_1}), \quad (1.5.13)$$

where $\rho(x)$ is a solution of the following equation

$$\rho(x) = 1 + z\int dy\left(e^{-\beta\Phi(x-y)}-1\right)\rho(y). \quad (1.5.14)$$

The sequence F (1.5.13) exists for $|z| < C^{-1}(\beta)$ and belongs to the space $L^\infty(\mathbb{R}^1)$ of the sequences $f = \{f_n(x_{-n_2},\ldots,x_{n_1})\}_{n_1+n_2>0}$ of those functions of n_1+n_2 variables, $f_n(x_{-n_2},\ldots,x_{n_1})$, which are zero on the set $\mathbb{W}_{n_1+n_2}$ of forbidden configurations (see subsec. 1.4.3) and continuous outside of them, with norm

$$\|f\| = \sup_{n=n_1+n_2>0}\xi^{-n}\sup_{x_{-n_2},\ldots,x_{n_1}}|f_n(x_{-n_2},\ldots,x_{n_1})|\exp\left\{\beta\sum_{i=-n_2}^{n_1}\frac{p_i^2}{2}\right\}.$$

The functions $F_{\Lambda.n}$ (1.5.12), $F_\Lambda \in L^1_\alpha(\Lambda) \cap L^\infty(\mathbb{R}^1)$ converge uniformly on each compact subset in $\mathbb{R}^n \times (\mathbb{R}^n \backslash \mathbb{W}_n)$ as $\Lambda \nearrow \mathbb{R}^1$, $\langle N\rangle \to \infty$, $\langle N\rangle/|\Lambda| = \frac{1}{v} < \infty$ to the limit functions F_n (1.5.13) for $|z| < C^{-1}(\beta)$ [PGM2].

1.5.4 The states close to equilibrium

It is natural to take as initial data for the BBGKY hierarchy distribution functions of an infinite system which are in a state close (in some sense) to equilibrium. We shall now describe one such class of distribution functions—the so–called locally perturbed equilibrium distribution functions [Pe1][PGM2].

We denote by $\delta\Psi_\Lambda = \{\delta\Psi_{\Lambda.n}(x_1,\ldots,x_n)\}_{n\geqslant 0}$ local perturbations of the functions Ψ_Λ defined according (1.5.2), upon which we impose the following conditions:

a) boundedness

$$\sup_{x_1,\ldots,x_n} |\delta\Psi_{\Lambda,n}(x_1,\ldots,x_n)| \leqslant c < \infty,$$

b) positivity

$$\delta\Psi_{\Lambda.n}(x_1,\ldots,x_n) \geqslant 0.$$

c) equivalence of thermodynamic properties of equilibrium and locally perturbed equilibrium systems of particles

$$\lim_{\Lambda\nearrow\mathbb{R}^\nu} \left(e^{\int\limits_\Lambda dx}\Psi_\Lambda\right)_0 \left(e^{\int\limits_\Lambda dx}\Psi_\Lambda\delta\Psi_\Lambda\right)_0^{-1} = \lim_{\Lambda\nearrow\mathbb{R}^\nu} C_1(\Lambda) = C_1 < \infty.$$

The sequence of locally perturbed Gibbs distribution functions F^0_Λ is defined by the sequence $\Psi_\Lambda\delta\Psi_\Lambda$ as follows

$$F^0_\Lambda = \left(e^{\int\limits_\Lambda dx}\Psi_\Lambda\delta\Psi_\Lambda\right)_0^{-1} e^{\int\limits_\Lambda dx}\Psi_\Lambda\delta\Psi_\Lambda \quad . \tag{1.5.15}$$

We also suppose that the sequence (1.5.15) F^0_Λ defined by a $\delta\Psi_\Lambda$ of general type exists for any Λ and converges as $\Lambda \nearrow \mathbb{R}^\nu$, $\langle N\rangle \to \infty$, $\langle N\rangle/|\Lambda| = \frac{1}{v} < \infty$ uniformly on each compact set to the limit distribution functions F^0. This limit sequence F^0 describes the corresponding state of an infinite system.

As an example of a local perturbations $\delta\Psi_\Lambda$, we can take $\delta\Psi_s(x_1,\ldots,x_s) = \exp\left\{-\beta\sum\limits_{i=1}^{s} h(q_i)\right\}$, where $h \geqslant 0$ is so small that we may neglect terms of order higher than first in h and write $\delta\Psi_s(x_1,\ldots,x_s) \cong 1-\beta\sum_{i=1}^{s} h(q_i)$. A direct calculation of the grand partition function for the locally perturbed equilibrium states by $h(q)$ yields

$$\left(e^{\int\limits_\Lambda dx}\Psi_\Lambda\delta\Psi_\Lambda\right)_0 = \left(e^{\int\limits_\Lambda dx}\Psi_\Lambda\right)_0 \left[1-\beta\int\limits_\Lambda dx_1 h(q_1)F_{\Lambda,1}(x_1)\right],$$

and thus

$$C_1(\Lambda) = 1+\beta\int\limits_\Lambda dx_1 h(q_1)F_{\Lambda,1}(x_1).$$

The function $h(q_1)$ is finite, so that

$$\lim_{\Lambda\nearrow\mathbb{R}^\nu} C_1(\Lambda) = 1+\beta\int dx_1 h(q_1)F_1(x_1) \equiv C_1 < \infty.$$

The definition (1.5.15) and the properties of the locally perturbed functions $\delta\Psi_\Lambda$ lead to the estimate

$$|F^0_{\Lambda.s}(x_1,\ldots,x_s)| \leqslant CC_1(\Lambda)F_{\Lambda,s}(x_1,\ldots,x_s). \tag{1.5.16}$$

It follows from (1.5.16) that the sequence of locally perturbed equilibrium distribution functions F^0_Λ (1.5.15) belongs to the spaces $L^1(\Lambda)$, L^∞ together with the sequence F_Λ of Gibbs distribution functions. We also suppose that locally perturbed equilibrium distribution functions are equal to zero in some ε–neighborhood of the set of forbidden configurations.

In the one–dimensional case, for a system of particles interacting with their nearest neighbours, locally perturbed Gibbs distribution functions are defined in a similar way.

We remark that another example of state close to equilibrium is a local equilibrium state, which is described by a local Gibbs distribution function [Zu].

APPENDIX I

The Infinitesimal Operator of the Group $S^N(t)$ [Kot].

Let us describe the domain $D(\mathcal{H}_N)$ of the infinitesimal operator $\mathcal{H}_N$. Denote $G = \Lambda \backslash \mathbb{W}_N$, $\Omega = \mathbb{R}^{\nu N} \times G$. Here we need not distinguish between the cases when Λ is a bounded domain, or $\Lambda = \mathbb{R}^{\nu N}$.

Suppose that $u \in L^1(\Omega)$. We will say that $\{u, H_N\} \in L^1(\Omega)$ where the Poisson bracket is understood in the weak sense, if there exists (of course, unique) a functional $f \in L^1(\Omega)$ such that

$$\int_{\Omega} u(x)\{\varphi, H_N\}(x)\,dx = -\int_{\Omega} f(x)\,\varphi(x)\,dx \tag{1.I.1}$$

for any continuously differentiable function $\varphi \in C^1(\Omega)$ with a compact support. Denote by D the set of all $u \in L^1(\Omega)$ for which $\{u, H_N\} \in L^1(\Omega)$ in the weak sense.

The boundary $\partial\Omega$ consists of a finite number of codimension one submanifolds intersecting along manifolds of lower dimensions. Let us say that a set $F \subset \partial\Omega$ has zero measure if for any chart $(V, \rho), \rho : V \to \mathbb{R}^{2\nu N-1}$, on each of the regular parts of the boundary $\partial\Omega$

$$\text{mes}\,\rho(F \cap V) = 0.$$

Now we may use the expression "almost everywhere on $\partial\Omega$". Of course, the set F_0 of singularities at the boundary $\partial\Omega$ has zero measure.

Let $F_1 \subset \partial\Omega \backslash F_0$ be the set of those points where the vector field R defined by the right–hand sides of the system (1.2.2) is tangent to the boundary, or vanishes. If $X = (P, Q) \in F_1$ then

$$R(X) = (0, \ldots, 0, P_1, \ldots, P_N)$$

where P is orthogonal to the unit normal $n(Q)$ of the surfuce ∂G. In other words, $\sum_{i=1}^{N} P_i \cdot n_i(Q) = 0$. This condition defines in each regular component of $\partial\Omega$ a codimension one submanifold. Hence $\text{mes}\,F_1 = 0$.

If $x_0 \in \partial\Omega \backslash F_0 \backslash F_1$ then there exists the trajectory $X(t) = X(t, x_0)$ in Ω intersecting the boundary at the point x_0 with a nonzero angle. This trajectory satisfies either $X(0) = x_0$ or $X(\Theta) = x_0$ for some $\Theta > 0$, depending on the direction of the vector field R at x_0. Denote the corresponding sets of points $x_0 \in \partial\Omega \setminus F_0 \setminus F_1$ by Γ_+ and Γ_- respectively.

Lemma 1. *Let $n \in D$. For almost every $x_0 \in \partial\Omega$ the function $z(t) = u(X(t, x_0))$ is (after correction on a subset of zero measure) absolutely continuous on some closed interval $[0, \delta(x_0)]$ (if $x_0 \in \Gamma_+$), or $[\Theta - \delta(x_0), \Theta]$ (if $x_0 \in \Gamma_-$), and*

$$z'(t) = f(X(t, x_0))$$

where $f = \{u, H_N\}$; the Poisson bracket is understood in the weak sense.

Proof. We shall consider the case when $x_0 \in \Gamma_+$; the case when $x_0 \in \Gamma_-$ can be dealt with in a similar way.

By the rectification theorem [Arn1] there exist a neighbourhood $U \subset \mathbb{R}^{2\nu N}$ of the point x_0 and a C^1–diffeomorphism Ψ defined on $\overline{U}$, with the following properties :

$$\Gamma' = \partial\Omega \cap U \subset \Gamma_+, \quad \Psi(X(t, x')) = \Psi(x') + et, \quad e = (1, 0, \ldots, 0),$$

for all $x' \in \Gamma'$ and sufficiently small t, say, for $t \in [0, \delta(x_0)]$; the diffeomorphism $y = \Psi(x)$ transforms the differential expression $\varphi \mapsto \{\varphi, H_N\}$ into the differentiation $\partial/\partial y_1$. We may assume that U is bounded and so small that $\Psi(\Gamma')$ can be projected one-to-one to the hyperlane orthogonal to the vector e. This projection induces on $\Psi(\Gamma)$ the measure $\nu(dy')$. It is sufficient to prove the assertion of our lemma for almost all points from Γ'.

Let $U_0 = U \cap \Omega$, $V_0 = \Psi(U_0)$. Consider the function

$$g(y) = \begin{cases} f(\Psi^{-1}(y))\,\mu(y), & y \in V_0, \\ 0, & y \notin V_0, \end{cases}$$

where μ is the Jacobian of the transformation Ψ^{-1}. Evidently, $g \in L^1(\mathbb{R}^{2\nu N})$. Denote by $\omega_\varepsilon \in C_0^\infty(\mathbb{R}^{2\nu N})$ an averaging kernel, $\omega_\varepsilon(x) = 0$ for $\|x\| \geqslant \varepsilon$ (see e.g. [MO] for information on Sobolev averaging). We have

$$\|g - \omega_\varepsilon * g\|_{L^1(\mathbb{R}^{2\nu N})} \to 0, \quad \varepsilon \to 0.$$

In particular,

$$\int_\Delta |g(y) - (\omega_\varepsilon * g)(y)| \to 0, \quad \varepsilon \to 0, \tag{1.I.2}$$

where $\Delta = \{ y = \Psi(x') + et,\ x' \in \Gamma',\ t \in [0, \delta(x_0)] \}$. By Fubini theorem the left–hand–part of (2) is equal to

$$\int_{\Psi(\Gamma')} \nu(dy') \int_0^{\delta(x_0)} | g(y' + et) - (\omega_\varepsilon * g)(y' + et) | \, dt.$$

The internal integral is finite for almost all y'. It determines an integrable function of y', L^1–tending to zero when $\varepsilon \to 0$. Thus there exists a sequence $\varepsilon_k \to 0$ such that for almost all $y' \in \Psi(\Gamma')$

$$\int_0^{\delta(x_0)} | g(y' + et) - (\omega_{\varepsilon_k} * g)(y' + et) | \, dt \to 0, \quad k \to \infty. \tag{1.I.3}$$

Taking $\sigma \in C_0^\infty(0, \delta(x_0))$, let us introduce the notation

$$\eta_{\varepsilon . x'}(s) = \int_0^{\delta(x_0)} \omega_\varepsilon\big(\Psi(x') + et - \Psi(s)\big)\sigma(t)\, dt, \quad s \in U_0,$$

where $\varepsilon > 0$, $x' \in \Gamma'$. It is clear that $\eta_{\varepsilon . x'} \in C^1(U_0)$. The function $\eta_{\varepsilon , x'}$ vanishes near ∂U_0. Indeed, suppose that

$$\operatorname{supp} \sigma \subset [\gamma, \delta(x_0) - \gamma], \quad 0 < \gamma < \delta(x_0).$$

The interval $\{ \Psi(x') + et,\ \gamma \leqslant t \leqslant \delta(x_0) - \gamma \}$ does not intersect the set $\Psi(\partial U_0)$ and its sufficiently small neighbourhood. Hence there exist such $\beta > 0$, $d > 0$ that the inequality dist $(s, \partial U_0) < \beta$ implies the inequality

$$| \Psi(x') + et - \Psi(s) | > d$$

so that $\omega_\varepsilon\big(\Psi(x') + et - \Psi(s)\big) = 0$ for $\varepsilon \leqslant d$, whence $\eta_{\varepsilon , x'}(s) = 0$. Now the function $\eta_{\varepsilon . x'}$ $(\varepsilon \leqslant d)$ can be extended to the function from $C^1(\Omega)$ with a compact support (for which we retain the same notation).

It follows from Fubini's theorem that $f\big(X(\cdot\,,x')\big) \in L^1\big(0,\delta(x_0)\big)$ for almost all $x' \in \Gamma'$. If $\varepsilon_k \leqslant d$, then

$$\int_\Omega f(s)\,\eta_{\varepsilon_k,x'}(s)\,ds - \int_0^{\delta(x_0)} f\big(X(t,x')\big)\sigma(t)\,dt$$

$$= \int_0^{\delta(x_0)} \sigma(t)\,dt\Big[\int_{U_0} f(s)\,\omega_{\varepsilon_k}\big(\Psi(x')+et-\Psi(s)\big)\,ds - f\big(X(t,x')\big)\Big]$$

$$= \text{const}\int_0^{\delta(x_0)} \sigma(t)\Big[(\omega_{\varepsilon_k} * g)\big(\Psi(x')+et\big) - g\big(\Psi(x')+et\big)\Big]\,dt \to 0,$$

when $k \to \infty$, for almost all $x' \in \Gamma'$. Here we have used (1.I.3) and the fact that $d\Psi(x') = d\Psi\big(X(t,x')\big) \circ d_{x'}X(t,x')$ whence (by Liouville's theorem) $\mu\big(\Psi(x')+et\big)$ does not depend on $t \in [\,0,\delta(x_0)\,]$. Thus

$$\int_\Omega f(s)\,\eta_{\varepsilon_k,x'}(s)\,ds \to \int_0^{\delta(x_0)} f\big(X(t,x')\big)\sigma(t)\,dt \tag{1.I.4}$$

when $k \to \infty$, for almost all $x' \in \Gamma'$.

By the definition of Ψ

$$\{\,\eta_{\varepsilon,x'}, H_N\,\}(s) = \int_0^{\delta(x_0)} \frac{\partial}{\partial \xi_1}\,\omega_\varepsilon\big(\Psi(x')+et-\xi\big)\Big|_{\xi=\Psi(s)}\sigma(t)\,dt$$

$$= -\int_0^{\delta(x_0)} \frac{\partial}{\partial t}\,\omega_\varepsilon\big(\Psi(x')+et-\Psi(s)\big)\sigma(t)\,dt$$

$$= \int_0^{\delta(x_0)} \omega_\varepsilon\big(\Psi(x')+et-\Psi(s)\big)\sigma'(t)\,dt.$$

Just as in (1.I.4), for almost all $x' \in \Gamma'$

$$\int_\Omega u(s)\,\{\,\eta_{\varepsilon_k,x'}, H_N\,\}(s)\,ds \to \int_0^{\delta(x_0)} u\big(X(t,x')\big)\sigma'(t)\,dt, \quad k \to \infty. \tag{1.I.5}$$

Put $\varphi = \eta_{\varepsilon_k, x'}$ in (1.I.1). Passing to the limit for $k \to \infty$ and taking into account (1.I.4), (1.I.5), we obtain that

$$\int_0^{\delta(x_0)} u\big(X(t, x')\big)\sigma'(t)\, dt = -\int_0^{\delta(x_0)} f\big(X(t, x')\big)\sigma(t)\, dt \tag{1.I.6}$$

for any $\sigma \in C_0^\infty(0, \delta(x_0))$, and for all $x' \in \Gamma'$ except a set of zero measure independent on σ. This implies [MO] the assertion of the lemma.

Let $\mathcal{M} \subset \partial\Omega$ be a set of zero measure, and $\mathfrak{M} \subset \Omega$ consist of those $x \in \Omega$ for which the trajectory $X(t, x)$, $0 < t \leqslant T$ (T is a fixed number) intersects $\partial\Omega$ at least once at a point from $\mathcal{M}$.

Lemma 2. *The Lebesgue measure of the set $\mathfrak{M}$ equals zero.*

Proof. It is sufficient to find, for any $x \in \mathfrak{M}$, a neighbourhood U such that $\operatorname{mes}(U \cap \mathfrak{M}) = 0$.

Let $t_1(x) < \cdots < t_m(x) \leqslant T$ be the instants when the trajectory $X(t, x)$ crosses $\partial\Omega$. Chose a neighbourhood U in such a way that the number m to be finite and same for all $x \in U$ (by Lemmas 1.2.2, 1.2.3 this can be done for almost all x). Assuming that U is sufficiently small, introduce in U the new coordinates $t_1(x)$, $b_1(x)$ where $b_1(x) = X\big(t_1(x), x\big)$. The mapping $x \mapsto \big(t_1(x), b_1(x)\big)$ is a C^1–diffeomorphism whence

$$\operatorname{mes}\big\{\, x \in U \;\big|\; b_1(x) \in \mathcal{M} \,\big\} = 0.$$

Denote by $\mathcal{X} : \partial\Omega \backslash F_0 \to \partial\Omega \backslash F_0$ the reflection mapping: $\mathcal{X}y = y^*$ (see Subsect. 1.2.2). Let $b_2 : \Gamma_+ \to \partial\Omega$ send a point $x_0 \in \Gamma_+$ to the point of the first intersection of the trajectory starting at x_0, with $\partial\Omega$. The mappings $\mathcal{X}$ and b_2 are local C^1–diffeomorphisms of the regular part of the boundary. Ignoring sets of zero measure we may assume that $X\big(t_2(x), x\big) = b_2 \circ \mathcal{X}\big(X(t_1, x), x\big)$ whence (due to what we have already proved)

$$\operatorname{mes}\big\{\, x \in U \;\big|\; X\big(t_2(x), x\big) \in \mathcal{M}) \,\big\} = 0.$$

Repeating these arguments $(m-2)$ times we finish the proof.

Now we can prove a "global" version of Lemma 1.

Lemma 3. *Let $u \in D$. For almost any $x \in \Omega$ the function $Z(t) = u(X(t,x))$ (after its correction on a set of zero measure) is absolutely continuous between instants when the trajectory $X(t,x)$ intersects the boundary, and $Z'(t) = f(X(t,x))$, $f = \{u, H_N\}$ (in the weak sense), $f(X(\cdot,x)) \in L^1_{\text{loc}}(0,\infty)$.*

Proof. Note first of all that the set $\mathfrak{M}_0 = \{x \in \Omega \mid R(x) = 0\}$ lies on the submanifold $\{(p,q) \in \Omega \mid p = 0\}$ of codimension $\nu N > 1$. Hence for almost every $x \in \Omega$ the trajectory $X(t,x)$ does not intersect $\mathfrak{M}_0$. In what follows we shall consider only trajectories with this property.

Let us fix an arbitrary $x_0 \in \Omega$ and prove the existence of a neighbourhood $U \ni x_0$ such that our assertion is true for almost all $x \in U$. Fix $T > 0$ and assume (discarding sets of zero measure) that the trajectory $X(t,x_0)$, $0 < t < T$ crosses the boundary on a finite number of time instants $t_1(x), \ldots, t_k(x)$ at regular points (the case when the trajectory does not cross the boundary is simpler and may be dealt with similarly).

Cover the trajectory's segment $X(t,x_0)$, $0 \leqslant t \leqslant t_1(x_0)$, by a finite number of balls $V_{1,1}, \ldots, V_{1,\ell_1}$, with centres on it, in such a way that a part of the trajectory in each of them is rectified by some diffeomorphism, and each ball intersect no more than two neighbouring ones. A similar covering $V_{\mu\nu}$ $(\mu = 1,\ldots,k; \nu = 1,\ldots \ell_\mu)$ is introduced for the next segments of the trajectory. The balls are indexed in the order of increase of t. Choose a neighbourhood U so that the whole tube $\mathbb{W} = \{X(t,x), x \in U, t \in [0,t_k]\}$ to be contained inside the covering.

Consider the last ball V_{k,ℓ_k}. The argument from the proof of Lemma 1 enables us to find a set $F_{k,\ell_k} \subset \partial V_{k,\ell_k} \cap \mathbb{W} \subset V_{k,\ell_k-1}$ of zero spherical measure such that the required assertion is valid for all parts of the trajectories going into the ball V_{k,ℓ_k} through the set

$$\mathcal{M}_{k,\ell_k} = (\partial V_{k,\ell_k} \cap \mathbb{W}) \setminus F_{k,\ell_k}.$$

For smaller t these trajectories go through the part $\mathbb{W}_{k,\ell_k-1}$ of the boundary $\partial V_{k,\ell_k-1}$ lying outside V_{k,ℓ_k}. Since the trajectory tube determines a C^1–diffeomorphism of spherical sets, we find a set $\mathcal{M}_{k,\ell_k-1} \subset \mathbb{W}_{k,\ell_k-1}$ whose complement in $\mathbb{W}_{k,\ell_k-1} \cap \mathbb{W}$ has zero measure, and for which our assertion is valid for trajectories coming from $\mathcal{M}_{k,\ell_k-1}$ into $V_{k,\ell_k-1} \cup V_{k,\ell_k}$. Repeating the procedure we obtain a set $\mathcal{M}_{k_1} \subset \partial V_{k_1}$, not inside the next balls, with the same property related to segments with $t_{k-1}(x) < t < t_k(x)$.

The inverse image of $\mathcal{M}_{k_1}$ with respect to the reflection $x \mapsto x^*$ has a full measure in the set cut by the tube out of the sphere $\partial V_{k-1.\ell_{k-1}}$. Repeating the whole reasoning we come at the end to a set $\mathcal{M}_{11}$ having a full measure in $\partial V_{11} \cap \mathbb{W}$, such that the required assertion is true for all the trajectories coming through $\mathcal{M}_{11}$. Now it remains to use rectifiability of trajectories in V_{11}, and the proof is over.

Let $u \in D$. By Lemma 1 for almost every $x \in \partial\Omega$ the function u has a limit value $u(x)$ along the trajectory going through x. Moreover, for almost all $x \in \partial\Omega$ there exist simultanuously the limit values $u(x)$, $u(x^*)$.

Theorem. *The domain $D(\mathcal{H}_N)$ consists of those functions $u \in D$ for which $u(x) = u(x^*)$ for almost all $x \in \partial\Omega$. If $u \in D(\mathcal{H}_N)$ then $\mathcal{H}_N u = \{u, H_N\}$ where the Poisson bracket is understood in the weak sense.*

Proof. Denote by $\mathcal{H}'_N$ the operator defined in the formulation of the theorem. We shall prove that $\mathcal{H}'_N = \mathcal{H}_N$.

Let $u \in D(\mathcal{H}'_N)$. By Lemma 1, for almost every $x \in \Omega$ the function $Z(t) = u(X(t,x))$ is continuous on $[0,\infty)$. Using Lemma 3 we obtain that for almost all $x \in \Omega$

$$u(X(t,x)) - u(x) = \int_0^t f(X(\tau,x))\, d\tau$$

where $f = \{u, H_N\} \in L^1(\Omega)$, $f(X(\cdot,x)) \in L^1_{\text{loc}}(0,\infty)$. Hence

$$\left\| \frac{1}{t}\left[u(X(t,\cdot)) - u\right] - f \right\|_{L^1(\Omega)} \leqslant \frac{1}{t}\int_0^t \| S^N(\tau) f - f \|_{L^1(\Omega)}\, d\tau. \qquad (1.I.7)$$

Since the group $S^N(\tau)$ is strongly continuous, the function $\| S^N(\tau) f - f \|$ is continuous with respect to τ and vanishes for $\tau = 0$, so that the right–hand side of (1.I.7) tends to 0 when $t \to 0$. This means that $u \in D(\mathcal{H}_N)$, $\mathcal{H}_N u = f$; thus $\mathcal{H}'_N \subset \mathcal{H}_N$.

It is clear that the operator $\mathcal{H}'_N$ is densely defined. Its domain is invariant with respect to the group $S^N(\tau)$. Indeed, the limit values along trajectories are the same for u and $S^N(\tau)u$. On the other hand, let $u \in D(\mathcal{H}'_N)$, $v_\tau = S^N(\tau)u$. Then

$$\frac{dv_\tau}{d\tau} = \mathcal{H}_N v_\tau = S^N(\tau)\mathcal{H}_N u = S^N(\tau)\mathcal{H}'_N u = S^N(\tau)\{u, H_N\}$$

whence, in particular, $\frac{dv_\tau}{d\tau} \in L^1(\Omega)$. For each real function $\varphi \in C^1(\Omega)$ with a compact support

$$\int_\Omega v_\tau(x)\{\varphi, H_N\}(x)\,dx = \int_\Omega v_\tau(x)\left[\frac{\partial}{\partial t}\left(S^N(t)\varphi\right)(x)\right]_{t=0} dx$$
$$= \frac{\partial}{\partial t}\left[\int_\Omega v_\tau(x)\left(S^N(t)\varphi\right)(x)\,dx\right]_{t=0}$$
$$= \frac{\partial}{\partial t}\left[\int_\Omega \left(S^N(-t)v_\tau\right)(x)\varphi(x)\,dx\right]_{t=0}$$
$$= -\frac{\partial}{\partial s}\left[\int_\Omega v_s(x)\varphi(x)\,dx\right]_{s=\tau}$$
$$= -\int_\Omega \frac{\partial v_\tau}{\partial \tau}(x)\varphi(x)\,dx.$$

so that $v_\tau \in D(\mathcal{H}'_N)$.

It is well known that the operator $\mathcal{H}_N$ is a closure of $\mathcal{H}'_N$. It remains to prove that $\mathcal{H}'_N$ is closed.

Let $u_m \in D(\mathcal{H}'_N)$, $f_m = \mathcal{H}'_N u_m$, $u_m \to u$. $f_m \to f$ in $L^1(\Omega)$. It follows from the definition of D that $u \in D$, $f \in \{u, H_N\}$ (in the weak sense). By Fubini's theorem (with notations from the proof of Lemma 1) we have

$$\| u_m - u \|_{L^1(\Omega)} \geqslant \| u_m - u \|_{L^1(U_0)} = \int_{V_0} \left| u_m\left(\Psi^{-1}(\xi)\right) - u\left(\Psi^{-1}(\xi)\right) \right| \mu(\xi)\,d\xi$$
$$\geqslant \gamma \int_{\Psi(\Gamma')} \nu(dy') \int_0^{\delta(x_0)} \Big| u_m\left(\Psi^{-1}(y'+et)\right)$$
$$- u\left(\Psi^{-1}(y'+et)\right) \Big| dt. \quad \gamma > 0.$$

This implies the existence of a subsequence $\{m_k\}$ such that for almost all $x' \in \Gamma'$

$$\int_0^{\delta(x_0)} \left| u_{m_k}(X(t,x')) - u(X(t,x'))\right| dt \to 0. \quad k \to \infty \tag{1.I.8}$$

Taking if necessary a subsequence of $\{m_k\}$ instead of $\{m_k\}$, we find also that

$$\int_0^{\delta(x_0)} \left| f_{m_k}(X(t,x')) - f(X(t,x')) \right| dt \to 0. \quad k \to \infty. \tag{1.I.9}$$

According to (6.I.1), for each m the function $f_m(X(t,x'))$ is a generalized derivative of the function $u_m(X(t,x'))$, and $f(X(t,x'))$ is a generalized derivative of $u(X(t,x'))$. The relations (8.I.1), (9.I.1) mean that $u_{m_k}(X(\cdot\,,x')) \to u(X(\cdot\,,x'))$ in the Sobolev space $\mathbb{W}_1^1(0,\delta(x_0))$. By the imbedding theorem [2] $u_{m_k}(X(t,x')) \to u(X(t,x'))$ uniformly with respect to $t \in [0,\delta(x_0)]$. In particular, $u_{m_k}(x') \to u(x')$ for almost all $x' \in \Gamma$. Taking a smaller neighbourhood U and a further subsequence of $\{m_k\}$, we also show that $u_{m_k}((x')^*) \to u((x')^*)$. Hence for almost all $x' \in \Gamma'$ or, in fact, for almost all $x \in \partial\Omega$ we find $u(x) = u(x^*)$, which concludes the proof.

APPENDIX II

The Formal Derivation of the Infinitesimal Operator [BB] of the Group $S^N(t)$.

In the proof of the Theorem 1.3.1 it has been calculated the derivative with respect to time of the functions $\left(S^N(t)f_N\right)(x)$ for $f_N \in L_0^1\left(\mathbb{R}^{\nu N} \times \left(\mathbb{R}^{\nu N} \backslash \mathbb{W}_N\right)\right)$. Here we will formally find such derivative for $f_N \in C^1\left(\mathbb{R}^{\nu N} \times \left(\mathbb{R}^{\nu N} \backslash \mathbb{W}_N\right)\right)$, where $C^1\left(\mathbb{R}^{\nu N} \times \left(\mathbb{R}^{\nu N} \backslash \mathbb{W}_N\right)\right)$ is the space of continuously differentiable functions that are equal to zero on the set of forbidden configurations $\mathbb{W}_N$.

For this problem, we will consider a typical example, that of a system of particles, which interact via a singular potential as hard spheres. To make our calculations more descriptive, we first examine the two–particle system. Let $f \in C^1\left(\mathbb{R}^{\nu 2} \times \left(\mathbb{R}^{\nu 2} \backslash \mathbb{W}_2\right)\right)$, then for each fixed point $x \equiv (x_1, x_2) \in \mathbb{R}^{2\nu} \times \left(\mathbb{R}^{2\nu} \backslash \mathbb{W}_2\right)$ we have

$$\begin{aligned}\left(\frac{d}{dt}S^2(t)f_2\right)(x_1,x_2) &= \lim_{\Delta t \to 0}\left(S^2(t)\frac{1}{\Delta t}\left(S^2(\Delta t) - I\right)f_2\right)(x_1,x_2)\\ &= \lim_{\Delta t \to 0}\left(S^2(t)\frac{1}{\Delta t}\left(S^2(\Delta t) - S_0^2(\Delta t)\right)f_2\right)(x_1,x_2)\\ &+ \lim_{\Delta t \to 0}\left(S^2(t)\frac{1}{\Delta t}\left(S_0^2(\Delta t) - I\right)f_2\right)(x_1,x_2). \qquad (1.\text{II}.1)\end{aligned}$$

where the operator $S^2(t)$ is defined by the formula (1.3.1) and $S_0^2(t)$ is a free motion operator:

$$\left(S_0^2(t)f_2\right)(x_1,x_2) = f_2\left(q_1 + p_1 t, p_1; q_2 + p_2 t, p_2\right).$$

Clearly, that expression $S^2(t)\left(S^2(\Delta t) - S_0^2(\Delta t)\right)f_2(x_1,x_2)$ in (1.II.1) differs from zero only when a collision between particles occurs during a time interval Δt. The collision takes place if the following inequalities are realized

$$|Q_2(t) - Q_1(t)| \geqslant a,$$

$$|Q_2(t) + P_2(t)\Delta t - Q_1(t) - P_1(t)\Delta t| < a. \qquad (1.\text{II}.2)$$

Denoting $(Q_2(t) - Q_1(t))/\,|Q_2(t) - Q_1(t)| = \eta$, we obtain for infinitesimal Δt

$$\begin{aligned} &\left| Q_2(t) - Q_1(t) + \Delta t\,(P_2(t) - P_1(t)) \right| \\ &\quad = \left| \eta\,|Q_2(t) - Q_1(t)| + \Delta t\,(P_2(t) - P_1(t)) \right| \\ &\quad = \Big((Q_2(t) - Q_1(t))^2 + 2\,|Q_2(t) - Q_1(t)| \\ &\qquad \times \langle \eta, (P_2(t) - P_1(t)) \rangle \Delta t + \Delta t^2\,(P_2(t) - P_1(t))^2 \Big)^{\frac{1}{2}} \\ &\quad = |Q_2(t) - Q_1(t)| + \langle \eta, (P_2(t) - P_1(t)) \rangle \Delta t + O\left(\Delta t^2\right), \end{aligned} \tag{1.II.3}$$

and hence for $\Delta t > 0$, in the domain defined by the inequalities (1.II.2), it holds

$$\langle \eta, (P_2(t) - P_1(t)) \rangle < 0.$$

In this domain we also have :

$$\Theta\Big(-\langle \eta, (P_2(t) - P_1(t)) \rangle \Big) = 1$$

$$\Theta\Big(|Q_2(t) - Q_1(t)| - a \Big) - \Theta\Big(|Q_2(t) - Q_1(t)| + \langle \eta, (P_2(t) - P_1(t)) \rangle \Delta t - a \Big) = 1, \tag{1.II.4}$$

where

$$\Theta(x) = \begin{cases} 1, & x > 0 \\ 0, & x \leqslant 0. \end{cases}$$

Since the considered expression $S^2(t)\left(S^2(\Delta t) - S_0^2(\Delta t)\right) f_2$ differs from zero only in the domain defined by (1.II.2), where relations (1.II.4) hold, we can represent, it taking into account (1.II.3), in the form

$$\begin{aligned} &S^2(t)\big(S^2(\Delta t) - S_0^2(\Delta t)\big) f_2(x_1, x_2) \\ &= \Theta\Big(-\langle \eta, (P_2(t) - P_1(t)) \rangle \Big) \Big\{ \Theta\Big(|Q_2(t) - Q_1(t)| - a \Big) \\ &\quad - \Theta\Big(|Q_2(t) - Q_1(t)| - a + \langle \eta, (P_2(t) - P_1(t)) \rangle \Delta t \Big) \Big\} \\ &\quad \times \Big\{ f_2\Big(Q_1(t + \Delta t), P_1(t + \Delta t); Q_2(t + \Delta t) . P_2(t + \Delta t) \Big) \\ &\quad - f_2\Big(Q_1(t) + P_1(t)\Delta t, P_1(t); Q_2(t) + P_2(t)\Delta t, P_2(t) \Big) \Big\}. \end{aligned} \tag{1.II.5}$$

Let us introduce one–dimensional δ–function

$$\delta(x) = \lim_{\varepsilon \to 0} \varepsilon^{-1}\big(\Theta(x) - \Theta(x - \varepsilon)\big).$$

Then, because $\langle \eta, (P_2(t) - P_1(t)) \rangle \Delta t < 0$, we have

$$\Theta\big(|Q_2(t) - Q_1(t)| - a\big) - \Theta\Big(|Q_2(t) - Q_1(t)| - a + \langle \eta, (P_2(t) - P_1(t)) \rangle \Delta t\Big)$$
$$= -\langle \eta, (P_2(t) - P_1(t)) \rangle \Delta t \, \delta\big(|Q_2(t) - Q_1(t)| - a\big).$$

In the domain (1.II.2) particles have the momenta after the collision :

$$P_1(t + \Delta t) \equiv P_1^*(t) = P_1(t) - \eta \langle \eta, (P_1(t) - P_2(t)) \rangle,$$
$$P_2(t + \Delta t) \equiv P_2^*(t) = P_2(t) + \eta \langle \eta, (P_1(t) - P_2(t)) \rangle.$$

Since the difference between $Q(t)$ and $Q(t+\Delta t)$ or $Q(t)+P(t)\Delta t$ is infinetely small, then, taking into account only the first order terms with respect to infinitesimal Δt, we find

$$S^2(t)\big(S^2(\Delta t) - S_0^2(\Delta t)\big) f_2(x_1, x_2)$$

$$= \Delta t \big| \langle \eta, (P_2(t) - P_1(t)) \rangle \big| \, \Theta\big(-\langle \eta, (P_2(t) - P_1(t)) \rangle\big) \, \delta\big(|Q_2(t) - Q_1(t)| - a\big)$$

$$\times \left\{ f_2\big(Q_1(t), P_1^*(t); Q_2(t), P_2^*(t)\big) - f_2\big(Q_1(t), P_1(t); Q_2(t), P_2(t)\big) \right\}. \qquad (1.II.6)$$

We remark that the right–hand side of (1.II.6) is a function of the form

$$\delta(|r| - a) \, f\left(\frac{r}{|r|}\right) = \int dR \, \delta(r - R) \, \delta(|R| - a) \, f\left(\frac{R}{|R|}\right).$$

Let us introduce a unit vector η in the direction of vector R, then $R = |R|\,\eta$ and $dR = R^2 \, d|R| \, d\eta$, where $d\eta$ is an element of solid angle. Thus, we find

$$\delta(|r| - a) \, f\left(\frac{r}{|r|}\right) = \int d|R| \, d\eta \, R^2 \, \delta(r - |R|\eta) \, \delta(|R| - a) f(\eta)$$
$$= a^2 \int \delta(r - a\eta) f(\eta) \, d\eta. \qquad (1.II.7)$$

Using the identity (1.II.7) we can write for the expression (1.II.6) the following representation

$$S^2(t)\left(S^2(\Delta t) - S_0^2(\Delta t)\right) f_2(x) = \Delta t\, a^2 \int d\eta\, \Theta\big(\langle \eta, (P_1(t) - P_2(t))\rangle\big) \times \big|\langle \eta, (P_1(t) - P_2(t))\rangle\big|\, \delta\big(Q_2(t) - Q_1(t) - a\eta\big) \times \Big\{ f_2\big(Q_1(t), P_1^*(t) ; Q_2(t), P_2^*(t)\big) - f_2\big(Q_1(t), P_1(t); Q_2(t), P_2(t)\big)\Big\}. \tag{1.II.8}$$

The expression $S^2(t)\left(S_0^2(\Delta t) - I\right) f_2(x_1, x_2)$ from (1.II.2) can be found by direct calculation

$$S^2(t)\left(S_0^2(\Delta t) - I\right) f_2(x_1, x_2) = \Delta t \sum_{i=1}^{2} \left\langle P_i(t), \frac{\partial}{\partial Q_i(t)} \right\rangle f_2\big(Q_1(t), P_1(t); Q_2(t), P_2(t)\big). \tag{1.II.9}$$

Finally, we have the following expression for the time derivative (1.II.2)

$$\left(\frac{d}{dt} S^2(t) f_2\right)(x_1, x_2) = \sum_{i=1}^{2} \left\langle P_i(t), \frac{\partial}{\partial Q_i(t)} \right\rangle f_2\big(Q_1(t), P_1(t); Q_2(t), P_2(t)\big) + a^2 \int d\eta\, \Theta\big(\langle \eta, (P_1(t) - P_2(t))\rangle\big) \big|\langle \eta, (P_1(t) - P_2(t))\rangle\big| \times \delta\big(Q_2(t) - Q_1(t) - a\eta\big) \times \Big\{ f_2\big(Q_1(t), P_1^*(t); Q_2(t), P_2^*(t)\big) - f_2\big(Q_1(t), P_1(t); Q_2(t), P_2(t)\big)\Big\}. \tag{1.II.10}$$

We define the function $f_2\big(Q_1(t) + P_1(t)\Delta t, P_1(t); Q_2(t) + P_2(t)\Delta t, P_2(t)\big)$ as

$$f_2\big(Q_1(t), P_1(t); Q_2(t), P_2(t)\big) + \Delta t \sum_{i=1}^{2} \left\langle P_i^*(t), \frac{\partial}{\partial Q_i(t)} \right\rangle \times f_2\big(Q_1(t), P_1(t); Q_2(t), P_2(t)\big)$$

even if the points $\big(Q_1(t) + P_1(t)\Delta t, P_1(t); Q_2(t) + P_2(t)\Delta t, P_2(t)\big)$ belong to forbidden configuration $\mathbb{W}_2$.

For a system of N particles we have the following result for each fixed point $x \equiv (x_1, \ldots, x_N) \in \mathbb{R}^{\nu N} \times \left(\mathbb{R}^{\nu N} \backslash \mathbb{W}_N\right) \backslash \mathcal{M}_N^0$

$$\frac{d}{dt} S^N(t) f_N = S^N(t) \left(\mathcal{H}_N^0 + T_N\right) f_N = \left(\mathcal{H}_N^0 + T_N\right) S^N(t) f_N,$$

where

$$\left(\mathcal{H}_N^0 f_N\right)(x) = \sum_{i=1}^{N} \left\langle p_i, \frac{\partial}{\partial q_i} \right\rangle f_N(x_1, \ldots, x_N)$$

and

$$\begin{aligned}(T_N f_N)(x) = a^2 \sum_{1=i<j}^{N} \int_{\mathbb{S}_+^{\nu-1}} d\eta \, \langle \eta, (p_i - p_j) \rangle \, \delta(q_j - q_i - a\eta) \\ \times \Big\{ f_N\Big(x_1, \ldots; q_i, p_i - \eta \langle \eta, (p_i - p_j) \rangle; \\ \ldots; q_j, p_j + \eta \langle \eta, (p_i - p_j) \rangle; \ldots, x_N\Big) \\ - f_N(x_1, \ldots, x_N) \Big\},\end{aligned} \tag{1.II.11}$$

where $\mathbb{S}_+^{\nu-1} = \left\{ \eta \in \mathbb{R}^\nu \mid |\eta| = 1, \langle \eta, (p_i - p_j) \rangle > 0 \right\}$. Remark that $(T_N f_N)(x) = 0$ for functions which satisfy the boundary condition $f_N(x) = f_N(x^*)$ on boundary $\partial \mathbb{W}_N$, and the infinitesimal operator of $S_N(t)$ coincides with $\mathcal{H}_N^0$ with boundary condition.

CHAPTER II

The Initial Value Problem for the BBGKY Hierarchy of a System of a Finite Number of Particles

2.1. Introduction

In this chapter we consider the BBGKY hierarchy as an abstract evolution equation in the space of sequences of summable functions L^1. The existence of solutions for a large class of singular interaction potentials and initial data in L^1 will be proved. From a physical point of view these solutions describe only states with a finite system of particles.

The results in this field are relatively new. Starting with 1972, the BBGKY hierarchy was first investigated as an abstract evolution equation in quantum statistical mechanics by Petrina [Pe2] [Pe3], who showed that the corresponding Cauchy problem has a unique solution in the space of sequences of trace class operators. In a series of papers of Petrina, Gerasimenko and Vidybida [PV][Pe1][PG1][Ger1] it was established that the Cauchy problem for the BBGKY hierarchy of different classical systems is soluble in the space of sequences of summable functions L^1. To be precise, the Cauchy problem for the BBGKY hierarchy in classical statistical mechanics was solved in L^1 for symmetric systems with smooth potentials with compact support by Petrina and Vidybida in 1972 [PV]. In these papers the existence of the group of evolution operators for the BBGKY hierarchy was established. These results were generalized for nonsymmetric systems by Gerasimenko and Petrina [Ger1][PG1][PG2] and the corresponding groups of evolution operators were constructed.

The Cauchy problem for the BBGKY hierarchy for systems of particles with singular interaction potentials (with hard core) was solved in L^1 by Gerasimenko and Petrina [GP1][PG1][PG2][PG3] and the existence of the group of evolution operators was also established.

Considerable difficulties connected with the fact that the operator defining the BBGKY hierarchy is unbounded and that the properties of its spectrum were not known were overcome. Moreover nothing was known about the resolvent of this operator. Therefore, it was impossible to construct an evolution operator for the BBGKY hierarchy by traditional methods, i. e. by using some information about the spectrum and the resolvent of the operator defining the BBGKY hierarchy.

Original methods for the construction of the evolution operator have been proposed. One of them is based on the fact that the evolution equation (the BBGKY hierarchy) can be converted into a form in which the operator in the right hand side of the BBGKY hierarchy is the derivative of a known operator, and in this form the equation can be integrated explicitly [PV].

Another method is based on the fact that starting out from the physical meaning of the BBGKY hierarchy one can write down the solution of the Cauchy problem immediately and define the BBGKY evolution operator for infinitely many particles in terms of the evolution operators for finitely many particles and of an operator analogous to the annihilation operator in quantum field theory [Pe1][PG1][PG2][PG3]. In [Ger1][BM1][MT] this method was extended to general BBGKY equations. The mathematical justification of the derivation of the BBGKY hierarchy as an evolution equation in the space L^1 of summable functions with the operator in right-hand side, which is the infinitesimal generator of a group of bounded evolution operator (and hence closed) will also be given in this chapter.

A third method of justification is based on giving a rigorous mathematical meaning to every term of the iteration series for the solutions of BBGKY hierarchy [PG3][GP2] and on the proof of convergence of the iteration series. The main difficulty in this problem is that the phase trajectories for a system with a singular interaction potential are defined almost everywhere on the phase space and initial distribution functions in the iteration series are concentrated on a lower dimension manifolds. It is necessary to check that the trajectories are defined on these manifolds. This problem was discussed in many papers [PG3][GP2][IP][CIP] and we shall present a complete solution of it.

The plan of this chapter is as follows. In Section 2 we introduce and describe the properties of the evolution operator for the BBGKY hierarchy. In Section 3 the BBGKY hierarchy is considered as an abstract evolution equation in L^1, and we formulate the initial value problem and state the existence of the solutions. The proof of these results is based on some papers of two of the authors and Vidybida [PV][Pe1][PG1][PG2] [PG3].

In Section 4 we introduce a new evolution operator for the BBGKY hierarchy of a one–dimensional system of particles interacting with their nearest neighbours. We also prove existence theorems for the solutions of the hierarchy in the case of these particle systems.

The connection between the iteration series for the BBGKY hierarchy and a functional series constructed in a previous section is discussed in Section 5. In that section wealso prove the existence of the iteration series.

2.2. The Evolution Operator for BBGKY Hierarchy

2.1.1. Definitions and basic result

In the present section we start the study of the Cauchy problem for the BBGKY hierarchy (1.4.7-9) in the space of sequences of summable functions L^1. First of all, in all this section we shall study the properties of the evolution operator for BBGKY hierarchy.

In Chapter 1 the Banach space L^1 whose elements are the infinite sequences $f = \{f_n(x_1, \dots, x_n)\}_{n \geqslant 0}$ of summable symmetric functions $f_n(x_1, \dots . x_n)$, defined on the phase space $\mathbb{R}^{\nu n} \times (\mathbb{R}^{\nu n} \backslash \mathrm{W}_n)$ of the n–particle system and equal to zero on the set W_n, was introduced. The norm of f in L^1 was introduced by the formula

$$\|f\| = \sum_{n=0}^{\infty} \int_{(\mathbb{R}^\nu \times \mathbb{R}^\nu)^n} dx_1 \dots dx_n \left| f_n(x_1, \dots . x_n) \right| = \sum_{n=0}^{\infty} \int_{(\mathbb{R}^\nu \times \mathbb{R}^\nu)^n} d(x)_n \left| f_n((x)_n) \right| .$$

The subset $L_0^1 \subset L^1$ consists of finite sequences of continuously differentiable functions with compact support and equal to zero in some ε–neighbourhood of the set W_n.

The isometric evolution operator $S(t)$ (1.3.1),(1.3.14) is defined on L^1 as a direct sum of isometric n–particle evolution operators $S_n(t)$ and also the operators

$$\left(e^{\pm \int dx} f\right)_k (x_1, \dots, x_k) =$$

$$\sum_{n=0}^{\infty} \frac{(\pm 1)^n}{n!} \int_{(\mathbb{R}^\nu \times \mathbb{R}^\nu)^n} dx_{k+1} \dots dx_{k+n} f_{k+n}(x_1, \dots, x_k, x_{k+1}, \dots, x_{k+n})$$

are defined (see Section 1.4).

Let us consider the following evolution operator on L^1

$$U(t) = e^{\int dx} S(-t) \, e^{-\int dx} . \tag{2.2.1}$$

The properties of this operator are described in the following theorem.

Theorem 2.2.1. *The operators* $U(t) = e^{\int dx} S(-t) e^{-\int dx}$, $t \in \mathbb{R}^1$ *are defined and bounded in* L^1,

$$\|U(t)\| \leqslant e^2,$$

are strongly continuous, and form a group. The infinitesimal generator $\mathcal{B}$ *of this group is closed.* $\mathcal{B}U(t) = U(t)\mathcal{B}$ *and its spectrum is concentrated on the imaginary axis. On the set* L_0^1, *everywhere dense in* L^1, $\mathcal{B}$ *coincides with the operator*

$$\mathcal{B} = -\mathcal{H} + \left[\mathcal{H}, \int dx\right] = -\mathcal{H} + a^{\nu-1}A,$$

or, componentwise

$$\left(\frac{d}{dt}U(t)\bigg|_{t=+0} f\right)_n (x_1,\ldots,x_n) = -(\mathcal{H}f)_n(x_1,\ldots,x_n)$$

$$+ \int\limits_{\mathbb{R}^\nu \times \mathbb{R}^\nu} dx_{n+1} \left\{ \sum_{i=1}^{n} \Phi(q_i - q_{n+1}), f_{n+1}(x_1,\ldots,x_{n+1}) \right\} \tag{2.2.2}$$

$$+a^{\nu-1} \sum_{i=1}^{n} \int\limits_{\mathbb{R}^\nu \times \mathbb{S}_+^{\nu-1}} d\eta dp_{n+1} \langle \eta, (p_i - p_{n+1})\rangle \left(f_{n+1}(x_1,\ldots; q_i, p_i - \eta\langle\eta,(p_i - p_{n+1})\rangle,\right.$$

$$\left.\ldots, x_n, q_i - a\eta, p_{n+1} + \eta\langle\eta,(p_i - p_{n+1})\rangle) - f_{n+1}(x_1,\ldots,x_n,; q_i + a\eta, p_{n+1})\right),$$

and, analogously,

$$\left(\frac{d}{dt}U(t)\bigg|_{t=-0} f\right)_n (x_1,\ldots,x_n) = -(\mathcal{H}f)_n(x_1,\ldots,x_n)$$

$$+ \int\limits_{\mathbb{R}^\nu \times \mathbb{R}^\nu} dx_{n+1} \left\{ \sum_{i=1}^{n} \Phi(q_i - q_{n+1}), f_{n+1}(x_1,\ldots,x_{n+1}) \right\} \tag{2.2.3}$$

$$+a^{\nu-1} \sum_{i=1}^{n} \int\limits_{\mathbb{R}^\nu \times \mathbb{S}_+^{\nu-1}} d\eta dp_{n+1} \langle \eta, (p_i - p_{n+1})\rangle \left(f_{n+1}(x_1,\ldots,x_n; q_i - a\eta, p_{n+1}) - \right.$$

$$\left. f_{n+1}(x_1,\ldots q_i, p_i - \eta\langle\eta,(p_i - p_{n+1})\rangle,\ldots,x_n; q_i + a\eta, p_{n+1} + \eta\langle\eta,(p_i - p_{n+1})\rangle)\right).$$

In the formulae (2.2.2) and (2.2.3) $\mathcal{H}$ is the infinitesimal generator of the group $S(t)$ (see Theorem 1.3.1), $\{.,.\}$ is the Poisson bracket (1.2.15) and $\mathbb{S}_+^{\nu-1} =$

$\{\eta \in \mathbb{R}^{\nu} \mid |\eta| = 1, \langle\eta, (p_i - p_{n+1})\rangle \ > 0\}$. In order to prove the theorem we need a few auxiliary statements.

2.2.2. Some auxiliary results

Let us calculate the commutator of the operators $\int dx$, given by (1.4.1), and $\mathcal{H}$, given by (1.3.15), in L^1.

Lemma 2.2.1. *For any $f \in L_0^1 \subset L^1$ we have*

$$\left(\left[\mathcal{H}, \int dx\right] f\right)_n (x_1, \ldots, x_n) =$$

$$\sum_{i=1}^{n} \int_{\mathbb{R}^{\nu}\times\mathbb{R}^{\nu}} dx_{n+1} \{\Phi(q_i - q_{n+1}), f_{n+1}(x_1, \ldots, x_{n+1})\}$$

$$+a^{\nu-1} \sum_{i=1}^{n} \int_{\mathbb{R}^{\nu}\times\mathbb{S}^{\nu-1}} d\eta dp_{n+1} \langle\eta, (p_i - p_{n+1})\rangle f_{n+1}(x_1, \ldots, x_n; q_i - a\eta, p_{n+1}) .$$

$$\textit{where } \mathbb{S}^{\nu-1} = \{\eta \in \mathbb{R}^{\nu} \mid |\eta| = 1\}, \nu \geqslant 2. \tag{2.2.4}$$

Proof. To prove (2.2.4), we consider in detail the projection of $\int dx \mathcal{H} f$ (where $f \in L_0^1$) onto subset $L^1(\mathbb{R}^{\nu n} \times (\mathbb{R}^{\nu n} \backslash \mathbb{W}_n))$

$$\left(\int dx \mathcal{H} f\right)_n (x_1, \ldots, x_n) = -\int_{\mathbb{R}^{\nu}\times\mathbb{R}^{\nu}} dx_{n+1} \left\{\sum_{i=1}^{n} \Phi(q_i - q_{n+1}), f_{n+1}(x_1, \ldots, x_{n+1})\right\}$$
$$+ \int_{\mathbb{R}^{\nu}\times\mathbb{R}^{\nu}} dx_{n+1} \mathcal{H}_n f_{n+1}(x_1, \ldots, x_{n+1})$$
$$+ \int_{\mathbb{R}^{\nu}\times\mathbb{R}^{\nu}} dx_{n+1} \langle p_{n+1}, \frac{\partial}{\partial q_{n+1}}\rangle f_{n+1}(x_1, \ldots, x_{n+1}) . \tag{2.2.5}$$

Since the integration over the configuration variables for $f \in L_0^1$ in (2.2.5) is in fact carried out over the allowed configurations, the boundary of the domain of integration depends on the variables $q_1, \ldots, q_n$. Taking this into account we will change the order of differentiation with respect to $q_1, \ldots, q_n$ and integration with respect to q_{n+1} in the second term, which is obviously possible because $f \in L_0^1$. We finally obtain

$$\int\limits_{\mathbb{R}^\nu \times \mathbb{R}^\nu} dx_{n+1} \mathcal{H}_n f_{n+1}(x_1, \ldots x_{n+1}) = \mathcal{H}_n \int\limits_{\mathbb{R}^\nu \times \mathbb{R}^\nu} dx_{n+1} f_{n+1}(x_1, \ldots, x_{n+1})$$

$$- \sum_{i=1}^{n} \int\limits_{\mathbb{R}^\nu} dp_{n+1} \int d\sigma \langle \eta, p_i \rangle f_{n+1}(x_1, \ldots, x_{n+1}), \tag{2.2.6}$$

where η is the inner normal to the surface $|q_{n+1} - q_i| = a$ with center at the point q_i, and $d\sigma$ is an element of this surface. Let us introduce in (2.2.6) new variables of integration by the formula

$$a\eta = q_i - q_{n+1},$$

then $d\sigma = a^{\nu-1} d\eta$, where $d\eta$ is the element of solid angle. The expression (2.2.6) in this case takes the form

$$\int\limits_{\mathbb{R}^\nu \times \mathbb{R}^\nu} dx_{n+1} \mathcal{H}_n f_{n+1}(x_1, \ldots, x_{n+1}) = \left(\mathcal{H} \int dx f \right)_n (x_1, \ldots, x_n)$$

$$-a^{\nu-1} \sum_{i=1}^{n} \int\limits_{\mathbb{R}^\nu} dp_{n+1} \int\limits_{\mathbb{S}^{\nu-1}} d\eta \langle \eta, p_i \rangle f_{n+1}(x_1, \ldots, x_n; q_i - a\eta, p_{n+1}). \tag{2.2.7}$$

The third term on the right–hand side of (2.2.5) can be transformed by using the Gauss theorem. For $f \in L_0^1$ we obtain

$$\int\limits_{\mathbb{R}^\nu \times \mathbb{R}^\nu} dx_{n+1} \langle p_{n+1}, \frac{\partial}{\partial q_{n+1}} \rangle f_{n+1}(x_1, \ldots, x_{n+1}) \tag{2.2.8}$$

$$= a^{\nu-1} \sum_{i=1}^{n} \int\limits_{\mathbb{R}^\nu} dp_{n+1} \int\limits_{\mathbb{S}^{\nu-1}} d\eta \langle \eta, p_{n+1} \rangle f_{n+1}(x_1, \ldots; q_i - a\eta, p_{n+1}).$$

Therefore it follows from (2.2.5–8) that

$$\left(\left[\mathcal{H}, \int dx \right] f \right)_n (x_1, \ldots, x_n) =$$

$$\sum_{i=1}^{n} \int\limits_{\mathbb{R}^\nu \times \mathbb{R}^\nu} dx_{n+1} \{ \Phi(q_i - q_{n+1}), f_{n+1}(x_1, \ldots, x_{n+1}) \}$$

$$+a^{\nu-1}\sum_{i=1}^{n}\int_{\mathbb{R}^\nu} dp_{n+1}\int_{\mathbb{S}^{\nu-1}} d\eta\langle\eta,(p_i-p_{n+1})\rangle f_{n+1}(x_1,\ldots,x_n,q_i-a\eta,p_{n+1}). \quad (2.2.9)$$

The expression (2.2.9) can be identically transformed in the following way. We split the second term of (2.2.9) into two parts: in the first part the integration is over the domain $\mathbb{R}^\nu\times\mathbb{S}_+^{\nu-1}$, where $\mathbb{S}_+^{\nu-1}=\{\eta\in\mathbb{R}^\nu|\,|\eta|=1,\langle\eta,(p_i-p_{n+1})\rangle\ >0\}$, while in the second part it is over the domain $\mathbb{R}^\nu\times\mathbb{S}_-^{\nu-1}$, where $\mathbb{S}_-^{\nu-1}=\{\eta\in\mathbb{R}^\nu|\,|\eta|=1,$ $\langle\eta,(p_i-p_{n+1})\rangle\ <0\}$. Making the change of variables $\eta\to-\eta$ in the second part and using the boundary condition for the Poisson bracket $\mathcal{H}$ (see Theorem 1.3.1) in the case $\Delta t\to+0$, the expression (2.2.9) is transformed as follows

$$\left(\left[\mathcal{H},\int dx\right]f\right)_n(x_1,\ldots,x_n)=$$

$$\sum_{i=1}^{n}\int_{\mathbb{R}^\nu\times\mathbb{R}^\nu} dx_{n+1}\{\Phi(q_i-q_{n+1}),f_{n+1}(x_1,\ldots,x_{n+1})\}$$

$$+a^{\nu-1}\sum_{i=1}^{n}\int_{\mathbb{R}^\nu\times\mathbb{S}_+^{\nu-1}} d\eta dp_{n+1}\langle\eta,(p_i-p_{n+1})\rangle\,(f_{n+1}(x_1,\ldots,q_i,p_i-\eta\langle\eta,(p_i-p_{n+1})\rangle,$$

$$\ldots,x_n;q_i-a\eta,p_{n+1}+\eta\langle\eta,(p_i-p_{n+1})\rangle)-f_{n+1}(x_1,\ldots,x_n;q_i+a\eta,p_{n+1})). \quad (2.2.10)$$

In a similar way the form (2.2.3) for expression (2.2.9) is determined in the case $\Delta t\to-0$. Lemma 2.2.1 was proved in a paper [Ce3] by Cercignani.

Using the same argument as above, and taking into account that the functions of the sequence $f\in L^1$ are symmetric, we immediately obtain

Lemma 2.2.2. *The following operator equalities hold for $f\in L_0^1$*

$$\left[\left(\int dx\right)^m,\mathcal{H}\right]f=m\left[\int dx,\mathcal{H}\right]\left(\int dx\right)^{m-1}f,\quad m\geqslant 2. \quad (2.2.11)$$

Proof. In fact, we have the following identity

$$\left[\left(\int dx\right)^m,\mathcal{H}\right]f=\sum_{k=1}^{m}\left(\int dx\right)^{k-1}\left[\int dx,\mathcal{H}\right]\left(\int dx\right)^{m-k}f$$

$$= \sum_{k=1}^{m} \left(\sum_{\ell=1}^{k-1} \left(\int dx \right)^{\ell-1} \left[\int dx, \left[\int dx, \mathcal{H} \right] \right] \left(\int dx \right)^{k-1-\ell} \right.$$

$$\left. + \left[\int dx, \mathcal{H} \right] \left(\int dx \right)^{k-1} \right) \left(\int dx \right)^{m-k} f \quad .$$

Then if we assume that for $f \in L_0^1$

$$\left[\int dx, \left[\int dx, \mathcal{H} \right] \right] f = 0 \quad , \tag{2.2.12}$$

we obtain (2.2.11).

Now let us show that the expression $\left[\int dx, \left[\int dx, \mathcal{H}\right]\right] f$ is zero. We consider the projection of this expression onto $L^1 (\mathbb{R}^{\nu n} \times (\mathbb{R}^{\nu n} \backslash \mathbb{W}_n))$, for example, in the case $\Delta t \to +0$. From (2.2.10) we have

$$\left(\left[\int dx, \left[\int dx, \mathcal{H} \right] \right] f \right)_n (x_1, \ldots, x_n) \tag{2.2.13}$$

$$= \int\limits_{\mathbb{R}^\nu \times \mathbb{R}^\nu} dx_{n+1} \int\limits_{\mathbb{R}^\nu \times \mathbb{S}_+^{\nu-1}} d\eta dp_{n+2} \langle \eta, (p_{n+1} - p_{n+2}) \rangle f_{n+2} (x_1, \ldots, x_{n+1}; q_{n+1} + a\eta, p_{n+2})$$

$$- \int\limits_{\mathbb{R}^\nu \times \mathbb{R}^\nu} dx_{n+1} \int\limits_{\mathbb{R}^\nu \times \mathbb{S}_+^{\nu-1}} d\eta dp_{n+2} \langle \eta, (p_{n+1} - p_{n+2}) \rangle f_{n+2} (x_1, \ldots, x_n; q_{n+1}, p_{n+1}$$

$$- \eta \langle \eta, (p_{n+1} - p_{n+2}) \rangle; q_{n+1} - a\eta, p_{n+2} + \eta \langle \eta, (p_{n+1} - p_{n+2}) \rangle) .$$

Let us show that the two terms in (2.2.13) coincide identically. For this purpose, we change the variables of integration in the second term by the formulae

$$p'_{n+1} = p_{n+1} - \eta \langle \eta, (p_{n+1} - p_{n+2}) \rangle,$$
$$p'_{n+2} = p_{n+2} + \eta \langle \eta, (p_{n+1} - p_{n+2}) \rangle.$$

The Jacobian of this transformation is unity, and the following equality holds

$$\langle \eta, (p_{n+1} - p_{n+2}) \rangle = - \langle \eta, (p'_{n+1} - p'_{n+2}) \rangle.$$

Changing the variables $\eta \to -\eta$, we obtain that, in the new variables, the second term in (2.2.13) coincides identically with the first term in (2.2.13), and thus (2.2.12) holds. The equality (2.2.12) was proved in [Ce1][Ce2][Ce3].

2.2.3. Properties of the BBGKY evolution operator

Now let us prove Theorem 2.2.1. The operator $U(t) = e^{\int dx} S(-t) e^{-\int dx}$ is defined as the products of three bounded operators in L^1; moreover

$$\|e^{\pm \int dx}\| \leqslant e \quad \text{and} \quad \|S(-t)\| = 1, \quad \text{thus} \quad \|U(t)\| \leqslant e^2.$$

The operators $U(t)$, $t \in \mathbb{R}^1$ possess the group property, for example

$$U(t_1) U(t_2) = e^{\int dx} S(-t_1) e^{-\int dx} e^{\int dx} S(-t_2) e^{-\int dx} = e^{\int dx} S(-t_1) S(-t_2) e^{-\int dx}$$
$$= e^{\int dx} S(-t_1 - t_2) e^{-\int dx} = U(t_1 + t_2)$$

for arbitrary $t_1, t_2 \in \mathbb{R}^1$.

The strong continuity of the group $U(t)$ defined in (2.2.1) follows from the strong continuity of $S(-t)$ (Theorem 1.3.1) and the boundedness of the operators $e^{\mp \int dx}$. It follows from the general properties of the groups of bounded operators [RS][Yo] that there exists an infinitesimal generator $\mathcal{B}$ of the group $U(t)$; $\mathcal{B}$ is closed and its spectrum is concentrated on the imaginary axis.

Let us show that on the subspace $L_0^1 \subset L^1$ this operator $\mathcal{B}$ coincides with the operator $-\mathcal{H} + [\mathcal{H}, \int dx]$ given by (2.2.2–3).

Taking into account the group properties of $U(t)$ and expanding the exponentials as power series, we obtain for $f \in L_0^1$ (in the sense of strong convergence in L^1):

$$\lim_{\Delta t \to +0} \frac{1}{\Delta t} (U(t + \Delta t) f - U(t) f) = \lim_{\Delta t \to +0} U(t) \frac{1}{\Delta t} (U(\Delta t) - I) f \tag{2.2.14}$$

$$= U(t) \lim_{\Delta t \to +0} \frac{1}{\Delta t} \Bigg[(S(-\Delta t) - I) f + \int dx \, (S(-\Delta t) - I) f - (S(-\Delta t) - I) \int dx f$$
$$+ \sum_{n=2}^{\infty} \sum_{k=0}^{n} \frac{(-1)^k}{k! \, (n-k)!} \left(\int dx \right)^{n-k} (S(-\Delta t) - I) \left(\int dx \right)^k f \Bigg].$$

Since the sequence $f \in L_0^1$ is finite, then $\left(\int dx\right)^n f = 0$ for sufficiently large n. Hence the sum over n in (2.2.14) contains for each $f \in L_0^1$ a finite number of terms. Moreover, $\int dx f$ is also a finite sequence of differentiable functions with compact support, and therefore, we can pass to the limit in (2.2.14) term by term. Using the fact that $\mathcal{H}$ is the infinitesimal generator of the group $S(t)$ (Theorem 1.3.1), the expression (2.2.14) can be transformed as follows

$$\lim_{\Delta t \to +0} U(t) \frac{1}{\Delta t} (U(\Delta t) - I) f = U(t) \left(-\mathcal{H} f + \left[\mathcal{H}, \int dx \right] f \right. \tag{2.2.15}$$

$$\left. - \sum_{n=2}^{\infty} \sum_{k=0}^{n} \frac{(-1)^k}{k!\,(n-k)!} \left(\int dx \right)^{n-k} \mathcal{H} \left(\left(\int dx \right)^k f \right) \right),$$

where the bracket $[.,.]$ is a commutator.

It follows from (2.2.11) that the series in the right–hand side of (2.2.15) can be transformed into

$$\sum_{n=2}^{\infty} \sum_{k=0}^{n} \frac{(-1)^k}{k!\,(n-k)!} \left(\int dx \right)^{n-k} \mathcal{H} \left(\int dx \right)^k f$$

$$= \sum_{n=2}^{\infty} \sum_{k=0}^{n} \frac{(-1)^k}{k!\,(n-k)!} \left(\left[\left(\int dx \right)^{n-k}, \mathcal{H} \right] \left(\int dx \right)^k f + \mathcal{H} \left(\int dx \right)^n f \right)$$

$$= \sum_{n=2}^{\infty} \sum_{k=0}^{n} \frac{(-1)^k}{k!\,(n-k)!} \left((n-k) \left[\int dx, \mathcal{H} \right] \left(\int dx \right)^{n-1} f + \mathcal{H} \left(\int dx \right)^n f \right).$$

This expression vanishes because the sum over k is identically zero.

Hence we have proved the equality

$$\lim_{\Delta t \to +0} U(t) \frac{1}{\Delta t} (U(\Delta t) - I) f = U(t) \left(-\mathcal{H} + \left[\mathcal{H}, \int dx \right] \right) f,$$

where the expression $\left[\mathcal{H}, \int dx \right] f$ is given by the formula (2.2.10). The following equality holds

$$\lim_{\Delta t \to +0} \frac{1}{\Delta t} (U(\Delta t) - I) U(t) f = \left(-\mathcal{H} + \left[\mathcal{H}, \int dx \right] \right) U(t) f.$$

This formula follows from the general theory of continuous groups of bounded operators [Ka][Kr][RS][Yo], but we prefer to give a detailed proof.

Let us first remark that the functions

$S^{s+n-k}(-t, x_1, \ldots, x_s, x_{s+1}, \ldots, x_{s+n-k}) f_{s+n}(0, x_1, \ldots, x_s, \ldots, x_{s+n-k}, \ldots, x_{s+n})$

are (possibly) different from zero in a neighborhood of the set of forbidden configurations where the functions $f_{s+n}(x_1, \ldots x_{s+n}) \in L_0^1$ vanish. From this it also follows that the functions $\left(U(t) f\right)_{s+1}(x_1, , \ldots x_{s+1})$, $f \in L_0^1$ are (possibly) different from zero in a neighborhood of forbidden configurations and on the hypersurfaces $|q_i - q_{s+1}| = a$, $i = (1, \ldots, s)$. This means that, in the formula

$$\lim_{\Delta t \to +0} \frac{1}{\Delta t} (U(\Delta t) - I) U(t) f = \left(-\mathcal{H} + a^{\nu-1} A\right) U(t) f \quad ,$$

the second term $a^{\nu-1} A U(t) f)_s(x_1, , \ldots x_s)$ is different from zero.

The proof of this equality is essentially the same as for the previous one. Indeed the finite sequences $U(t)f \in L^1$ consist of functions $(U(t)f)_s(x_1, \ldots x_s)$ with compact support which are continuously differentiable with respect to x almost everywhere and belong to L^1 together with their first derivatives. This property is a direct consequence of the representation of $(U(t)f)_s(x_1, , \ldots x_s)$ in terms of $f_1(x_1), \ldots, f_{n_0}(x_1, \ldots x_{n_0}), f_n = 0, n > n_0$:

$$\left(U(t)f\right)_s(x_1, \ldots, x_s) = \sum_{n=0}^{n_0 - s} \sum_{k=0}^{n} \frac{(-1)^k}{k!(n-k)!} \int dx_{s+1} \ldots dx_{s+n} \times$$
$$\times\, S^{s+n-k}(-t, x_1, \ldots, x_s, x_{s+1}, \ldots, x_{s+n-k}) \times$$
$$\times\, f_{s+n}(x_1, \ldots, x_{s+n-k}, \ldots, x_{s+n})$$

and of the properties of the sequence $S(-t)f$ for $f \in L_0^1$ (see section 1.3).

The operator $S(-\Delta t)$ is defined on $U(t)f$ according to the formula

$$S^{s-l}(-\Delta t, x_1, \ldots, x_{s-l}) \left(U(t)f\right)_s(x_1, \ldots, x_s) = \sum_{n=0}^{n_0 - s} \sum_{k=0}^{n} \frac{(-1)^k}{k!(n-k)!} \int dx_{s+1} \ldots dx_{s+n} \times$$
$$f_{s+n}(X^{(s+n-k)}(-t, X^{(s-l)}(-\Delta t, x_1, \ldots, x_{s-l}), x_{s-l+1}, \ldots, x_{s+n-k}), x_{s+n-k+1}, \ldots x_{s+n})$$
$$0 \le l < s \; , \quad X^{(s+n-k)} = (X_1, \ldots, X_{s+n-k}) \; .$$

Hence it follows that there exists

$$\lim_{\Delta t \to +0} \frac{1}{\Delta t} (S(-\Delta t) - I) U(t) f = (-\mathcal{H}) U(t) f \quad ,$$

and $U(t)f \in D(\mathcal{H})$ if $f \in L_0^1$ because $\mathcal{H} U(t) f \in L^1$ (see the proof of Theorem 1.3.1).

The functions $(U(t)f)_{s+1}(x_1, \ldots x_{s+1})$ are defined on the hypersurfaces $q_i - q_{s+1} = a\eta$, $|\eta| = 1$, together with $\left(S(-t)f\right)_n(x_1, \ldots, x_n)$ and are almost everywhere continuous functions of $x_1, \ldots x_s, p_{s+1}, \eta$ with compact support (for more details, see Theorem 2.5.1 in Section 2.5). From this it follows that the functions

$\left(U(t)f\right)_s(x_1, \ldots, x_s)$ belong to L^1 and are almost everywhere continuous functions of $(x_1, \ldots, x_s)$ with compact support.

Thus the operators $\mathcal{H}$ and A are defined on $U(t)f$, $f \in L_0^1$, $\mathcal{H}U(t)f \in L^1$, $AU(t)f \in L^1$. The calculation of the commutator $[\int dx, \mathcal{H}]$ on $U(t)f$ is the same as on f.

The proof of the theorem for $\Delta t \to -0$ is obtained in analogous way. The theorem is proved.

Remark 1. As shown in Theorem 2.2.1, the infinitesimal generator $\mathcal{B}$ of the group $U(t)$ possesses different boundary conditions for $t > 0$ and $t < 0$, because the infinitesimal generator $\mathcal{H}$ of the group $S(t)$ possesses different boundary conditions for $t > 0$ and $t < 0$. This means that the group $U(t)$, $-\infty < t < \infty$ in fact consists of two semigroups : 1) $U(t), t \in [0. \infty[$ 2) $U(t), t \in]-\infty, 0]$.

Remark 2. Let us consider the following operator in $L^1(\Lambda)$:

$$U_\Lambda(t) = e^{\int\limits_\Lambda dx} S_\Lambda(-t)\, e^{-\int\limits_\Lambda dx}, \tag{2.2.16}$$

for a system of particles in a bounded domain Λ. Theorem 2.2.1 is also valid for the evolution operator (2.2.16). In fact, all the steps of the proof of the theorem hold for functions concentrated on compact sets. The only difference here is that configurational variables belong to the domains Λ_δ.

The infinitesimal generator $\mathcal{B}_\Lambda$ of the group (2.2.16) coincides with the operator $-\mathcal{H}_\Lambda + \left[\mathcal{H}_\Lambda, \int\limits_\Lambda dx\right]$. If the domain $\Lambda \subset \mathbb{R}^\nu$ is a ball of finite radius and on $\partial\Lambda$ the boundary condition of the elastic collision (1.3.2′) is defined, then $\left[\mathcal{H}_\Lambda, \int\limits_\Lambda dx\right] = \left[\mathcal{H}, \int\limits_\Lambda dx\right]$. In othere words in this case the term in the expression $\left[\mathcal{H}_\Lambda, \int\limits_\Lambda dx\right]$, describing the interaction of particles with boundary of domain Λ, is equal to zero [Ce1][Ce2][Ce3].

Remark 3. It is an open question how to identify the entire domain of definition $\mathcal{D}\left(-\mathcal{H} + \left[\mathcal{H}, \int dx\right]\right)$ of the operator $-\mathcal{H} + \left[\mathcal{H}, \int dx\right]$. But in what follows we shall use the infinitesimal generator of the group (2.2.1) only on the set $L_0^1 \subset \mathcal{D}\left(-\mathcal{H} + \left[\mathcal{H}, \int dx\right]\right)$, where it is well defined. Let us stress that Theorem 2.2.1 and Lemmas 2.2.1, 2.2.2 hold for the subset $L_0^1 \subset L^1$ which consists of finite sequences of continuously differentiable functions with compact support satisfying the boundary condition $f_n(x) = f_n(x^*)$ on the boundary of forbidden configurations, $f_n \subset \mathcal{D}(\mathcal{H}_N)$. It is obvious that these finite sequences belong to $\mathcal{D}\left(-\mathcal{H} + \left[\mathcal{H}, \int dx\right]\right)$.

2.3. Existence of Solutions of the BBGKY Hierarchy

2.3.1. The BBGKY hierarchy

The BBGKY hierarchy is an evolution equation for an infinite sequence of probability distribution functions $F(t) = (F_1(t, x_1), \quad \dots, F_n(t, x_1, \dots, x_n), \dots)$. This equation reads as follows

$$\frac{d}{dt} F(t) = -\mathcal{H} F(t) + \left[\mathcal{H}, \int dx\right] F(t), \tag{2.3.1}$$

where $-\mathcal{H} + \left[\mathcal{H}, \int dx\right]$ is the infinitesimal generator of the group (2.2.1). For the system of particles under consideration (see Section 1.2) the BBGKY hierarchy (2.3.1) in componentwise form was written as (1.4.8) and (1.4.9).

Let us consider the BBGKY hierarchy as an abstract evolution equation in the Banach space L^1 with the infinitesimal generator of the group $U(t)$ in the right–hand side and with initial data

$$F(0) = F^0 \subset L^1. \tag{2.3.1'}$$

It follows from Theorem 2.2.1 that the sequence

$$F(t) = U(t) F^0 = e^{\int dx} S(-t) e^{-\int dx} F^0, \tag{2.3.2}$$

is the strong solution of the Cauchy problem for the BBGKY hierarchy (2.3.1) with initial data $F^0 \in L^1_0$. The formula (2.3.2) can be written componentwise

$$F_s(t, x_1, \dots, x_s) = \left(U(t) F^0\right)_s (x_1, \dots, x_s) \tag{2.3.3}$$

$$= \sum_{n=0}^{\infty} \sum_{k=0}^{n} \frac{(-1)^k}{k!\,(n-k)!} \int\limits_{(\mathbb{R}^\nu \times \mathbb{R}^\nu)^n} dx_{s+1} \dots dx_{s+n} \times$$

$$\times S^{s+n-k}(-t, x_1, \dots, x_s, x_{s+1}, \dots, x_{s+n-k}) F^0_{s+n}(x_1, \dots, x_{s+n}).$$

By using the symmetry of the distribution functions F^0_{s+n}, we can rewrite (2.3.3) identically in the following form

$$F_s(t, x_1, \dots, x_s) = \sum_{n=0}^{\infty} \sum_{k=0}^{n} \frac{(-1)^k}{n!} \int\limits_{(\mathbb{R}^\nu \times \mathbb{R}^\nu)^n} dx_{s+1} \dots dx_{s+n} \times \tag{2.3.3}$$

$$\times \sum_{\{i_1,\dots,i_{n-k}\} \in \{s+1,\dots,s+n\}} S^{s+n-k}\left(-t.\, x_1, \dots, x_s, x_{i_1}, \dots, x_{i_{n-k}}\right) F^0_{s+n}(x_1, \dots, x_{s+n}).$$

Here $S^{s+n-k}\left(-t, x_1, \dots, x_s, x_{i_1}, \dots, x_{i_{n-k}}\right)$ denotes the evolution operator for $s+n-k$ particles with initial data $x_1, \dots, x_s, x_{i_1}, \dots, x_{i_{n-k}}$, and the summation is over all choices of the $n-k$ numbers $\{i_1, \dots, i_{n-k}\} \in \{s+1, \dots, s+n\}$.

Note that every term of the functional series (2.3.3) is correctly defined for $F(0) \in L^1_0$, because the integrands are defined everywhere outside the hypersurfaces $\mathcal{M}^0_{s+n-k}$, and the series (2.3.3) converges in the norm of the space L^1 for $t \in \mathbb{R}^1$. Also we have proven existence of strong solutions of the BBGKY hierarchy for initial data $F^0 \in L^1_0$.

2.3.2. The existence theorem

We are now able to prove the main result about existence of solutions in the entire space L^1. The result is contained in the following

Theorem 2.3.1. *The Cauchy problem for the BBGKY hierarchy (2.3.1) has a unique solution in L^1, given by the formula (2.3.2). For initial data belonging to $L^1_0 \subset L^1$ this solution is strong, while for arbitrary initial data it is generalized (in the sense to be explained below) and for a smooth potential without hard core it is a weak solution.*

Proof. According to well–known theorems of functional analysis [Ka][Kr][RS] [Yo], the Cauchy problem (2.3.1) for the evolution equation, with evolution operator given by the infinitesimal generator of a strongly continuous group of bounded operators $U(t)$, has a unique solution in L^1 given by the formula (2.3.2). We have already proved in Theorem 2.2.1 that for $F^0 \in L^1_0$ this solution is strong.

Please note that the generalized solutions in the case of hard spheres interacting via a smooth potential exist in the following sense [Kr]. The strong solutions exists for $F^0 \in L^1_0$ and are represented by the formula (2.3.2). The set L^1_0 is everywhere dense in L^1 and for an arbitrary $F^0 \in L^1$ there exists a sequence $F^{0(i)} \in L^1_0$ which strongly converges to F^0. From the boundedness of $U(t)$ it follows that the sequence $U(t)F^{0(i)}$ also converges to $U(t)F^0$ and in this sense $F(t) = U(t)F^0$ is a generalized solution of the BBGKY hirarchy. We do not know the properties of differentiability of these generalized solutions.

We shall now show that in the general case $F^0 \in L^1$ the formula (2.3.2) gives a weak solution of the initial value problem for the BBGKY hierarchy (2.3.1). We are able to give the proof only for the case when the interaction potential is a smooth one without a hard core and F^0 is finite. (We hope that the general case can be proven in a similar way.)

Consider the functional

$$(\varphi, F(t)) = \left(e^{\int dx}\varphi F(t)\right)_0,$$

or, componentwise,

$$\left(e^{\int dx}\varphi F(t)\right)_0 = \sum_{n=0}^{\infty} \frac{1}{n!} \int\limits_{(\mathbb{R}^\nu \times \mathbb{R}^\nu)^n} dx_1 \dots dx_n \varphi_n(x_1, \dots, x_n) F_n(t, x_1, \dots, x_n),$$

where φ is a finite sequence of continuously differentiable symmetric functions $\varphi(x_1, \dots, x_n)$ with compact supports. As φ and $F(t)$ are respectively sequences of bounded and summable functions, the functional $(\varphi, F(t))$ exists.

Using the formula (2.3.2), we can transform the functional $(\varphi, F(t))$ as follows

$$(\varphi, F(t)) = (\varphi, U(t) F^0) = (U^*(t)\varphi, F^0). \tag{2.3.4}$$

In the last equality we used the following notations.

$$U^*(t) = e^{-\int dx^*} S(t) e^{\int dx^*}, \tag{2.3.5}$$

where the operator $\int dx^*$ (analogous to a creation operator in quantum field theory) is the adjoint of the operator (1.4.1) $\int dx$

$$\left(\int dx^* f\right)_n (x_1, \dots, x_n) = \sum_{i=1}^{n} f_{n-1}(x_1, \dots, x_{i-1}, x_{i+1}, \dots, x_n). \tag{2.3.6}$$

The operator $\int dx^*$ multiplies the function $f_{n-1}(x_1,\ldots,x_{n-1})$ by a function which depends on x_n and is equal to unity and then the resulting function is symmetrized. Rewrite (2.3.5) componentwise

$$\left(U^*(t)\varphi\right)_s(x_1,\ldots,x_s)=\sum_{n=0}^{s-1}\sum_{k=0}^{n}\frac{(-1)^k}{k!\,(n-k)!}\sum_{\substack{i_1,\ldots,i_k=1\\ i_1\neq\cdots\neq i_k}}^{s}\;\sum_{\substack{j_1,\ldots,j_{n-k}\\ j_1\neq\cdots\neq j_{n-k}\\ \{j_1,\ldots,j_{n-k}\}\neq\{i_1,\ldots,i_k\}}}^{s-k}\times$$

$$\times S^{s-k}\left(t,x_1,\overset{i_1}{\vee}\!\!\cdots\!\!\overset{i_k}{\vee},x_s\right)\varphi_{s-n}\left(x_1,\overset{i_1}{\vee}\!\!\cdots\!\!\overset{i_k}{\vee},\overset{j_1}{\vee}\!\!\cdots\!\!\overset{j_{n-k}}{\vee},x_s\right). \tag{2.3.7}$$

The function $\left(U^*(t)\varphi\right)_s(x_1,\ldots,x_s)$ is differentiable with respect to t; moreover, the expression

$$\left|\frac{1}{\Delta t}\left(U^*(t+\Delta t)\varphi-U^*(t)\varphi\right)_s-\left(\mathcal{B}^*U^*(t)\varphi\right)_s\right|$$

converges to zero uniformly with respect to $x_1,\ldots,x_s$ on compact sets as $\Delta t\to 0$. In this expression $\mathcal{B}^*$ is the infinitesimal generator of the group $U^*(t)$ (2.3.5). The operator $\mathcal{B}^*$ coincides with $\mathcal{H}+\left[\mathcal{H},\int dx^*\right]$

$$\left(\left(\mathcal{H}+\left[\mathcal{H},\int dx^*\right]\right)\varphi\right)_s(x_1,\ldots,x_s)=\mathcal{H}_s\varphi_s(x_1,\ldots,x_s) \tag{2.3.8}$$

$$-\sum_{\substack{i,k=1\\ i\neq k}}^{s}\left\langle\frac{\partial}{\partial q_k}\Phi(q_k-q_i),\frac{\partial}{\partial p_k}\right\rangle\varphi_{s-1}(x_1,\ldots,x_{i-1},x_{i+1},\ldots,x_s).$$

Since the function $\left(\mathcal{B}^*U^*(t)\varphi\right)_s$ is bounded, the function $\left(\mathcal{B}^*U^*(t)\varphi\right)_sF_s^0$ is summable and the following functional exists : $\left(\mathcal{B}^*U^*(t)\varphi,F^0\right)$. Hence the equality (2.3.4) can be differentiated with respect to time under the integral sign, and we get finally

$$\frac{d}{dt}(\varphi,F(t))=\left(\left(\mathcal{H}+\left[\mathcal{H},\int dx^*\right]\right)U^*(t)\varphi,F^0\right)$$

$$=\left(U^*(t)\left(\mathcal{H}+\left[\mathcal{H},\int dx^*\right]\right)\varphi,F^0\right)=\left(\left(\mathcal{H}+\left[\mathcal{H},\int dx^*\right]\right)\varphi,F(t)\right).$$

These equalities mean that the sequence of functions (2.3.2) for $F^0\in L^1$ is a weak solution of the Cauchy problem for the BBGKY hierarchy.

Theorem 2.3.1 is also valid for the BBGKY hierarchy of a system of particles in a bounded domain $\Lambda \subset \mathbb{R}^\nu$ (1.4.11). The solution of the initial value problem is given, in this case, by the following formula:

$$F_\Lambda(t) = U_\Lambda(t) F_\Lambda^0 = e^{\int\limits_\Lambda dx} S_\Lambda(-t) e^{-\int\limits_\Lambda dx} F_\Lambda^0. \tag{2.3.9}$$

In order to prove the existence of weak solutions, we have restricted ourselves to smooth potentials because for the terms connected with a hard core, we have

$$\int \varphi_s(x_1,\ldots,x_s) \sum_{i=1}^{s} \int d\eta dp_{s+1} \langle \eta, (p_i - p_{s+1}) \rangle F_{s+1}(t, x_1, \ldots, x_s, q_i - a\eta, p_{s+1})$$

$$= \int \varphi_s(x_1,\ldots,x_s) \sum_{i=1}^{s} \int d\eta dq_{s+1} dp_{s+1} \langle \eta, (p_i - p_{s+1}) \rangle \delta(q_{s+1} - q_i + a\eta) \times$$

$$\times F_{s+1}(t, x_1, \ldots, x_s, q_i - a\eta, p_{s+1}) =$$

$$\int \Big[\varphi_s(x_1,\ldots,x_s) \sum_{i=1}^{s} \int d\eta \delta(q_{s+1} - q_i + a\eta) \langle \eta, (p_i - p_{s+1}) \rangle \Big] \times$$

$$\times F_{s+1}(t, x_1, \ldots, x_{s+1}) \, dx_1, \ldots, dx_{s+1}$$

and it is necessary to prove that the product of $\varphi_s(x_1,\ldots,x_s)\{\sum_{i=1}^{s} \int d\eta \delta(q_{s+1} - q_i + a\eta)$ and $F_{s+1}(t, x_1, \ldots, x_{s+1})$ is integrable. This problem is open at present.

Remark 4. In Section 1.4 the formal derivation of the BBGKY hierarchy from the particle dynamics was presented. In the present section this problem has been solved in another manner. The main idea is based on the existence of a closed operator, that is the infinitesimal generator of the evolution group. Theorem 2.2.1 gives the rigorous justification of the mathematical derivation of the BBGKY hierarchy, as an evolution equation in the space of sequences of summable functions L^1 with the infinitesimal generator $-\mathcal{H} + \left[\mathcal{H}, \int dx\right]$ of the group $U(t)$ defined by (2.2.1).

Remark 5. It is easy to see that the series (2.3.3) for $F_s(t, x_1, \ldots, x_s)$ are convergent in the norm of $L^1(\mathbb{R}^{\nu s}, \mathbb{R}^{\nu s} \backslash \mathbb{W}_s)$ if the initial data F^0 satisfy the following conditions

$$\sum_{n=0}^{\infty} \frac{2^n}{n!} \int dx_1 \ldots dx_n \left| F_{s+n}^0(x_1, \ldots, x_{s+n}) \right| < \infty \quad , \quad s = 1, 2, \ldots .$$

For such initial data all functions $F_s(t, x_1, \ldots, x_s)$ exist and are elements of spaces $L^1(\mathbb{R}^{\nu s} \times \mathbb{R}^{\nu s} \backslash \mathbb{W}_s)$ but the sequence $F(t)$ does not belong, in general, to the direct sum of the spaces $L^1(\mathbb{R}^{\nu s} \times \mathbb{R}^{\nu s} \backslash \mathbb{W}_s)$, i.e. to L^1.

Remark 6. Let us consider a sequence of functions $G(t) = \big(G_1(t_1, x_1), \ldots, G_n(t.\ x_1, \ldots, x_n), \ldots\big)$ and the following chain of equations

$$\frac{d}{dt} G(t) = \mathcal{H} G(t) + \left[\mathcal{H}, \int dx^*\right] G(t), \tag{2.3.10}$$

where $\mathcal{H} + \left[\mathcal{H}, \int dx^*\right]$ is the infinitesimal generator of the group $U^*(t)$ (2.3.5). Equation (2.3.10) is called the conjugated (dual) BBGKY hierarchy. This hierarchy in componentwise form is written as (2.3.8).

From a physical point of view the evolution of the observables of a system of particles is described by the solutions of the hierarchy (2.3.10). This interpretation follows from the formula for average values

$$\langle G\rangle(t) = \left(e^{\int dx} G(0) F(t)\right)_0 = \left(e^{\int dx} G(t) F(0)\right)_0. \tag{2.3.11}$$

As shown above, the solution of the Cauchy problem for the conjugated BBGKY hierarchy is given by the formula

$$G(t) = e^{-\int dx^*} S(t)\, e^{\int dx^*} G(0), \tag{2.3.12}$$

which componentwise has the form (2.3.7).

2.4. Existence of Solutions for a One–dimensional BBGKY Hierarchy

2.4.1. Properties of the evolution operator

In the present section we will prove the existence of solutions to the Cauchy problem for the BBGKY hierarchy of a one–dimensional system of particles interacting with their nearest heighbours (1.4.12). To start with, we shall now recall some notions.

In subsection 1.4.3 the Banach space L^1_α of sequences of summable functions with the norm (1.4.13) was introduced. In this space L^1_α the evolution operator $S(t)$ and the operators $(1 - \int dx^\pm)$, $(1 - \int dx^\pm)^{-1}$ were also defined

$$\left(\left(1 - \int dx^+\right)^{-1} \left(1 - \int dx^-\right)^{-1} f\right)_s (x_{-s_2}, \ldots, x_{s_1}) = \sum_{n=0}^{\infty} \sum_{\substack{n_1, n_2 \geqslant 0 \\ n_1 + n_2 = n}} \int dx_{-(n_2+s_2)} \times$$

$$\times \ldots dx_{-(s_2+1)} dx_{s_1+1} \ldots dx_{s_1+n_1} f_{s+n}\left(x_{-(n_2+s_2)}, \ldots, x_{n_1+s_1}\right).$$

Let us now consider, in L^1_α, the operator

$$U(t) = \left(1 - \int dx^+\right)^{-1} \left(1 - \int dx^-\right)^{-1} S(-t) \left(1 - \int dx^+\right) \left(1 - \int dx^-\right) \tag{2.4.1}$$

The properties of the evolution operator (2.4.1) in L^1_α are described in the following theorem (similar to Theorem 2.2.1).

Theorem 2.4.1. *The operators* $U(t) = \left(1 - \int dx^+\right)^{-1} \left(1 - \int dx^-\right)^{-1} S(-t) (1- - \int dx^+) \left(1 - \int dx^-\right)$, $t \in \mathbb{R}^1$ *are defined, bounded*

$$\|U(t)\| \leqslant (\alpha+1)^2 \Big/ (\alpha-1)^2$$

strongly continuous in L^1_α, *and form a group. The infinitesimal generator* $\mathcal{B}$ *of this group is closed,* $\mathcal{B}U(t) = U(t)\mathcal{B}$ *and its spectrum is concentrated on the imaginary axis. On the everywhere dense subspace* L^0_α *of* L^1_α *it coincides with the operator*

$$\mathcal{B} = -\mathcal{H} + \left[\mathcal{H}, \int dx^+\right] + \left[\mathcal{H}, \int dx^-\right],$$

which componentwise has the following form ($n = n_1 + n_2$):

$$\begin{aligned}
&\left(\frac{d}{dt} U(t) \big|_{t=+0} f\right)_n (x_{-n_2}, \ldots, x_{n_1}) = -(\mathcal{H}f)_n (x_{-n_2}, \ldots, x_{n_1}) \\
&+ \int\limits_{\mathbb{R}^1 \times \mathbb{R}^1} dx_{n_1+1} \left\{\Phi(q_{n_1} - q_{n_1+1}), f_{n+1}(x_{-n_2}, \ldots, x_{n_1+1})\right\} \\
&+ \int\limits_{\mathbb{R}^1 \times \mathbb{R}^1} dx_{-(n_2+1)} \left\{\Phi\left(q_{-n_2} - q_{-(n_2+1)}\right), f_{n+1}\left(x_{-(n_2+1)}, \ldots, x_{n_1}\right)\right\} \\
&+ \int_0^\infty dP P \big[f_{n+1}(x_{-n_2}, \ldots, x_{n_1-1}; q_{n_1}, p_{n_1} + P; q_{n_1} + a, p_{n_1}) \\
&- f_{n+1}(x_{-n_2}, \ldots. x_{n_1}; q_{n_1} + a, p_{n_1} - P)\big] \\
&+ \int_0^\infty dP P \big[f_{n+1}(p_{-n_2}, q_{-n_2} - a; p_{-n_2} - P, q_{-n_2}; x_{-(n_2-1)}, \ldots, x_{n_1}) \\
&- f_{n+1}(p_{-n_2} + P. q_{-n_2} - a; x_{-n_2}, \ldots, x_{n_1})\big],
\end{aligned} \tag{2.4.2}$$

and, analogously,

$$\left(\frac{d}{dt}U(t)\mid_{t=-0} f\right)_n (x_{-n_2},\dots,x_{n_1}) = -(\mathcal{H}f)_n (x_{-n_2},\dots,x_{n_1})$$

$$+ \int\limits_{\Re^1\times\Re^1} dx_{n_1+1}\left\{\Phi\left(q_{n_1}-q_{n_1+1}\right), f_{n+1}\left(x_{-n_2},\dots,x_{n_1+1}\right)\right\}$$

$$+ \int\limits_{\Re^1\times\Re^1} dx_{-(n_2+1)}\left\{\Phi\left(q_{-n_2}-q_{-(n_2+1)}\right), f_{n+1}\left(x_{-(n_2+1)},\dots,x_{n_1}\right)\right\}$$

$$+ \int\limits_0^\infty dPP\big[f_{n+1}\left(x_{-n_2},\dots,x_{n_1}; q_{n_1}+a, p_{n_1}+P\right)$$

$$- f_{n+1}\left(x_{-n_2},\dots,x_{n_1-1}; q_{n_1}, p_{n_1}-P; q_{n_1}+a, p_{n_1}\right)\big]$$

$$+ \int\limits_0^\infty dPP\big[f_{n+1}\left(p_{-n_2}-P, q_{-n_2}-a; x_{-n_2},\dots,x_{n_1}\right)$$

$$- f_{n+1}\left(p_{-n_2}, q_{-n_2}-a; p_{n_2}+P, q_{-n_2}; x_{-(n_2-1)},\dots,x_{n_1}\right)\big], \tag{2.4.3}$$

In these formulae $\mathcal{H}$ is the infinitesimal generator of the group $S(t)$ (see subsection 1.4) , $\{.,.\}$ is the Poisson bracket.

Proof. The proof of this theorem is analogous to Theorem 2.2.1 for the higher–dimensional case.

The operator $U(t)$, given by (2.4.1), is well defined and bounded in L^1_α, being the product of operators that are bounded in L^1_α; moreover

$$\|S(-t)\| = 1,$$

$$\left\|\left(1-\int dx^\pm\right)\right\| \leqslant \frac{\alpha+1}{\alpha},$$

$$\left\|\left(1-\int dx^\pm\right)^{-1}\right\| \leqslant \frac{\alpha}{\alpha-1},$$

and hence

$$\|U(t)\| \leqslant \frac{(\alpha+1)^2}{(\alpha-1)^2}.$$

Strong continuity with respect to t of the group (2.4.1) follows from strong continuity of $S(-t)$ and boundness of the operators $\left(1-\int dx^\pm\right)$, $\left(1-\int dx^\pm\right)^{-1}$.

The group property of (3.4.1) follows from the corresponding properties of the operator $S(-t)$; for example,

$$U(t_1)U(t_2) = \left(1-\int dx^+\right)^{-1}\left(1-\int dx^-\right)^{-1} S(-t_1)\left(1-\int dx^+\right)$$
$$\times\left(1-\int dx^-\right)\left(1-\int dx^+\right)^{-1}\left(1-\int dx^-\right)^{-1} S(-t_2)\left(1-\int dx^+\right)\left(1-\int dx^-\right)$$
$$= \left(1-\int dx^+\right)^{-1}\left(1-\int dx^-\right)^{-1} S(-t_1)\,S(-t_2)\left(1-\int dx^+\right)\left(1-\int dx^-\right)$$
$$= \left(1-\int dx^+\right)^{-1}\left(1-\int dx^-\right)^{-1} S(-t_1-t_2)\left(1-\int dx^+\right)\left(1-\int dx^-\right)$$
$$= U(t_1+t_2)$$

for arbitrary t_1 and t_2.

The facts that the infinitesimal generator $\mathcal{B}$ of $U(t)$ is closed and that its spectrum is concentrated on the imaginary axis follow from general properties of generators of groups of bounded operators [Ka][Kr][RS][Yo].

Let us now construct the infinitesimal generator of the group (2.4.1). For $f \in L^0_\alpha \subset L^1_\alpha$ we obtain in the sense of strong convergence in L^1_α

$$\lim_{\Delta t\to 0}\frac{1}{\Delta t}\left(U(t+\Delta t)f - U(t)f\right) = \lim_{\Delta t\to 0} U(t)\frac{1}{\Delta t}\left(U(\Delta t)-I\right)f$$

$$= U(t)\lim_{\Delta t\to 0}\frac{1}{\Delta t}\left(\sum_{n=0}^{\infty}\sum_{\substack{n_1,n_2\geqslant 0\\ n_1+n_2=n}}\sum_{k_1=0}^{\min(1,n_1)}\sum_{k_2=0}^{\min(1,n_2)}(-1)^{k_1+k_2}\times\right.$$
$$\left.\times\left(\int dx^+\right)^{n_1-k_1}\left(\int dx^-\right)^{n_2-k_2} S(-\Delta t)\left(\int dx^+\right)^{k_1}\left(\int dx^-\right)^{k_2} - I\right)f$$

$$= U(t)\lim_{\Delta t\to 0}\frac{1}{\Delta t}\left(\left(S(-\Delta t)-I\right)+\left[\int dx^+,\left(S(-\Delta t)-I\right)\right]+\left[\int dx^-,\left(S(-\Delta t)-I\right)\right]\right.$$
$$+\sum_{n=2}^{\infty}\sum_{\substack{n_1,n_2\geqslant 0\\ n_1+n_2=n}}\sum_{k_1=0}^{\min(1,n_1)}\sum_{k_2=0}^{\min(1,n_2)}(-1)^{k_1+k_2}\left(\int dx^+\right)^{n_1-k_1}\left(\int dx^-\right)^{n_2-k_2}\times$$

$$\times \left(S\left(-\Delta t\right)-I\right)\left(\int dx^{+}\right)^{k_1}\left(\int dx^{-}\right)^{k_2}\Bigg)\, f, \tag{2.4.4}$$

where I is the unit operator and the bracket $[.,.]$ is a commutator. The sequence $f \in L^0_\alpha$ is finite. Hence $\left(\int dx^+\right)^{n_1}\left(\int dx^-\right)^{n_2} f = 0$ for sufficiently large values of $n_1 + n_2 = n$ and the sum over n in the formula (2.4.4) contains for each $f \in L^0_\alpha$ a finite number of terms. Moreover, the sequences $\int dx^\pm f$ are also finite sequences of differentiable functions with compact supports. Therefore the limit can be taken term by term in (2.4.4).

Using the fact that $\mathcal{H}$ is the infinitesimal generator of the group $S(t)$, the expression (2.2.4) can be transformed as follows

$$\lim_{\Delta t\to 0} U(t)\,\frac{1}{\Delta t}\left(U(\Delta t)-I\right)f = U(t)\left(-\mathcal{H}f + \left[\mathcal{H}, \int dx^+ + \int dx^-\right] f\right.$$

$$-\sum_{n=2}^{\infty}\sum_{\substack{n_1,n_2\geqslant 0\\ n_1+n_2=n}}\sum_{k_1=0}^{\min(1,n_1)}\sum_{k_2=0}^{\min(1,n_2)}(-1)^{k_1+k_2}\left(\int dx^+\right)^{n_1-k_1}\left(\int dx^-\right)^{n_2-k_2}\mathcal{H}\times$$

$$\left.\times\left(\int dx^{+}\right)^{k_1}\left(\int dx^{-}\right)^{k_2} f\right) \tag{2.4.5}$$

Using the same argument as in Lemma 2.2.2, we can easily obtain the following equality

$$\left[\mathcal{H}, \left(\int dx^+\right)^{m_1}\left(\int dx^-\right)^{m_2}\right] = \left[\mathcal{H}, \int dx^+\right]\left(\int dx^+\right)^{m_1-1}\left(\int dx^-\right)^{m_2} +$$

$$\left[\mathcal{H}, \int dx^-\right]\left(\int dx^+\right)^{m_1}\left(\int dx^-\right)^{m_2-1} \tag{2.4.6}$$

From this equality it follows that the series on the right–hand side of (2.4.5) can be transformed into

$$\sum_{n=2}^{\infty}\sum_{\substack{n_1,n_2\geqslant 0\\ n_1+n_2=n}}\sum_{k_1=0}^{\min(1,n_1)}\sum_{k_2=0}^{\min(1,n_2)}(-1)^{k_1+k_2}\left(\int dx^+\right)^{n_1-k_1}\left(\int dx^-\right)^{n_2-k_2}\times$$

$$\times\mathcal{H}\left(\int dx^{+}\right)^{k_1}\left(\int dx^{-}\right)^{k_2} f$$

$$=\sum_{n=2}^{\infty}\sum_{\substack{n_1,n_2\geqslant 0\\ n_1+n_2=n}}\sum_{k_1=0}^{\min(1,n_1)}\sum_{k_2=0}^{\min(1,n_2)}(-1)^{k_1+k_2}\left(\left[\left(\int dx^{+}\right)^{n_1-k_1}\left(\int dx^{-}\right)^{n_2-k_2},\mathcal{H}\right]\times\right.$$

$$\left.\times\left(\int dx^{+}\right)^{k_1}\left(\int dx^{-}\right)^{k_2}+\mathcal{H}\left(\int dx^{+}\right)^{n_1}\left(\int dx^{-}\right)^{n_2}\right)f$$

$$=\sum_{n=2}^{\infty}\sum_{\substack{n_1,n_2\geqslant 0\\ n_1+n_2=n}}\sum_{k_1=0}^{\min(1,n_1)}\sum_{k_2=0}^{\min(1,n_2)}(-1)^{k_1+k_2}\left(\left[\int dx^{+},\mathcal{H}\right]\left(\int dx^{+}\right)^{n_1-1}\left(\int dx^{-}\right)^{n_2}\right.$$

$$\left.+\left[\int dx^{-},\mathcal{H}\right]\left(\int dx^{+}\right)^{n_1}\left(\int dx^{-}\right)^{n_2-1}+\mathcal{H}\left(\int dx^{+}\right)^{n_1}\left(\int dx^{-}\right)^{n_2}\right)f.$$

This expression is zero because it contains the coefficient

$$\sum_{k_1=0}^{\min(1,n_1)}\sum_{k_2=0}^{\min(1,n_2)}(-1)^{k_1+k_2},\quad n_1+n_2=n\geqslant 2,$$

which is identically zero.

Hence we have proved the equality

$$\lim_{\Delta t\to 0}U(t)\frac{1}{\Delta t}\left(U(\Delta t)-I\right)f=U(t)\left(-\mathcal{H}+\left[\mathcal{H},\int dx^{+}\right]+\left[\mathcal{H},\int dx^{-}\right]\right)f.$$

The following equality can be proved similarly

$$\lim_{\Delta t\to 0}\frac{1}{\Delta t}\left(U(\Delta t)-I\right)U(t)f=\left(-\mathcal{H}+\left[\mathcal{H},\int dx^{+}\right]+\left[\mathcal{H},\int dx^{-}\right]\right)U(t)f.$$

Our further step is to calculate the commutators $\left[\mathcal{H},\int dx^{\pm}\right]$. Let us consider, for example, the projection of $\int dx^{+}\mathcal{H}$ onto subspace $L^0_\alpha(\mathbb{R}^n\times(\mathbb{R}^n\backslash\mathbb{W}_n))$

$$\left(\int dx^{+}\mathcal{H}f\right)_n(x_{-n_2},\ldots,x_{n_1})=\int_{\mathbb{R}^1\times\mathbb{R}^1}dx_{n_1+1}\mathcal{H}_n f_{n+1}(x_{-n_2},\ldots,x_{n_1+1})$$

$$-\int_{\mathbb{R}^1\times\mathbb{R}^1}dx_{n_1+1}\left\{\Phi\left(q_{n_1}-q_{n_1+1}\right),f_{n+1}\left(x_{-n_2},\ldots,x_{n_1+1}\right)\right\}$$

$$+\int\limits_{\Re^1\times\Re^1} dx_{n_1+1}p_{n_1+1}\frac{\partial}{\partial q_{n_1+1}}f_{n+1}(x_{-n_2},\ldots,x_{n_1+1}), \tag{2.4.7}$$

where $\{.,.\}$ is a Poisson bracket. The integration over the configuration variable q_{n_1+1} for $f \in L^0_\alpha$ is in fact carried out over an interval, whose extremes depend on q_{n_1}, therefore changing the order of differentiation with respect to q_{n_1} and integration with respect to q_{n_1+1}, we obtain

$$\int\limits_{\Re^1\times\Re^1} dx_{n_1+1}\mathcal{H}_n f_{n+1}(x_{-n_2},\ldots,x_{n_1+1}) =$$

$$(\mathcal{H}\int dx^+ f)_n(x_{-n_2},\ldots,x_{n_1+1}) + \int\limits_{\Re^1} dp_{n_1}p_{n_1}f_{n+1}(x_{-n_2},\ldots,x_{n_1};q_{n_1}+a,p_{n_1+1}) \tag{2.4.8}$$

Analogously, the last term in (2.4.7) can be transformed into

$$\int\limits_{\Re^1\times\Re^1} dx_{n_1+1}p_{n_1+1}\frac{\partial}{\partial q_{n_1+1}}f_{n+1}(x_{-n_2},\ldots,x_{n_1+1})$$

$$= -\int\limits_{\Re^1} dp_{n_1+1}p_{n_1+1}f_{n+1}(x_{-n_2},\ldots,x_{n_1};q_{n_1}+a,p_{n_1+1}). \tag{2.4.9}$$

Therefore it follows from (2.4.7–9) that

$$\left(\left[\mathcal{H},\int dx^+\right]f\right)_n(x_{-n_2},\ldots,x_{n_1})$$

$$= \int\limits_{\Re^1\times\Re^1} dx_{n_1+1}\{\Phi(q_{n_1}-q_{n_1+1}), f_{n+1}(x_{-n_2},\ldots,x_{n_1+1})\}$$

$$-\int\limits_{\Re^1} dp_{n_1+1}(p_{n_1}-p_{n_1+1})f_{n+1}(x_{-n_2},\ldots,x_{n_1};q_{n_1}+a,p_{n_1+1}). \tag{2.4.10}$$

The expression (2.4.10) can be transformed in the following way. We represent the second term of (2.4.10) as the sum of two expressions, in which integration is over the interval $]-\infty, p_{n_1}]$ and $[p_{n_1}, +\infty[$, respectively. Using the boundary condition for the Poisson bracket $\mathcal{H}$ (see subsection 1.4), for example, in the case $\Delta t \to +0$, we obtain

$$\int_{\mathbb{R}^1} dp_{n_1+1}\,(p_{n_1} - p_{n_1+1})\, f_{n+1}\,(x_{-n_2},\ldots,x_{n_1}; q_{n_1} + a, p_{n_1+1}) \tag{2.4.11}$$

$$= \int_{-\infty}^{p_{n_1}} dp_{n_1+1}\,(p_{n_1} - p_{n_1+1})\, f_{n+1}\,(x_{-n_2},\ldots,x_{n_1-1}, x_{n_1}; q_{n_1} + a, p_{n_1+1})$$

$$+ \int_{p_{n_1}}^{\infty} dp_{n_1+1}\,(p_{n_1} - p_{n_1+1})\, f_{n+1}\,(x_{-n_2},\ldots,x_{n_1-1}; q_{n_1}, p_{n_1+1}; q_{n_1} + a, p_{n_1})\,.$$

Making the changes of variables $p_{n_1+1} = p_{n_1} - P$ in the first part on the right–hand side of (2.4.11) and $p_{n_1+1} = p_{n_1} + P$ in the second part of (2.4.11), respectively,, the expression $\left(\left[\mathcal{H}, \int dx^+\right] f\right)_n$ is transformed into

$$\left(\left[\mathcal{H}, \int dx^+\right] f\right)_n (x_{-n_2},\ldots,x_{n_1})$$

$$= \int_{\mathbb{R}^1\times\mathbb{R}^1} dx_{n_1+1}\{\Phi\,(q_{n_1} - q_{n_1+1})\,, f_{n+1}\,(x_{-n_2},\ldots,x_{n_1+1})\}$$

$$+ \int_0^{\infty} dP P\big(f_{n+1}\,(x_{-n_2},\ldots,x_{n_1-1}; q_{n_1}, p_{n_1} + P; q_{n_1} + a, p_{n_1})$$

$$- f_{n+1}\,(x_{-n_2},\ldots,x_{n_1}; q_{n_1} + a, p_{n_1} - P)\big). \tag{2.4.12}$$

In a similar way the following equality ($\Delta t \to +0$) is proved

$$\left(\left[\mathcal{H}, \int dx^-\right] f\right)_n (x_{-n_2},\ldots,x_{n_1})$$

$$= \int_{\mathbb{R}^1\times\mathbb{R}^1} dx_{-(n_2+1)}\{\Phi\,\big(q_{-n_2} - q_{-(n_2+1)}\big)\,, f_{n+1}\,\big(x_{-(n_2+1)},\ldots,x_{n_1}\big)\}$$

$$+ \int_0^{\infty} dP P\big(f_{n+1}\,(p_{-n_2}, q_{-n_2} - a; p_{-n_2} - P, q_{-n_2}; x_{-(n_2-1)},\ldots,x_{n_1})$$

$$- f_{n+1}\,(p_{n_1} + P. q_{-n_2} - a; x_{-n_2},\ldots,x_{n_1})\big). \tag{2.4.13}$$

Analogously, one can calculate the commutators $\left[\mathcal{H}, \int dx^{\pm}\right]$ in the case $\Delta t \to -0$ to obtain (2.4.3).

Hence, we have proven the statement of the Theorem 2.4.1. It is quite obvious that Theorem 2.4.1 is also valid for the evolution operator of a system of particles in a bounded interval $\Lambda \subset \mathbb{R}^1$

$$U_\Lambda (t) = \left(1 - \int_\Lambda dx^+\right)^{-1} \left(1 - \int_\Lambda dx^-\right)^{-1} S_\Lambda (-t) \left(1 - \int_\Lambda dx^+\right) \left(1 - \int_\Lambda dx^-\right) \tag{2.4.14}$$

in the space $L^1_\alpha (\Lambda)$. The infinitesimal generator of the group (2.4.14) coincides with the operator : $-\mathcal{H}_\Lambda + \left[\mathcal{H}_\Lambda, \int_\Lambda dx^+\right] + \left[\mathcal{H}_\Lambda, \int_\Lambda dx^-\right]$ on L^0_α.

2.4.2. The formula for the solution of the Cauchy problem

The BBGKY hierarchy for a one–dimensional system of particles interacting with their nearest neighbours is an evolution equation for an infinite sequence of distribution functions $F(t) = \left\{F_n (x_{-n_2}, \ldots, x_{n_1})\right\}_{n=n_1+n_2\geqslant 1}$ and reads as follows:

$$\frac{d}{dt} F(t) = -\mathcal{H} F(t) + \left[\mathcal{H}, \int dx^+\right] F(t) + \left[\mathcal{H}, \int dx^-\right] F(t), \tag{2.4.15}$$

where $-\mathcal{H} + \left[\mathcal{H}, \int dx^+\right] + \left[\mathcal{H}, \int dx^-\right]$ is the infinitesimal generator of the group (2.4.1) (see Theorem 2.4.1).

Let us consider the BBGKY hierarchy (2.4.15) as an abstract evolution equation in the Banach space L^1_α with initial data

$$F(0) = F^0 .$$

It follows from Theorem 2.4.1 that the Cauchy problem for (2.4.15) has a unique solution in L^1_α, which is given by the formula

$$F(t) = \left(1 - \int dx^+\right)^{-1} \left(1 - \int dx^-\right)^{-1} S(-t) \left(1 - \int dx^+\right) \left(1 - \int dx^-\right) F^0, \tag{2.4.16}$$

or. componentwise,

$$F_s\left(t, x_{-s_2}, \ldots, x_{s_1}\right) = \sum_{n=0}^{\infty} \sum_{\substack{n_1, n_2 \geqslant 0 \\ n_1+n_2=n}} \sum_{k_1=0}^{\min(1,n_1)} \sum_{k_2=0}^{\min(1.n_2)} (-1)^{k_1+k_2} \times$$

$$\times \int_{(\mathbb{R}^1\times\mathbb{R}^1)^n} dx_{-(n_2+s_2)} \ldots dx_{-(s_2+1)} dx_{s_1+1} \ldots dx_{s_1+n_1} S^{s+n-k}(-t) \times$$

$$\times F^0_{s+n}\left(x_{-(s_2+n_2)}, \ldots, x_{s_1+n_1}\right), \qquad s = s_1 + s_2, n = n_1 + n_2. \tag{2.4.17}$$

The operator $S^{s+n-k}(-t)$ acts on F^0_{s+n} as follows

$$S^{s+n-k}(-t) F^0_{s+n}\left(x_{-(s_2+n_2)}, \ldots, x_{s_1+n_1}\right) = F^0_{s+n}\big(x_{-(s_2+n_2)}, \ldots, x_{-(s_2+n_2-k_2+1)},$$

$$X_{-(s_2+n_2-k_2)}(-t), \ldots, X_{s_1+n_1-k_1}(-t), x_{s_1+n_1-k_1+1}, \ldots, x_{s_1+n_1}\big),$$

where the functions defined by $X_i(-t) \equiv X_i\left(-t, x_{-(s_2+n_2-k_2)}, \ldots, x_{s_1+n_1-k_1}\right)$, for $i \in \left(-(s_2+n_2-k_2), \ldots, s_1+n_1-k_1\right)$ give the solution of the corresponding Cauchy problem for the Hamilton equations.

Every term of the functional series (2.4.17) is correctly defined, because the integrands are defined everywhere, outside of $\mathcal{M}^0_{s+n}$, and the series (2.4.17) converges in the norm of the space L^1_α.

2.4.3. The existence theorem

The following theorem holds.

Theorem 2.4.2. *The Cauchy problem for the BBGKY hierarchy (2.4.15) has a unique solution in the space L^1_α given by the formula (2.4.16). For initial data $F^0 \in L^0_\alpha \subset L^1_\alpha$ this solution is strong, and for arbitrary initial data $F^0 \in L^0_\alpha$ and smooth potentials it is a weak solution.*

Proof. In fact, we have already proved in Theorem 2.4.1 this statement for $F^0 \in L^0_\alpha \subset L^1_\alpha$ and for a smooth potential with hard core. Now, let us show that in the general case ($F^0 \in L^1_\alpha$ and smooth potentials) the formula (2.4.16) gives a weak solution of the initial value problem (2.4.15).

Let us consider the following functional

$$(\varphi, F(t)) = \left(\left(1 - \int dx^+\right)^{-1}\left(1 - \int dx^-\right)^{-1} \varphi F(t)\right)_0, \tag{2.4.18}$$

or. componentwise.

$$\left(\left(1-\int dx^+\right)^{-1}\left(1-\int dx^-\right)^{-1}\varphi F(t)\right)_0=\sum_{n=0}^{\infty}\sum_{\substack{n_1,n_2\geqslant 0\\ n_1+n_2=n}}\int dx_{-n_2}\dots dx_{n_1}\times$$

$$\times\varphi\left(x_{-n_2},\dots,x_{n_1}\right)F_n\left(t,x_{-n_2},\dots,x_{n_1}\right),$$

where we have denoted by φ a finite sequence of continuously differentiable functions $\varphi_n(x_{-n_2},\dots,x_{n_1})$ with compact supports. As $F(t)\in L^1_\alpha$ and φ is sequence of bounded functions, then the functional $(\varphi,F(t))$ exists.

Using the formula (2.4.16), we can transform the functional $(\varphi,F(t))$ as follows

$$(\varphi,F(t))=(\varphi,U(t)F^0)=(U^*(t)\varphi,F^0). \tag{2.4.19}$$

In (2.4.19) it was introduced the notation

$$U^*(t)=\left(1-\int d\overset{*}{x}{}^+\right)\left(1-\int d\overset{*}{x}{}^-\right)S(t)\left(1-\int d\overset{*}{x}{}^+\right)^{-1}\left(1-\int d\overset{*}{x}{}^-\right)^{-1}, \tag{2.4.20}$$

where the operators $\int d\overset{*}{x}{}^\pm$ are the adjoints of the operators $\int dx^\pm$ (2.4.14)

$$\left(\int d\overset{*}{x}{}^+f\right)_n(x_{-n_2},\dots,x_{n_1})=f_{n-1}\left(x_{-n_2},\dots,x_{n_1-1}\right)$$
$$\left(\int d\overset{*}{x}{}^-f\right)_n(x_{-n_2},\dots,x_{n_1})=f_{n-1}\left(x_{-(n_2-1)},\dots,x_{n_1}\right). \tag{2.4.21}$$

The operator (2.4.20) acts on $\varphi=\left\{\varphi_n\left(x_{-n_2},\dots,x_{n_1}\right)\right\}_{n\geqslant 1}$ by the formula

$$\left(U^*(t)\varphi\right)_s\left(x_{-s_2},\dots,x_{s_1}\right) \tag{2.4.22}$$
$$=\sum_{n_1=0}^{s_1}\sum_{n_2=0}^{s_2}\sum_{k_1=0}^{\min(1,n_1)}\sum_{k_2=0}^{\min(1,n_2)}(-1)^{k_1+k_2}S^{s-k_1-k_2}(t)\varphi_{s-n}\left(x_{-(s_2-n_2)},\dots,x_{s_1-n_1}\right).$$

Note. that the group (2.4.20) defines the solution of the initial value problem for a conjugated BBGKY hierarchy.

The function $\left(U^*(t)\varphi\right)_s\left(x_{-s_2},\dots,x_{s_1}\right)$ is differentiable with respect to t, and in the sense of uniform convergence on compact we have

$$\lim_{\Delta t\to 0}\left|\frac{1}{\Delta t}\left(U^*(t+\Delta t)\varphi-U^*(t)\varphi\right)_s-\left(\mathcal{B}^*U^*(t)\varphi\right)_s\right|=0,$$

where $\mathcal{B}^*$ is the infinitesimal generator of the group $U^*(t)$ given by (2.4.20). The operator $\mathcal{B}^*$ coincides with $\mathcal{H} + \left[\mathcal{H}, \int d\overset{*}{x}^+\right] + \left[\mathcal{H}, \int d\overset{*}{x}^-\right]$.

Since the functions $\left(\mathcal{B}^* U^*(t)\varphi\right)_s$ are bounded, the function $(\mathcal{B}^* U^*(t)\varphi)_s F_s^0$ is summable and the functional $\left(\mathcal{B}^* U^*(t)\varphi, F^0\right)$ exists for finite F^0.

Hence the equality (2.4.19) can be differentiated with respect to time t under the integral sign and we have

$$\begin{aligned}\frac{d}{dt}(\varphi, F(t)) &= \left(\left(\mathcal{H} + \left[\mathcal{H}, \int d\overset{*}{x}^+\right] + \left[\mathcal{H}, \int d\overset{*}{x}^-\right]\right) U^*(t)\varphi, F^0\right) \\ &= \left(U^*(t)\left(\mathcal{H} + \left[\mathcal{H}, \int d\overset{*}{x}^+\right] + \left[\mathcal{H}, \int d\overset{*}{x}^-\right]\right)\varphi, F^0\right) \\ &= \left(\left(\mathcal{H} + \left[\mathcal{H}, \int d\overset{*}{x}^+\right] + \left[\mathcal{H}, \int d\overset{*}{x}^-\right]\right)\varphi, F(t)\right).\end{aligned}$$

These equalities mean that for $F^0 \in L^1_\alpha$ the sequence (2.4.16) is a weak solution of the Cauchy problem (2.4.15).

We note that Theorem 2.4.2 remains valid for the BBGKY hierarchy of a system of particles in a bounded interval $\Lambda \subset \mathbb{R}^1$.

$$\frac{d}{dt}F_\Lambda(t) = -\mathcal{H}_\Lambda F_\Lambda(t) + \left[\mathcal{H}_\Lambda, \int_\Lambda dx^+\right] F_\Lambda(t) + \left[\mathcal{H}_\Lambda, \int_\Lambda dx^-\right] F_\Lambda(t), \qquad (2.4.23)$$

where $-\mathcal{H}_\Lambda + \left[\mathcal{H}_\Lambda, \int_\Lambda dx^+\right] + \left[\mathcal{H}_\Lambda, \int_\Lambda dx^-\right]$ is the infinitesimal generator of the group $U_\Lambda(t)$ (2.4.14). The solution of initial value problem in this case is given the formula : $F_\Lambda(t) = U_\Lambda(t) F^0_\Lambda$.

2.4.4. A hard rod particle system

Let us consider an example of BBGKY hierarchy for a one–dimensional system of particles interacting with their nearest neighbours, which will be studied later. It is a hard rod particle system.

We denote by q_i the position of hard rod's center, by v_i its velocity, $x_i \equiv (q_i, v_i) \in \mathbb{R}^1 \times \mathbb{R}^1$, and let us consider the case of particles with arbitrary length $2a_i > 0$ and mass $m_i > 0$.

In this case the BBGKY hierarchy for $t \in \mathbb{R}^1_+$ has the following form [Ger2]

$$\frac{\partial}{\partial t}F(t) = -\mathcal{H}F(t) + \left[\mathcal{H}, \int dx^+\right] F(t) + \left[\mathcal{H}, \int dx^-\right] F(t), \qquad (2.4.24)$$

where

$$\left(\left[\mathcal{H}, \int dx^{+}\right] F(t)\right)_n (x_{-n_2}, \dots, x_{n_1}) = \int_0^{\infty} dV V \Big(F_{n+1}(t, x_{-n_2}, \dots, x_{n_1-1};$$

$$q_{n_1}, v^*_{n_1}(v_{n_1}, V); q_{n_1} + (a_{n_1} + a_{n_1+1}), v^*_{n_1+1}(v_{n_1}, V))$$

$$- F_{n+1}(t, x_{-n_2}, \dots, x_{n_1}; q_{n_1} + (a_{n_1} + a_{n_1+1}), v_{n_1} - V) \Big),$$

$$\left(\left[\mathcal{H}, \int dx^{-}\right] F(t)\right)_n (x_{-n_2}, \dots, x_{n_1}) = \int_0^{\infty} dV V \Big(F_{n+1}(t, v^*_{-(n_2+1)}(v_{-n_2}, V),$$

$$q_{-n_2} - (a_{-n_2} + a_{-(n_2+1)}); v^*_{-n_2}(v_{-n_2}, V), q_{-n_2}; x_{-(n_2-1)}, \dots, x_{n_1})$$

$$- F_{n+1}(t, v_{-n_2} + V, q_{-n_2} - (a_{-n_2} + a_{-(n_2+1)}); x_{-n_2}, \dots, x_{n_1}) \Big).$$

Here we denote by $v^*_i(v_i, V)$, $v^*_{i+1}(v_i, V)$ the expressions

$$v^*_i(v_i, V) = v_i + \frac{2m_{i+1}}{m_i + m_{i+1}} V,$$

$$v^*_{i+1}(v_i, V) = v_i + \frac{m_{i+1} - m_i}{m_i + m_{i+1}} V, \qquad (2.4.25)$$

$$v^*_{-(i+1)}(v_{-i}, V) = v_{-i} - \frac{m_{-(i+1)} - m_{-i}}{m_{-(i+1)} + m_{-i}} V,$$

$$v^*_{-i}(v_{-i}, V) = v_{-i} - \frac{2m_{-i}}{m_{-(i+1)} + m_{-i}} V.$$

In (2.4.24) $\mathcal{H}$ is the Liouville operator : $\left(\mathcal{H}F(t)\right)_n = \left(\mathcal{H}^0 F(t)\right)_n \equiv \sum_{i=-n_2}^{n_1} v_i \frac{\partial}{\partial q_i} F_n(t)$ with boundary conditions on $\partial(\mathbb{R}^n \backslash \mathbb{W}_n)$. This boundary conditions are defined as follows. If $|q_i - q_{i+1}| = a_i + a_{i+1}$ for some pair $(i, i+1)$, then at $v_i > v_{i+1}$ one must replace in the Poisson bracket $\mathcal{H}^0$ the velocities $(v_i, v_{i+1}) \to (v^*_i, v^*_{i+1})$ according to the expression

$$v^*_i = \frac{m_i - m_{i+1}}{m_i + m_{i+1}} v_i + \frac{2m_{i+1}}{m_i + m_{i+1}} v_{i+1},$$

$$v^*_{i+1} = \frac{m_i - m_{i+1}}{m_i + m_{i+1}} v_{i+1} + \frac{2m_i}{m_i + m_{i+1}} v_i \,. \qquad (2.4.25)$$

The operator $\mathcal{H}$ is a closed infinitesimal generator of the isometric group $S(t)$ of evolution operators of the hard rod particle system.

Now, repeating the proof given in the previous subsection, we can conclude that Cauchy problem for BBGKY hierarchy (2.4.24) has a unique solution in L^1_α, wich is given by the functional series (2.4.17) with evolution operator $S(-t)$

2.5. The Iteration Series

2.5.1. Construction of the evolution operator for the BBGKY hierarchy by iteration

The solution of the Cauchy problem for the BBGKY hierarchy can be represented formally by the iteration series of this hierarchy.

In fact, the BBGKY hierarchy (2.3.1) can be written in an integral form, i.e. by the following integral equation

$$F(t) = S(-t)F^0 + \int_0^t d\tau S(\tau - t) \left[\mathcal{H}, \int dx\right] F(\tau). \tag{2.5.1}$$

The formal iteration series for (2.5.1) has the following form:

$$F(t) = U(t)F^0 \tag{2.5.2}$$

$$= \sum_{n=0}^{\infty} \int_0^t dt_1 \cdots \int_0^{t_{n-1}} dt_n S(-t)S(t_1) \left[\mathcal{H}, \int dx\right] S(-t_1) \ldots S(t_n) \left[\mathcal{H}, \int dx\right] S(-t_n)F^0.$$

It will be shown in the next chapter that the sequence $F(t)$ in (2.5.2) is a weak solution of the BBGKY hierarchy (2.3.1). We have obtained two different representations of the solution $F(t)$ of the BBGKY hierarchy, namely the series (2.3.3) and (2.5.2). We want to show that each term in (2.5.2) can be directly defined through a corresponding term of the functional series (2.3.3), and that the series (2.5.2) and (2.3.3) coincide term–by–term. This follows from the fact that the expression (2.5.2) after integrations with respect to time can be identically transformed into the series (2.3.3).

Indeed, as the Poisson bracket is invariant under the action of the evolution operator $S(t) : [\mathcal{H}, S(t)] = 0$, we obtain, according to Theorem 1.3.1

$$S(t)\left[\mathcal{H}, \int dx\right] S(-t) = \left[\mathcal{H}, S(t)\int dx\right] S(-t)$$
$$= \frac{d}{dt}\left(S(t)\int dx S(-t)\right). \qquad (2.5.3)$$

From the definitions of the operators $S(t)$ given by (1.3.14) and $\int dx$ given by (1.4.1) it follows that the following operator equalities hold

$$\frac{d}{dt}\left(S(t)\int dx S(-t)\right)\left(S(t)\int dx S(-t)\right)^{k-1} = \frac{1}{k}\frac{d}{dt}\left(S(t)\int dx S(-t)\right)^{k},$$

$$\left[\frac{d}{dt}\left(S(t)\int dx S(-t)\right), S(t)\int dx S(-t)\right] = 0. \qquad (2.5.4)$$

Integrating (2.2.5) with respect to time, and using (2.5.4) we get

$$F(t) = S(-t)F^0 + \sum_{n=1}^{\infty} S(-t)\sum_{k=0}^{n} \frac{(-1)^k}{k!(n-k)!}\left(S(t)\int dx S(-t)\right)^{n-k}\left(\int dx\right)^k F^0$$
$$= e^{\int dx} S(-t) e^{-\int dx} F^0 ;$$

hence, we have formally obtained the required formula (2.3.2) and have defined the terms of the series (2.5.2) by the terms of the series (2.3.3).

We now consider another approach to the problem of defining the terms of the iteration series (2.5.2). We shall need in what follows the independent, direct definition of every terms in (2.5.2). The main difficulty in this problem is that the functions $(S(t)F^0)_n(x_1, \ldots, x_n)$ are defined everywhere on the phase space outside of $\mathcal{M}_n^0$ and initial distribution functions F_n^0 from iteration series (2.5.2) are concentrated on manifolds of lower dimension. It is necessary to verify that these manifolds belong to the domain of definition of the functions $(S(t)F^0)_n(x_1, \ldots, x_n)$.

2.5.2. The explicit form of the iteration series

Let us consider the iteration series (2.5.2) for the BBGKY hierarchy of the hard spheres particle system and give the exact mathematical meaning to every term of this series.

Let us rewrite (2.5.2) in componentwise form. To this purpose, it is convenient to represent the n–th term of the iteration series

$$\mathcal{I}_n = \int_0^t dt_1 \cdots \int_0^{t_{n-1}} dt_n S(t_1 - t) \left[\mathcal{H}, \int dx\right] \ldots S(t_n - t_{n-1}) \left[\mathcal{H}, \int dx\right] S(-t_n) F^0,$$

where $\left[\mathcal{H}, \int dx\right]$ is defined by (2.2.10) with $\Phi = 0$, through the following operators $\left(t \in \mathbb{R}^1_+\right)$

$$\left(S(t_k - t_{k-1}) \left[\mathcal{H}, \int dx\right] f\right)_{s+k-1} (x_1, \ldots, x_{s+k-1})$$

$$= a^{\nu-1} \sum_{i_k=1}^{s+k-1} ((a^*_{i_k} - a_{i_k}) f)_{s+k-1} (x_1, \ldots, x_{s+k-1}), \tag{2.5.5}$$

$$(a_{i_k} f)_{s+k-1} (x_1, \ldots, x_{s+k-1}) = \int\limits_{\mathbb{R}^\nu \times \mathbb{S}^{\nu-1}_+(t_k - t_{k-1})} d\eta_k dp_{s+k} \langle \eta_k, (P_{i_k}(t_k - t_{k-1}) - p_{s+k})\rangle$$

$$\times f_{s+k}\big(X_1(t_k - t_{k-1}), \ldots, X_{s+k-1}(t_k - t_{k-1}); Q_{i_k}(t_k - t_{k-1}) + a\eta_k, p_{s+k}\big),$$

$$(a^*_{i_k} f)_{s+k-1} (x_1, \ldots, x_{s+k-1}) = \int\limits_{\mathbb{R}^\nu \times \mathbb{S}^{\nu-1}_+(t_k - t_{k-1})} d\eta_k dp_{s+k} \langle \eta_k, (P_{i_k}(t_k - t_{k-1}) - p_{s+k})\rangle$$

$$\times f_{s+k}\big(X_1(t_k - t_{k-1}), \ldots; Q_{i_k}(t_k - t_{k-1}), P^*_{i_k}(t_k - t_{k-1}); \ldots, X_{s+k-1}(t_k - t_{k-1});$$
$$Q_{i_k}(t_k - t_{k-1}) - a\eta_k, p^*_{s+k}\big),$$

where $\mathbb{S}^{\nu-1}_+(t_k - t_{k-1}) = \{\eta_k \in \mathbb{R}^\nu \mid |\eta_k| = 1, \ \langle \eta_k, (P_{i_k}(t_k - t_{k-1}) - p_{s+k})\rangle > 0\}$. $X_i(t_k - t_{k-1}) \equiv X_i(t_k - t_{k-1}, x_1, \ldots, x_{s+k-1})$ and $P^*_{i_k}(t_k - t_{k-1})$, p^*_{s+k} denote the expressions

$$P^*_{i_k}(t_k - t_{k-1}) = P_{i_k}(t_k - t_{k-1}) - \eta_k \langle \eta_k, (P_{i_k}(t_k - t_{k-1}) - p_{s+k})\rangle,$$
$$p^*_{s+k} = p_{s+k} + \eta_k \langle \eta_k, (P_{i_k}(t_k - t_{k-1}) - p_{s+k})\rangle.$$

Thus, we have

$$\left(\mathcal{I}_n\right)_s (x_1, \ldots, x_s) = \int_0^t dt_1 \cdots \int_0^{t_{n-1}} dt_n \sum_{i_1=1}^{s} \cdots \sum_{i_n=1}^{s+n-1} a^{(\nu-1)n}$$

$$\times\left(\prod_{k=1}^{n}(a_{i_k}^* - a_{i_k})S(-t_n)F^0\right)_s(x_1,\dots,x_s) \tag{2.5.6}$$

$$= \sum_{m=0}^{n}(-1)^{n-m}\sum_{i_1=1}^{s}\cdots\sum_{i_n=1}^{s+n-1}\sum_{1\leqslant j_1<\dots<j_m}^{n} a^{(\nu-1)n}\mathcal{I}^m_{\{i_1,\dots,i_n\}}(x_1,\dots,x_s),$$

where

$$\mathcal{I}^m_{\{i_1,\dots,i_n\}}(x_1,\dots,x_s) = \int_0^t dt_1\cdots\int_0^{t_{n-1}}dt_n$$

$$\times(a_{i_1}\dots a_{i_{j_1-1}}a^*_{i_{j_1}}a_{i_{j_1+1}}\dots a_{i_{j_m-1}}a^*_{i_{j_m}}a_{i_{j_m+1}}\dots a_{i_n}S(-t_n)F^0)_s(x_1,\dots,x_s).$$

This formula allows us to classify all possible elements out of which the n-th term $\mathcal{I}_n$ of the series (2.5.2) is constructed.

Let us write the typical expression $\mathcal{I}^m_{\{i_1,\dots,i_n\}}$ in component form. Using for phase states of a system of the $s+k$ particles the following notation : $X^{(s+k)} \equiv (X_1,\dots,X_{s+k})$, from the definition (2.5.5) of the operator a_{i_k}, we have for $\mathcal{I}^0_{\{i_1,\dots,i_n\}}$

$$\mathcal{I}^0_{\{i_1,\dots,i_n\}}(x_1,\dots,x_s) = \int_0^t dt_1\cdots\int_0^{t_{n-1}}dt_n\int_{\mathbb{R}^\nu\times\mathbb{S}^{\nu-1}_+(t_1-t)}d\eta_1 dp_{s+1}$$

$$\dots\int_{\mathbb{R}^\nu\times\mathbb{S}^{\nu-1}_+(t_n-t_{n-1})}d\eta_n dp_{s+n}\langle\eta_1,(P_{i_1}(t_1-t)-p_{s+1})\rangle\dots\langle\eta_n,(P_{i_n}(t_n-t_{n-1})-p_{s+n})\rangle$$

$$\times F^0_{s+n}\Bigg(X^{(s+n)}\Bigg(-t_n,X^{(s+n-1)}\Bigg(t_n-t_{n-1},X^{(s+n-2)}\big(t_{n-1}-t_{n-2},\dots X^{(s+1)}(t_2-t_1,$$

$$X^{(s)}(t_1-t,x^{(s)});Q_{i_1}(t_1-t,x^{(s)})+a\eta_1,p_{s+1})\dots\big);Q_{i_{n-1}}(t_{n-1}-t_{n-2},\dots)+a\eta_{n-1},$$

$$p_{s+n-1}\Bigg);Q_{i_n}(t_n-t_{n-1},\dots)+a\eta_n,p_{s+n}\Bigg)\Bigg). \tag{2.5.7}$$

Let us use now the notation $(X_1;\dots;Q_{i_k},\overset{\boxed{i_k}}{P}_{i_k};\dots;X_{s+k-1}) \equiv \overset{\boxed{i_k}}{X}{}^{(s+k-1)}$ (where $\overset{\boxed{i_k}}{P}_{i_k}$ and correspondingly $\overset{\boxed{i_k}}{p}_{s+k}$ are the functions of P_{i_k}, p_{s+k} and η_k, as in

(2.5.5)). Then according to the definition (2.5.5) of the operator $a^*_{i_k}$, the expression $\mathcal{I}^n_{\{i_1,\ldots,i_n\}}$ have the form

$$\mathcal{I}^n_{\{i_1,\ldots,i_n\}}(x_1,\ldots,x_s) = \int_0^t dt_1 \cdots \int_0^{t_{n-1}} dt_n \int_{\mathbb{R}^\nu \times \mathbb{S}_+^{\nu-1}(t_1-t)} d\eta_1 dp_{s+1}$$

$$\cdots \int_{\mathbb{R}^\nu \times \mathbb{S}_+^{\nu-1}(t_n-t_{n-1})} d\eta_n dp_{s+1} \langle \eta_1, (P_{i_1}(t_1-t)-p_{s+1})\rangle \ldots \langle \eta_n, (P_{i_n}(t_n-t_{n-1})-p_{s+1})\rangle$$

$$\times F^0_{s+n}\Bigg(X^{(s+n)}\Bigg(-t_n, \overset{\boxed{i_n}}{X}{}^{(s+n-1)}\Big(t_n - t_{n-1}, \overset{\boxed{i_{n-1}}}{X}{}^{(s+n-2)}\big(t_{n-1} - t_{n-2}, \ldots$$

$$\overset{\boxed{i_2}}{X}{}^{(s+1)}(t_2 - t_1, \overset{\boxed{i_1}}{X}{}^{(s)}(t_1 - t, x^{(s)}); Q_{i_1}(t_1 - t, x^{(s)}) - a\eta_1, \overset{\boxed{i_1}}{p}{}_{s+1}) \ldots);$$

$$Q_{i_{n-1}}(t_{n-1} - t_{n-2}, \ldots) - a\eta_{n-1}, \overset{\boxed{i_{n-1}}}{p}{}_{s+n-1}\Big);$$

$$Q_{i_n}(t_n - t_{n-1}, \ldots) - a\eta_n, \overset{\boxed{i_n}}{p}{}_{s+n}\Bigg)\Bigg). \qquad (2.5.8)$$

The other expressions in (2.5.6) for $m = 1, \ldots, n-1$ have a structure similar to (2.5.7) and (2.5.8).

2.5.3. Existence of the iteration series

The following statement shows that the integrals with respect to $p_{s+1}, \ldots, p_{s+n}; \eta_1, \ldots, \eta_n$ and $t_1, \ldots, t_n$ in the expressions $\mathcal{I}^m_{\{i_1,\ldots,i_n\}}$ from (2.5.6) have a quite definite sense. In what follows their convergence will also be proved.

Theorem 2.5.1. *Let $F^0_{s+n} \in L^1_0(\mathbb{R}^{\nu(s+n)} \times (\mathbb{R}^{\nu(s+n)} \backslash \mathbb{W}_{s+n}))$. Then the integrand in the expression $(\mathcal{I}_n)_s(x_1, \ldots, x_s)$ (2.5.6) is well defined, continuous with respect to $t \in \mathbb{R}^1$, $t_1 \in [0, t], \ldots, t_n \in [0, t_{n-1}]$, and continuously differentiable with respect to $t, t_1, \ldots, t_n$ in the time intervals between collision instants and with respect to $x_1, \ldots, x_s; \eta_1, \ldots, \eta_n; p_{s+1}, \ldots, p_{s+n}$ almost everywhere in the set of admissible configurations (or outside $\mathcal{M}^0_{s+n}$).*

Proof. Let us consider the expression (2.5.7) $\mathcal{I}^0_{\{i_1,\dots,i_n\}}(x_1,\dots,x_s)$. According to Theorem 1.2.1 , the phase trajectory $X^{(s)}(t_1-t,x^{(s)})$ is a continuously differentiable function with respect to $x^{(s)} \equiv (x_1,\dots,x_s)$ outside the set $\mathcal{M}^0_s$ (see Section 1.2) and with respect to $t_1 \in [0,t]$ in the time intervals between the collision instants in a subsystem of s particles. Similary, the phase trajectories $X^{(s+1)}(t_2-t_1, X^{(s)}(t_1-t,x^{(s)}); Q_{i_1}(t_1-t,x^{(s)})+a\eta_1, p_{s+1})$, $i_1 \in (1,\dots,s)$ are continuously differentiable functions with respect to $X^{(s)}(t_1-t,x^{(s)})$, $Q_{i_1}(t_1-t,x^{(s)})+a\eta_1$ and p_{s+1} outside the set $\mathcal{M}^0_{s+1}$,and with respect to $t_2 \in [0,t_1]$ on every time interval between two subsequent collision instants in the system of $s+1$ particles. If in a system of $s+1$ particles at $t_2 \in [0,t_1]$ only the particles labelled $1,\dots,s$ interact with each other with initial data $X^{(s)}(t_1-t,x^{(s)})$, then it means that in the interval $t_2 \in [0,t]$ the evolution of the particles $1,\dots,s$ takes place according to the s-particle dynamics and therefore $X^{(s)}(t_2-t,x^{(s)})$ is a continuously differentiable function with respect to $t_2 \in [0,t]$ outside the collision instants and with respect to $x^{(s)}$ outside the set $\mathcal{M}^0_s$. In the system of $s+1$ particles at $t_2 \in [0,t_1]$ the particles $1,\dots s$ can also interact with the $(s+1)$-th particle, in accordance with the initial phase states $X^{(s)}(t_1-t,x^{(s)})$, $\left(Q_{i_1}(t_1-t,x^{(s)})+a\eta_1, p_{s+1}\right)$. These states belong to a $(6\nu(s+1)-1)$–dimensional manifold, as $Q_{s+1}(t_1-t) = Q_{i_1}(t_1-t,x^{(s)})+a\eta_1$, $|\eta_1|=1$. Let us show that on the manifold

$$\left\{x^{(s)} \in \mathbb{R}^{\nu s} \times (\mathbb{R}^{\nu s}\backslash \mathrm{W}_s)\ ,\ q_{s+1} = q_{i_1} + a\eta_1, |\eta_1| = 1, p_{s+1} \in \mathbb{R}^{\nu}\right\}$$

the set $\mathcal{M}^0_{s+1}$ is also of zero Lebesgue measure.

The equations of the hypersurface, which define the set $\mathcal{M}^0_{s+1}$, for example, the triple collision of $(s+1), j$ and l particles ($j \neq l \neq i_1 \in (1,\dots,s)$) have the following form:

$$\begin{cases} \left|Q_{s+1}(t_2-t_1, X^{(s+1)}(t_1-t,\dots)) - Q_j(t_2-t_1, X^{(s+1)}(t_1-t,\dots))\right| & = a, \\ \left|Q_j(t_2-t_1, X^{(s+1)}(t_1-t,\dots)) - Q_l(t_2-t_1, X^{(s+1)}(t_1-t,\dots))\right| & = a, \end{cases} \tag{2.5.9}$$

At first, let us consider the case when in a system of $s+1$ particles the first collision occurs. Then Eqs. (2.5.9) take the form

$$\left\{\begin{aligned} &\Big|Q_{i_1}(t_1-t,x^{(s)})+a\eta_1+p_{s+1}(t_2-t_1)-Q_j(t_1-t,x^{(s)})\\ &\qquad\qquad -P_j(t_1-t,x^{(s)})(t_2-t_1)\Big|=a\\ &\Big|Q_j(t_1-t,x^{(s)})+P_j(t_1-t,x^{(s)})(t_2-t_1)-Q_l(t_1-t,x^{(s)})\\ &\qquad\qquad -P_l(t_1-t,x^{(s)})(t_2-t_1)\Big|=a \end{aligned}\right. \tag{2.5.10}$$

Let us show that Eqs. (2.5.10) are solvable with respect to t_2 and $p^{\alpha}_{s+1}, \alpha=1,\dots,\nu$.

From the first of Eqs. (2.5.10) it follows that all partial derivatives of the expression in the left hand side, with respect to $p^{\alpha}_{s+1}, \alpha=1,\dots,\nu$ are equal to zero, if the vector $Q_{i_1}(t_1-t,x^{(s)})+a\eta_1+p_{s+1}(t_2-t_1)-Q_j(t_1-t,x^{(s)})-P_j(t_1-t,x^{(s)})(t_2-t_1)$ is zero. But this contradicts the equation itself, according to which the magnitude of the vector under consideration is a. Therefore all partial derivatives of the expression in the left hand side with respect to p^{α}_{s+1} cannot simultaneously vanish and hence, on manifold $\{x^{(s)};\eta_1,p_{s+1}\}$ this equation defines a set of lower dimension at fixed value $t_2\in[0,t_1]$.

From the second of Eqs. (2.5.10) it follows that the partial derivative of the expression in the left hand side with respect to t_2 is equal to zero, if the following equation holds

$$\begin{aligned} &\Big\langle\big(Q_j(t_1-t,x^{(s)})+P_j(t_1-t,x^{(s)})(t_2-t_1)-Q_l(t_1-t,x^{(s)})\\ &-P_l(t_1-t,x^{(s)})(t_2-t_1)\big),\big(P_j(t_1-t,x^{(s)})-P_l(t_1-t,x^{(s)})\big)\Big\rangle=0 \end{aligned}\tag{2.5.11}$$

This equation means that the relative momentum of the j-th and l-th particles at a collision instant is perpendicular to the vector joining their centers and defines a two–dimensional surface in the three–dimensional space of relative momenta. If one observes, that the hypersurface in terms of the variables $X^{(s)}$ according to Liouville's theorem for the mapping $x^{(s)}\to X^{(s)}(t_1-t,x^{(s)})$, is a hypersurface, also in variables $x^{(s)}$, then Eq. (2.5.11) can be valid only on a set of zero Lebesgue measure in the space described by the $x^{(s)}\in(\mathbb{R}^{\nu s}\times(\mathbb{R}^{\nu s}\backslash\mathbb{W}_s))$. Thus Eqs. (2.5.10) are solvable with respect to t_2 and $p^{\alpha}_{s+1},\alpha=1,\dots,\nu$ and they define a set of lower dimension on the manifold $\{x^{(s)},\eta_1,p_{s+1}\}$, i.e. a set of zero Lebesgue measure.

We consider now the general case (2.5.9). Let only pair collisions take place in the system till an instant when a triple collision occurs. All results obtained above are valid in this case as well. The only difference is that the partial derivatives of the expression in the left–hand side of the first of Eqs. (2.5.9) with respect to $P^{\alpha}_{s+1}, \alpha = 1, \ldots, \nu$ must be calculated after the last pair collision , and one must take into account that P_{s+1} linearly depends upon the momenta of the particles, which interact with the $s+1$-th particle. The Jacobian of this transformation $p_{s+1} \to P_{s+1}$ is not zero and therefore the manifold of lower dimension in terms of the variables P_{s+1} is a manifold of lower dimension also in terms of p_{s+1} as well.

Thus, the phase trajectories $X^{(s+1)}(t_2 - t_1, X^{(s)}(t_1 - t, x^{(s)}); Q_{i_1}(t_1 - t, x^{(s)}) + a\eta_1, p_{s+1})$, $i_1 \in (1, \ldots, s)$ are continuously differentiable functions with respect to $x^{(s)} \in (\mathbb{R}^{\nu s} \times (\mathbb{R}^{\nu s} \backslash \mathbb{W}_s))$, $p_{s+1} \in \mathbb{R}^{\nu}, \eta_1(|\eta_1| = 1)$ outside a set of zero Lebesgue measure and with respect to $t_2 \in [0, t_1]$ in the time intervals between the collision instants in the system of $(s+1)$ particles. According to the relation (1.2.11), the phase trajectory $X^{(s+1)}(t_2 - t_1, X^{(s)}(t_1 - t, x^{(s)}); Q_{i_1}(t_1 - t, x^{(s)}) + a\eta_1, p_{s+1})$ is continuous with respect to t_1 in the instants of pair collisions in the system of s particles.

Continuing, we obtain that the phase trajectories $X^{(s+n)}(-t_n, X^{(s+n-1)}(t_n - t_{n-1}, \ldots, X^{(s)}(t_1 - t, x^{(s)}), Q_{i_1}(t_1 - t, \ldots) + a\eta_1, p_{s+1}) \ldots; Q_{i_n}(t_n - t_{n-1}, \ldots) + a\eta_n, p_{s+n})$ are continuously differentiable functions with respect to $t_n \in [0, t_{n-1}]$ in the time intervals between collision instants in a system of $(s+n)$ particles and almost everywhere with respect to $x^{(s)}; \eta_1, \ldots, \eta_n; p_{s+1}, \ldots, p_{s+n}$, and also that they are continuous with respect to $t. t_1 \in [0, t], \ldots, t_{n-1} \in [0, t_{n-2}]$ and continuously differentiable functions with respect to $t, t_1, \ldots, t_{n-1}$ with the exception of the instants of pair collisions.

The continuity of the function $F^0_{s+n}(X^{(s+n)}(-t_n, \ldots), Q_{i_n}(t_n - t_{n-1}, \ldots) + a\eta_n, p_{s+n})$, $F^0_{s+n} \in L^1_0(\mathbb{R}^{\nu(s+n)} \times (\mathbb{R}^{\nu(s+n)} \backslash \mathbb{W}_{s+n}))$ with respect to t_n at the collision instants in the system of $(s+n)$ particles follows from the boundary conditions (2.4.4), which are a consequence of the relation (1.2.11). The integrand contains also multipliers which depend on differences of momenta. They have the same properties, as the phase points.

Thus, the expressions $\mathcal{I}^0_{\{i_1,\ldots,i_n\}}(x_1, \ldots, x_s)$ make sense and the statement of Theorem 2.5.1 is true.

Now, let us consider the expressions $\mathcal{I}^n_{\{i_1,\ldots,i_n\}}(x_1, \ldots, x_s)$ (2.5.8). If one notes, that the functions $\overset{\boxed{i_1}}{P}_{i_1}(t_1 - t, x^{(s)}), \ldots; \overset{\boxed{i_1}}{p}_{s+1}, \ldots, \overset{\boxed{i_1}}{p}_{s+n}$ are continuous almost everywhere with respect to the initial data for arbitrary times with the exception of instants of pair collisions, and that a set of zero Lebesgue measure in terms of variables $(P_{i_1}^{\boxed{i_1}}, p_{s+1}^{\boxed{i_1}}), \ldots, (P_{i_n}^{\boxed{i_n}}, p_{s+n}^{\boxed{i_n}})$ is a set of zero Lebesgue measure also in terms of the variables $(P_{i_1}, p_{s+1}), \ldots (P_{i_n}, p_{s+n})$ (as the Jacobians of the transformations $(P_{i_k}^{\boxed{i_k}}, p_{s+k}^{\boxed{i_k}}) \to (P_{i_k}, p_{s+k})$, $k = 1, \ldots, n$ are unity), then our theorem is valid in this case as well. The analysis of other expressions from (2.5.6) reduces to the cases considered above.

Thus, the integrand in (2.5.6) is well defined and the theorem is proven. Let us stress that the integrand is a continuous function of t_1 in intervals between pair collisions in a system of s particles with initial states $(x_1, \ldots, x_s)$.

It is important to note that the surfaces $t = t_i(x^{(s+n)})$, where $t_i(x^{(s+n)})$ are pair collision instants, is a set of zero Lebesgue measure in the space described by the variables $p_{s+1}, \ldots, p_{s+n}; \eta_1, \ldots, \eta_n; t_1, \ldots, t_n$ in formula (2.5.6).

Remark 7. Theorem 2.5.1 is also true if the initial functions $F_{s+n}^0(x_1, \ldots$ x_{s+n} are continuous on the set $\mathbb{R}^{\nu(s+n)} \times (\mathbb{R}^{\nu(s+n)} \backslash \mathbb{W}_{(s+n)})$ and satisfy boundary conditions (2.4.4) and are different from zero in some neighbourhood of the set of forbidden configurations $\mathbb{W}_{s+n}$.

2.5.4. Existence theorem for initial data in some subspace in L^1

Let us consider the BBGKY hierarchy for a hard sphere particle systems and let us show that the iteration series (2.5.2),(2.5.6) is uniformly convergent for some subset of initial data in L^1 and represents the solution of the Cauchy problem for this hierarchy.

We consider initial data in the space $X_{\xi,\beta} \subset L^1$ with the norm

$$\|f\| = \sup_{n \geqslant 0} \xi^{-n} \sup_{x_1, \ldots, x_n} |f_n(x_1, \ldots, x_n)| \exp\left\{ \frac{\beta}{2} \sum_{i=1}^{n} (p_i^2 + q_i^2) \right\},$$

where ξ and β are certain positive parameters. In what follows it will be convenient to use a normalization condition for $F(t)$ such that a factor $za^{\nu-1} = \lambda^{-1}$ appears before the expression $\left[\mathcal{H}, \int dx\right] F(t)$ in the BBGKY hierarchy, and the notation : $(T)_n(x_1, \ldots, x_n) = \frac{1}{2}\sum_{i=1}^{n} p_i^2$, $(I)_n(x_1, \ldots, x_n) = \frac{1}{2}\sum_{i=1}^{n} q_i^2$.

The proof of the uniform convergence of the iteration series (2.5.6) in this case has been obtained in [IP] and is based on the following inequality

$$(S^n(t) I_n)(x_1, \ldots x_n) \geqslant (S_0^n(t) I_n)(x_1, \ldots, x_n), \tag{2.5.12}$$

where $S_0^n(t)$ is the evolution operator of the system of n noninteracting particles. This estimate follows from the definition of a phase trajectory $X(t,s)$ for hard spheres (Chapter 1). Indeed, let $0 \leqslant t_1 < \cdots < t_k \leqslant t$ be the instants of pair collisions of particles, p_i and p_i' the incoming momenta in the instant t_i and η_i the unit vector directed along the line connecting the centers of particles. Then, according to energy and momentum conservation and the definition of a phase trajectory we obtain

$$\left(S^n(t)I_n\right)(x_1,\dots,x_n) = \left(S_0^n(t)I_n\right)(x_1,\dots,x_n) + a\sum_{i=1}^{k}(t-t_i)\langle\eta_i,(p_i-p_i')\rangle. \tag{2.5.13}$$

As $\langle\eta_i,(p_i-p_i')\rangle \geqslant 0$ at $t \geqslant 0$, the inequality (2.5.12) follows from the above equality. Let us construct the majorant for iteration series (2.5.2),(2.5.6). From energy conservation we have

$$\left(T\right)_{s+n}\left(X^{(s+n)}(-t_n,\dots,Q_{i_n}+a\eta_n,p_{s+n})\right)$$
$$= \left(T\right)_{s+n}(x_1\dots,x_s,\dots,q_{i_n}+a\eta_n,p_{s+n})$$

According to (2.5.12) the following inequality is valid

$$\left(I\right)_{s+n}\left(X^{(s+n)}(-t_n,\dots,Q_{i_n}+a\eta_n,p_{s+n})\right) \geqslant \left(S_0^s(-t)I_s\right)(x_1,\dots,x_s)$$
$$+\sum_{k=1}^{n}\left(Q_{t_k}+a\eta_k-p_{s+k}t_k\right)^2.$$

The initial distribution functions satisfy the inequality

$$\left|F_{s+n}^0(x_1,\dots,x_{s+n})\right| \leqslant \exp\left\{-\beta\left(\frac{1}{2}\sum_{i=1}^{n+s}(q_i^2+p_i^2)\right)\right\}\xi^{s+n}$$

Using these inequalities we obtain the following estimate for the n–th term of the iteration series

$$|\mathcal{I}_n(x_1,\dots,x_s)| \leqslant \text{const}\,\xi^{s+n}\lambda^{-n}\exp\left\{-\beta\left(S_0^s(-t)I_s\right)(x_1,\dots,x_s)\right.$$
$$\left.-\beta\left(T\right)_s(x_1,\dots,x_s)\right\}\int_0^t dt_1\cdots\int_0^{t_{n-1}}dt_n\sup_{y_1,\dots,y_n}\sum_{i_1=1}^{s}\sum_{i_n=1}^{s+n-1}\int dp_{s+1}\cdots\int dp_{s+n}$$

$$\times \prod_{i=1}^{n} \left(|P_{i_k}(t_k - t_{k-1}, \dots)| + |p_{s+k}| \right) \exp\left\{ -\beta \sum_{j=1}^{n} (y_j - p_{s+j} t_j)^2 - \beta \frac{1}{2} \sum_{j=1}^{n} p_{s+j}^2 \right\}.$$

Using the inequalities

$$|p| \exp\{-\alpha p^2\} \leqslant \text{const}\, \alpha^{-1/2}$$

$$\sup_y \int dp \exp\left\{ -(y - pt)^2 - \frac{1}{2} p^2 \right\} \leqslant \text{const}\,(1 + t^3)^{-1},$$

we obtain

$$|\mathcal{I}_n(x_1, \dots, x_n)| \leqslant \text{const}\, \xi^s \exp\left\{ -\beta \left(S_0^s(-t) I_s + T_s \right) (x_1, \dots, x_n) \right\}$$

$$\times \frac{(s+n)^n}{n!} \left(\xi \lambda^{-1} \text{const}\,(\beta) \left(\int_0^\infty d\tau (1 + \tau^3)^{-1} \right) \right)^n . \qquad (2.5.13)$$

The convergence of the series (2.5.5) and (2.5.6) follows directly from these estimates. Thus the following statement is proven:

Proposition 2.5.1. If $F^0 \in \mathbf{X}_{\xi,\beta} \subset L^1$, then for sufficiently small $\xi\lambda^{-1}$ the iteration series (2.5.2) (2.5.6) converges almost everywhere, uniformly with respect to $(x_1, \dots, x_s)$ on every compact set for $t \in \mathbb{R}^1$.

In Chapter III for more general situations it will be proven that the iteration series is a weak solution of the BBGKY hierarchy.

As was mentioned above, the initial data $F^0 \in \mathbf{X}_{\xi,\beta}$ are sequences of summable functions. The existence of global solutions for the BBGKY hierarchy in this case has been proven in Theorem 2.3.1, using the functional series (2.3.3). Indeed, the series (2.3.3) for the functions $F_s(t)$ for $t \in \mathbb{R}^1$ converge in the norm of $L^1(\mathbb{R}^{\nu s} \times (\mathbb{R}^{\nu s} \setminus \mathbb{W}_s))$, and, hence, almost everywhere, if the initial data F^0 satisfy the following conditions

$$\sum_{n=0}^{\infty} \frac{2^n z^n}{n!} \int dx_1 \dots dx_{s+n} \left| F_{s+n}^0(x_1, \dots, x_{s+n}) \right| < \infty \quad , \quad z a^{\nu - 1} = \lambda^{-1}$$

for arbitrary $s = 1, 2, \dots$.(see Remark 5). It is easy to see that these conditions are fulfilled for $F^0 \in \mathbf{X}_{\xi,\beta}$.

Remark 8. We have considered two different representations, (2.3.3) and (2.5.2), of solutions of the hierarchy for initial data $F^0 \in L^1$. In the first represen- ation we used the group $U(t)$ of evolution operators, in the second one the solution was represented as the sum of an iteration series. If we prove that both expressions are solutions of the hierarchy, then, since the solution of the Cauchy problem is unique, it follows that they coincide.

From formulas (2.5.3-4) it also follows that it is enough to prove than an arbitrary term in the iteration series (2.5.2) exists and the integration with respect to the time variables can be carried out. Then, in fact, every term of the iteration series (2.5.2) coincides with the corresponding term of the series (2.3.3), representing the solution as the result of the action of the group of evolution operators $U(t)$ on the initial data F^0. An independent proof of the convergence of the series is not necessary in this case because the convergence of the iteration series follows directly from the existence of the group of evolution operators and the convergence of the series (2.3.3) in L^1. From theorem 2.5.1 it follows that for $F^0 \in L_0^1$ the terms of the iteration series (2.5.2) possess the sufficient properties with respect to the variables $t_1, \ldots, t_n$ and thus the integration with respect to $t_1, \ldots, t_n$ can be performed. In this case the representations of the solutions by formulas (2.3.3) and (2.5.2) coincide term by term.

2.5.5. The iteration series for one–dimensional hard spheres system

The solution of the Cauchy problem for the BBGKY hierarchy for non–symmetric systems (2.4.15) can be represented by the iteration series of this hierarchy

$$F(t) = \sum_{n=0}^{\infty} \int_0^t dt_1 \cdots \int_0^{t_{n-1}} dt_n S(-t)S(-t_1) \left(\left[\mathcal{H}, \int dx^+ \right] + \left[\mathcal{H}, \int dx^- \right] \right)$$
$$\times S(-t_1) \ldots S(t_n) \left(\left[\mathcal{H}, \int dx^+ \right] + \left[\mathcal{H}, \int dx^- \right] \right) S(-t_n) F^0 \qquad (2.5.14)$$

where $S(t)$ is the evolution operator (see Section 1.4). Let us show that the functional series (2.4.17) can be obtained after integration with respect to time from the iteration series (2.5.14).

Using the equality

$$S(t) \left[\mathcal{H}, \int dx^{\pm} \right] S(-t) = \frac{d}{dt} \left(S(t) \int dx^{\pm} S(-t) \right) \qquad (2.5.15)$$

one obtains from the definitions of the operators $S(t)$ and $\int dx^{\pm}$:

$$\left(\frac{d}{dt}\left(S(t)\int dx^{\pm}S(-t)\right)\right)\left(S(t)\int dx^{\pm}S(-t)\right)^{k-1}$$

$$= k^{-1}\frac{d}{dt}\left(S(t)\int dx^{\pm}S(-t)\right)^{k}. \tag{2.5.16}$$

Integrating (2.5.14) with respect to time, we get from (2.5.15–16)

$$F(t) = S(-t)F^0 + \sum_{n=1}^{\infty} S(-t)\sum_{k=0}^{1}(-1)^k$$

$$\times\left(S(t)\left(\int dx^+ + \int dx^-\right)S(-t)\right)^{n-k}\left(\int dx^+ + \int dx^-\right)F^0$$

$$=\sum_{n=0}^{\infty}\sum_{\substack{n_1,n_2\geqslant 0\\ n_1+n_2=n}}\sum_{k_1=0}^{\min(1,n_1)}\sum_{k_2=0}^{\min(1.n_2)}(-1)^{k_1+k_2}\left(\int dx^+\right)^{n_1-k_1}\left(\int dx^-\right)^{n_2-k_2}$$

$$\times S(-t)\left(\int dx^+\right)^{k_1}\left(\int dx^-\right)^{k_2}F^0. \tag{2.5.17}$$

Using the definitions of the operators $\left(1-\int dx^{\pm}\right)^{-1}$ and $\left(1-\int dx^{\pm}\right)$, from (2.5.17) we obtain the formula (2.4.16).

We note, that as in the multi–dimensional case, that every term of the (2.5.14) is rigorously defined and the statement similar to Theorem 2.5.1 holds.

CHAPTER III

The Initial Value Problem for L^∞ Data: Thermodynamic Limit

3.1. Introduction

In the previous chapter the solutions of the initial value problem for the BBGKY hierarchy have been constructed for systems with a finite average number of particles (or zero density in the entire space $\mathbb{R}^\nu$). To describe the evolution of states of infinite systems we must construct the solutions in a functional space different from L^1, which contains the sequences of distribution functions, describing states of infinite systems. Such functional space is the space L^∞_ξ of the sequences of functions bounded with respect to the configuration variables and decreasing with respect to the momentum ones (this space was defined in Section 1.5). As is well-known, the equilibrium distribution functions belong to this space.

This chapter deals with the problem of existence and uniqueness of solutions of the Cauchy problem for the BBGKY hierarchy in the space L^∞_ξ. In order to solve these problems the method of the thermodynamic limit will be developed. It implies the extension of the BBGKY evolution operator from the space L^1 to the space L^∞_ξ or its subspaces.

In fact, our problem is to give a meaning to the functional series (2.3.3, 2.4.17) or the iteration series (2.5.2) for initial data from the space L^∞_ξ. We will first consider as initial data the sequences of functions $F^0_\Lambda \in L^\infty_\xi \cap L^1(\Lambda)$ for which the solutions of the Cauchy problem for the BBGKY hierarchy exist and are meaningful. Then, as in the equilibrium case, we define the distribution functions $F_s(t, x_1, \ldots, x_s)$ of a system with an infinite number of particles as a limit of the sequences $F_{\Lambda,s}(t, x_1, \ldots, x_s)$ ($F_\Lambda(t) = U_\Lambda(t)F^0_\Lambda$) for every $s = 1, 2\ldots$ in the sense of uniform convergence with respect to $x_1, \ldots, x_s$ on each compact when $\Lambda \nearrow \mathbb{R}^\nu$, assuming that F^0_Λ converges to F^0 in the same sense. After this we show that the limiting distribution functions $F(t, x_1, \ldots, x_s)$ satisfy the BBGKY hierarchy.

This Chapter is organized as follows. In Sections 3.2 and 3.3 we investigate the solution of initial value problem for the BBGKY hierarchy for hard spheres in the space L^∞_ξ. The solutions are represented through an iteration series. In Section 3.2 we prove a local existence theorem for arbitrary initial data from E^∞_ξ and the existence of the thermodynamic limit for these solutions. In Section 3.3 we prove a global existence theorem for initial data close to equilibrium distribution functions.

We also establish there a global existence theorem for the BBGKY hierarchy of hard spheres in the one-dimensional case. Let us emphasize that it is valid for arbitrary initial data $F^0 \in L^\infty_\xi$ and not only for initial data close to an equilibrium state. In these sections we follow the papers [Pe1][PG1][PG2][PG3][PGM1][PGM2][GP1] [Ger2] [MP1][Ma].

In sections 3.4 we consider the solutions of the Cauchy problem for the BBGKY hierarchy, which are represented through the functional series (2.3.3) and (2.4.17). We develop the concept of interaction region, which allows us to show that for systems with short range potentials the integration over the configuration variables in formula (2.3.3) (or (2.4.17)) is carried out over a finite domain and obtain an estimate for the BBGKY evolution operator in L^∞_ξ. This method was formulated in Petrina's paper [Pe1], and was further used to prove existence theorems for three–dimensional systems [MP][PGM1][PGM2] and for non–symmetrical systems in [Ger1].

We emphasize that in this chapter we use a purely functional approach to the considered problems. The BBGKY hierarchy is interpreted as an abstract evolution equation and for the initial value problem we prove existence and uniqueness theorems by the thermodynamic limit method.

There is another approach to the one-dimensional BBGKY hierarchy [GLL] [SS][H1][H2][Ze1] as well as the results of other authors [Lan3][U][Sh] for three-dimensional systems.

3.2 A local existence theorem for the BBGKY hierarchy of hard spheres

3.2.1 Formulation of problem and existence theorem Let us consider the initial value problem for the BBGKY hierarchy of hard spheres for $\nu = 3$:

$$\frac{d}{dt}F(t) = -\mathcal{H}F(t) + \left[\mathcal{H}, \int dx\right] F(t), \tag{3.2.1}$$

$$F(0) = F^0,$$

where for $t > 0$:

$$\left(\left[\mathcal{H}, \int dx\right] F(t)\right)_s (x_1, \dots, x_s) = a^2 \sum_{i=1}^{s} \int\limits_{\mathbb{R}^3 \times \mathbb{S}^2_+} d\eta dp_{s+1} \langle \eta, (p_i, -p_{s+1})\rangle \times$$

$$\times \big(F_{s+1}(t, x_1, \dots ; q_i, p_i - \eta\langle\eta, (p_i - p_{s+1})\rangle; \dots ; x_s; q_i - a\eta, p_{s+1} + \eta\langle\eta, (p_i - p_{s+1})\rangle\big) -$$

$$-F_{s+1}(t, x_1, \ldots, x_s; q_i + a\eta, p_{s+1})\big).$$

The operator $-\mathcal{H} + \left[\mathcal{H}, \int dx\right]$ was defined as the infinitesimal generator of the group $U(t)$ (2.2.1) of the evolution operators for hard spheres $U(t)$ in space L^1 (Theorem 2.2.1). In the previous chapter it was shown that the evolution operator $U(t)$ exists in L^1 and an existence theorem for the solutions of the initial value problem for the BBGKY hierarchy was proved. Now we want to prove that a solution exists also for initial data in L^∞_ξ. We recall that L^∞_ξ ($L^\infty_{\xi,\beta}$ when we want to stress the role of the constant β which appears in the definition) is the space of sequences of continuous symmetric functions on the admissible configurations (see Section 1.5) with the norm

$$\|f\| = \sup_{n\geqslant 0} \xi^{-n} \sup_{x_1,\ldots,x_n} \left|f_n(x_1, \ldots, x_n)\right| \exp\left\{\beta \sum_{i=1}^{n} \frac{p_i^2}{2}\right\}.$$

One cannot use directly the formula $F(t) = U(t)F^0$ for $F^0 \in L^\infty_\xi$ because the operator $U(t)$ was not defined on L^∞_ξ. To solve this problem we shall use the following approach.

The solution of the Cauchy problem (3.2.1) for $F^0 \in L^\infty_\xi$ is defined at first formally by a series of iterations of the hierarchy (3.2.1)

$$F(t) = U(t)F^0 = \tag{3.2.2}$$

$$\sum_{n=0}^{\infty} \int_0^t dt_1 \cdots \int_0^{t_{n-1}} dt_n S(-t)S(t_1) \left[\mathcal{H}, \int dx\right] S(-t_1) \ldots S(t_n) \left[\mathcal{H}, \int dx\right] S(-t_n)F^0.$$

The explicit form and a mathematical meaning of the individual terms of this series was obtained in Theorem 2.5.1. Then the uniform convergence of the series (3.2.2) will be proved and at last we will show that the series (3.2.2) is a solution of the BBGKY hierarchy (3.2.1).

The procedure of solution of this problem is subdivided into several steps. First, we shall prove the convergence of the series (3.2.2) for $F_s(t, x_1, \ldots, x_s)$ ($s = 1, 2, \ldots$) and the existence of a solution local in time.

Let $L^{\infty,0}_\xi \subset L^\infty_\xi$ denote the set of sequences of functions f_n which are equal to zero in some ε–neighborhood of the set $\mathbb{W}_n$ or satisfy boundary conditions (2.4.4) on $\partial\left(\mathbb{R}^{\nu n} \backslash \mathbb{W}_n\right)$. The following result holds.

Theorem 3.2.1. *If $F^0 \in L^{\infty,0}_{\xi}$, then the sequence $F(t)$ (3.2.2) exists and the iteration series (3.2.2) for the functions $F_s(t, x_1, \dots, x_s)$ converges uniformly on each compact of the set $\mathbb{R}^{\nu s} \times (\mathbb{R}^{\nu s} \backslash \mathbb{W}_s)$ almost evrywhere for $t \in [-t_0, t_0]$, where $t_0 = t_0(\xi, \beta, a)$ is a finite number defined according to (3.2.10). The sequence of functions $F_s(t)$ is a weak unique solution of the Cauchy problem for the BBGKY hierarchy (3.2.1).*

Let us proceed to the proof of this existence theorem.

3.2.2 Basic estimate First of all, we prove the uniform convergence of series (3.2.2) for $F_s(t, x_1, \dots, x_s)$ $(s = 1, 2, \dots)$ on each compact for arbitrary initial data $F^0 \in L^{\infty,0}_{\xi}$.

Using the representation (2.5.6) for n–th term I_n of the series (3.2.2), taking into account Theorem 2.5.1 and the invariance of the Maxwellian distribution under the action of both the evolution operators $S(\pm t)$ and the mapping (1.2.5), the following inequality can be obtained

$$|I_n(x_1, \dots, x_s)| \leqslant \sum_{m=0}^{n} \sum_{i_1=1}^{s} \cdots \sum_{i_n=1}^{s+n-1} \sum_{1 \leqslant j_1 < \dots < j_m}^{n} \int_0^t dt_1$$

$$\cdots \int_0^{t_{n-1}} dt_n \int_{\mathbb{R}^3 \times \mathbb{S}^2_+} d\eta_1 dp_{s+1} \cdots \int_{\mathbb{R}^3 \times \mathbb{S}^2_+} d\eta_n dp_{s+n} \prod_{k=1}^{n} \sum_{i_k=1}^{s+k-1} \left(|P_{i_k}(t_k - t_{k-1}, \dots)| + |p_{s+k}|\right)$$

$$\times \xi^{s+n} \exp\left\{-\beta \sum_{i=1}^{s+n} \frac{p_i^2}{2}\right\} \|F^0_{s+n}\|, \tag{3.2.3}$$

where the functions $P_{i_k}(t_k - t_{k-1}, \dots)$ depend on the corresponding initial data (2.5.6). Using the estimate

$$|P_{i_k}(t_k - t_{k-1}, \dots)| \leqslant |P_{i_k}(t_k - t_{k-1}, \dots)|^2 + 1,$$

and taking into account that at the collision instants the kinetic energy of a finite number of hard spheres is conserved we have, for example,

$$\sum_{i_k=1}^{s+k-1} P^2_{i_k}\left(t_k - t_{k-1}, X^{(s+k-2)}(t_{k-1} - t_{k-2}, \dots); Q_{i_{k-1}}(t_{k-1} - t_{k-2}, \dots)\right)$$

$$+a\eta_{k-1}, p_{s+k-1}\big) = p^2_{s+k-1} + \sum_{i_k=1}^{s+k-2} P^2_{i_k}\left(t_{k-1}-t_{k-2},\ldots\right) = \cdots = \sum_{i_k=1}^{s+k-1} p^2_{i_k},$$

we obtain

$$\sum_{i_k=1}^{s+k-1}\left(\left|P_{i_k}\left(t_k-t_{k-1},\ldots\right)\right| + \left|p_{s+k}\right|\right) \leqslant \sum_{i_k=1}^{s+k-1}\left(\left|p_{i_k}\right|^2 + 1 + \left|p_{s+k}\right|\right), \quad k=1,\ldots,n.$$

We then have

$$\sum_{m=0}^{n} \sum_{1\leqslant j_1<\cdots<j_m}^{n} 1 = 2^n,$$

and we can perform the integration over the variables $t_1,\ldots,t_n$, and $\eta_1,\ldots,\eta_n$ in the majorant (3.2.3) explicitly. As a result we obtain for $n \geqslant 1$

$$\left|I_n\left(x_1,\ldots,x_s\right)\right| \leqslant \|F^0\|\xi^s \exp\left\{-\beta\sum_{i=1}^{s}\frac{p_i^2}{2}\right\}\frac{t^n}{n!}\left(2^3\pi\xi\right)^n$$

$$\times \int_{\mathbb{R}^{3n}} dp_{s+1}\ldots dp_{s+n}\prod_{k=1}^{n}\sum_{i_k=1}^{s+k-1}\left(\left|p_{i_k}\right|^2+1+\left|p_{s+k}\right|\right)\exp\left\{-\beta\sum_{i=s+1}^{s+n}\frac{p_i^2}{2}\right\}. \qquad (3.2.4)$$

Let us estimate the integrand in (3.2.4)

$$\prod_{k=1}^{n}\sum_{i_k=1}^{s+k-1}\left(\left|p_{i_k}\right|^2+1+\left|p_{s+k}\right|\right) \leqslant \prod_{k=1}^{n}\left(\sum_{i=1}^{s+n}\left|p_i\right|^2 + (s+k)\left(1+\left|p_{s+k}\right|\right)\right)$$

$$= \sum_{k=0}^{n}\sum_{1=j_1<\cdots<j_k}^{n}(s+j_1)\ldots(s+j_k)\left(\sum_{i=1}^{s+n}\left|p_i\right|^2\right)^{n-k}$$

$$\times\sum_{r=0}^{k}\sum_{1\leqslant i_1<\cdots<i_r}^{k}\left|p_{s+j_{i_1}}\right|\ldots\left|p_{s+j_{i_r}}\right|$$

$$= \sum_{k=0}^{n}\sum_{1\leqslant j_1<\cdots<j_k}^{n}(s+j_1)\ldots(s+j_k)\sum_{r=0}^{k}\sum_{1\leqslant i_1<\cdots<i_r}\left|p_{s+j_{i_1}}\right|\ldots\left|p_{s+j_{i_r}}\right|$$

$$\times\sum_{l=0}^{n-k}\frac{(n-k)!}{l!}\left(\sum_{i=1}^{s}\left|p_i\right|^2\right)^l$$

$$\times\sum_{\substack{m_1,\ldots,m_n=0\\ m_1+\cdots+m_n=n-k-l}}^{n-k-l}\frac{1}{m_1!\ldots m_n!}\left|p_{s+1}\right|^{2m_1}\ldots\left|p_{s+n}\right|^{2m_n}.$$

(3.2.5)

Using these inequalities, we calculate in (3.2.4) the integrals over the momentum variables

$$\int_{\Re^{3n}} dp_{s+1}\ldots dp_{s+n}\exp\left\{-\beta\sum_{i=s+1}^{s+n}\frac{p_i^2}{2}\right\}|p_{s+1}|^{2m_1}\ldots|p_{s+n}|^{2m_n}\left|p_{s+j_{i_1}}\right|\ldots\left|p_{s+j_{i_r}}\right|$$

where $i_1,\ldots,i_r\in(1,\ldots,k)$; $i_1\neq\cdots\neq i_r$; $j_1,\ldots,j_k\in(1,\ldots,n)$; $j_1\neq\cdots\neq j_k$. Since the result of the integration is independent of the choice of the values of the indices $j_1,\ldots,j_n$, we put $i_1=1,\ldots,i_r=r$ and $j_1=1,\ldots,j_k=k$. Then we obtain

$$\int_{\Re^{3n}} dp_{s+1}\ldots dp_{s+n}\exp\left\{-\beta\sum_{i=s+1}^{s+n}\frac{p_i^2}{2}\right\}|p_{s+1}|^{2m_1+1}\ldots|p_{s+r}|^{2m_r+1}|p_{s+r+1}|^{2m_{r+1}}$$

$$\ldots|p_{s+n}|^{2m_n}=(2\pi)^n\left(\frac{2}{\beta}\right)^{m_1+\cdots+m_n+\frac{1}{2}(3n+r)}\Gamma(m_1+2)\ldots\Gamma(m_r+2)$$

$$\times\Gamma\left(m_{r+1}+\frac{3}{2}\right)\ldots\Gamma\left(m_n+\frac{3}{2}\right), \qquad (3.2.6)$$

where $\Gamma(x)$ is the gamma–function. Hence, using the inequality (3.2.5) and equality (3.2.6), the estimate (3.2.4) for $I_n(x_1,\ldots,x_s)$ can be converted to the form

$$|I_n(x_1,\ldots,x_s)|\leqslant\|F^0\|\xi^s\exp\left\{-\beta\sum_{i=1}^{s}\frac{p_i^2}{2}\right\}$$

$$\times\frac{t^n}{n!}\left(2^4\pi^2\xi\right)^n\sum_{k=0}^{n}\sum_{1\leqslant j_1<\cdots<j_k}^{n}(s+j_1)\ldots(s+j_k)\sum_{r=0}^{k}\sum_{1\leqslant i_1<\cdots<i_r}^{k}$$

$$\times\sum_{l=0}^{n-k}\frac{(n-k)!}{l!}\left(\sum_{i=1}^{s}p_i^2\right)^l\sum_{\substack{m_1,\ldots,m_n=0\\ m_1+\cdots+m_n=n-k-l}}^{n-k-l}\left(\frac{2}{\beta}\right)^{\frac{5}{2}n-k-l+\frac{r}{2}}$$

$$\times\frac{\Gamma(m_1+2)}{m_1!}\ldots\frac{\Gamma(m_r+2)}{m_r!}\frac{\Gamma\left(m_{r+1}+\frac{3}{2}\right)}{m_{r+1}!}\ldots\frac{\Gamma\left(m_n+\frac{3}{2}\right)}{m_n!}. \qquad (3.2.7)$$

The following inequalities hold

$$\frac{1}{m_i!}\Gamma(m_i+2)\leqslant 2^{m_i+1}, \quad \frac{1}{m_j!}\Gamma\left(m_j+\frac{3}{2}\right)\leqslant\sqrt{\pi}2^{m_j+1},$$

$$(s+j_1)\dots(s+j_k) \leqslant (s+n-k)\dots(s+n) = \frac{(s+n)!}{(s+n-k)!},$$

Taking into account that

$$\sum_{1\leqslant j_1<\dots j_k}^{n} 1 = \frac{n!}{k!\,(n-k)!}, \quad \sum_{1\leqslant i_1<\dots i_r}^{k} 1 = \frac{k!}{r!\,(k-r)!},$$

$$\sum_{\substack{m_1,\dots,m_n=0\\ m_1+\dots+m_n=n-k-l}}^{n-k-l} 1 = C_{2n-k-l-1}^{n-k-l} \leqslant 2^{2n-k-l},$$

the estimate (3.2.7) can be transformed in the following way

$$|I_n(x_1,\dots,x_s)| \leqslant \|F^0\|\xi^s \exp\left\{-\beta\sum_{i=1}^{s}\frac{p_i^2}{2}\right\}\left(2^4\pi^2\xi t\right)^n$$

$$\times\sum_{k=0}^{n}\sum_{r=0}^{k}\frac{(s+n)!}{k!\,(s+n-k)!}\,\frac{k!}{r!\,(k-r)!}\sum_{l=0}^{n-k}\frac{1}{l!}\left(\sum_{i=1}^{s}p_i^2\right)^l 4^{2n-k-l}$$

$$\times\pi^{\frac{1}{2}(n-r)}\left(\frac{2}{\beta}\right)^{\frac{5}{2}n-k-l+\frac{1}{2}r}. \tag{3.2.8}$$

In this expression, we first estimate the sum over the index l

$$\sum_{l=0}^{n-k}\frac{1}{l!}\left(\sum_{i=1}^{s}p_i^2\right)^l 2^{-l}\left(\frac{2}{\beta}\right)^{-l} \leqslant \exp\left\{\frac{1}{2}\beta\sum_{i=1}^{s}\frac{p_i^2}{2}\right\}.$$

Then, using the inequality

$$\pi^{\frac{1}{2}(n-r)}4^{2n-k}2^{\frac{5}{2}n-k+\frac{1}{2}r} \leqslant \pi^{\frac{n}{2}}2^{\frac{13}{2}n-\frac{5}{2}k},$$

$$\beta^{-\frac{5}{2}n+k-\frac{1}{2}r} \leqslant (l(\beta))^n,$$

where $l(\beta) \equiv \max\left(\beta^{-\frac{3}{2}},\beta^{-\frac{5}{2}}\right)$, $0<\beta<\infty$, we finally estimate the sums over indices r and k from (3.2.8)

$$2^{-\frac{5}{2}k}\sum_{r=0}^{k}\frac{k!}{r!\,(k-r)!} = 2^{-\frac{3}{2}k}2^k \leqslant 1,$$

$$\sum_{k=0}^{n} \frac{(s+n)!}{k!\,(s+n-k)!} \leqslant \sum_{k=0}^{s+n} \frac{(s+n)!}{k!\,(s+n-k)!} = 2^{s+n}.$$

Using these inequalities, we obtain

$$|I_n(x_1,\ldots,x_s)| \leqslant \|F^0\|\,(2\xi)^s \exp\left\{-\frac{1}{2}\beta\sum_{i=1}^{s}\frac{p_i^2}{2}\right\}\left(2^{12}\pi^{\frac{5}{2}}\xi a^2 l(\beta)\,t\right)^n \tag{3.2.9}$$

or

$$|F_s(t,x_1,\ldots,x_s)| \leqslant \|F^0\|\xi^s \exp\left\{-\beta\sum_{i=1}^{s}\frac{p_i^2}{2}\right\} + \|F^0\|\,(2\xi^s)$$

$$\times \exp\left\{-\frac{1}{2}\beta\sum_{i=1}^{s}\frac{p_i^2}{2}\right\}\sum_{n=1}^{\infty}\left(2^{12}\pi^{\frac{5}{2}}\xi a^2 l(\beta)\,t\right)^n. \tag{3.2.9'}$$

For any $t \in [-t_0, +t_0]$, where

$$t_0 \equiv \left(2^{12}\pi^{\frac{5}{2}}\xi l(\beta)\,a^2\right)^{-1} \tag{3.2.10}$$

is an average time of the free motion of hard spheres in a system with a fixed density at $t=0$, we finally obtain the estimates

$$|F_s(t,x_1,\ldots,x_s)| \leqslant \|F^0\| 2\,(2\xi)^s \exp\left\{-\frac{1}{2}\beta\sum_{i=1}^{s}\frac{p_i^2}{2}\right\}, \quad s \geqslant 1. \tag{3.2.11}$$

Hence we have defined the evolution operator $U(t)$ which appears in (3.2.2) on the sequences $F^0 \in L_\xi^{\infty,0}$ in the sense that the series (3.2.2) converges uniformly with respect to $x_1,\ldots,x_s$ on each compact almost everywhere and with respect to $t \in [-t_0, t_0]$.

From this result it follows that functions $F_s(t,x_1,\ldots,x_s)$ are continuous with respect to $t \in [-t_0, t_0]$ outside of instants of collision of systems of s particles and, as functions of $x_1,\ldots,x_s$, are defined on compacts almost everywhere according to the properties of I_n (Theorem 2.5.1).

Remark 1. The integrand of the n-th term I_n of the series (3.2.2) is, according to Theorem 2.5.1, defined almost everywhere with respect to $x_1,\ldots,x_s$, $\eta_1,\ldots,\eta_n$, $p_{s+1},\ldots,p_{s+n}$ $t_1,\ldots,t_n$ and absolutely integrable with respect to all these variables, according to the above estimates (on arbitrary compacts with respect to $q_1,\ldots,q_s$). Then, thanks to Fubini's theorem, the function $F_s(t,x_1,\ldots,x_s)$ is also defined almost everywhere with respect to $x_1,\ldots,x_s$, since it is a sum of the uniformly convergent series (3.2.2). Henceforth, for the sake of simplicity, we shall omit the words "almost everywhere" with the understanding that they should appear in all the statements we make.

To finish the proof of the local existence theorem let us show that the sequence $F(t)$ (3.2.2) is a weak solution of the BBGKY hierarchy (3.2.1). To this aim, let us consider the functional

$$(\varphi, F(t)) = \left(e^{\int dx}\varphi F(t)\right)_0 ,$$

where

$$\left(e^{\int dx}\varphi F(t)\right)_0 = \sum_{n=0}^{\infty} \frac{1}{n!} \int\limits_{\mathbb{R}^{3n}\times\mathbb{R}^{3n}} dx_1 \ldots dx_n \varphi_n(x_1, \ldots, x_n) F_n(t, x_1, \ldots, x_n) ,$$

and $\varphi = \{\varphi_n(x_1, \ldots, x_n)\}_{n \geqslant 0}$ is a finite sequence of continuously differentiable functions with compact supports, equal to zero in the ε–neighbourhood of the forbidden configurations $\mathbb{W}_n$. According to the inequalities (3.2.11) this functional $(\varphi, F(t))$ exists.

The integrands in every term of the iteration series (3.2.2) (Theorem 2.5.1) are continuous functions with respect to t_1 within time intervals between pair collisions of s–particle systems with initial data $x_1, \ldots, x_s$ and therefore $(S^s(t) I_n)(x_1, \ldots, x_s)$ are continuously differentiable function with respect to t from intervals between collisions and their derivatives satisfy inequality (3.2.9). The series of derivatives of the functions $(S^s(t) I_n)(x_1, \ldots, x_s)$ absolutely converge with respect to t on the finite time interval $[-t_0, t_0]$; therefore the function $\left(S^s(t) F_s(t)\right)(x_1, \ldots, x_s)$ is continuously differentiable with respect to t in intervals between collisions and $\frac{d}{dt}(S^s(t) F_s(t))(x_1, \ldots, x_s) = S^s(t)\left(\left[\mathcal{H}, \int dx\right] F(t)\right)_s \times (x_1, \ldots, x_s)$. The functions $(S^s(t)\varphi_s)(x_1, \ldots, x_s)$ are also continuously differentiable with respect to t, and the following equality holds

$$\frac{d}{dt}(\varphi, F(t)) = \frac{d}{dt}\Big(S(t)\varphi, \sum_{n=0}^{\infty} \int_0^t dt_1 \cdots \int_0^{t_{n-1}} dt_n S(t_1)\left[\mathcal{H}, \int dx\right] S(-t_1)$$

$$\ldots S(t_n)\left[\mathcal{H}, \int dx\right] S(-t_n) F^0\Big)$$

$$= (-\mathcal{H}S(t)\varphi, S(t) F(t)) + \left(S(t)\varphi, S(t)\left[\mathcal{H}, \int dx\right] F(t)\right)$$

$$= (\mathcal{H}\varphi, F(t)) + \left(\varphi, \left[\mathcal{H}, \int dx\right] F(t)\right).$$

The functionals appearing in the right–hand side exist because $(\mathcal{H}\varphi)_s(x_1,\dots,x_s)$ and $\varphi_s(x_1,\dots,x_s)$ are continuous functions with supports on compacts, and the functions $F_s(t,x_1,\dots,x_s)$, $\left(\left[\mathcal{H},\int dx\right]F(t)\right)_s(x_1,\dots,x_s)$ are bounded according to (3.2.11).

Thus, if $F^0 \in L_\xi^{\infty,0}$ the formula (3.2.2) describes the weak solution of the initial value problem (3.2.1). Obviously, the estimates (3.2.11) do not depend on the volume of the domain Λ, which contains the particles and therefore the result (Theorem 3.2.1) is also valid for the evolution operator (3.2.2), defined for the bounded domain Λ.

3.2.3 Other methods of construction of the basic estimate. Estimates similar to (3.2.11) can be obtained in different ways. Let us give some examples.

Firstly this problem was discussed in Lanford's paper [Lan3], where the BBGKY hierarchy was considered in the framework of the canonical ensemble and the normalization condition used for $F(t)$ was such that in front of the expression $\left[\mathcal{H},\int dx\right]F(t)$ in the BBGKY hierarchy there is a factor Na^2, where N is the number of particles. The space considered there was $L_{\xi'}^{\infty}$, with $\xi' = \left(\frac{\beta}{2\pi}\right)^{-\frac{3}{2}}\xi$, where ξ is a parameter of the norm L_ξ^∞ defined above. In this paper the following statement was proved

Proposition 3.2.1. Let $\beta > \beta' > 0$ and $\xi'' > \left(\frac{\beta'}{\beta}\right)^{\frac{3}{2}}\xi'$, then there exists a constant c, depending only on $\frac{\beta'}{\beta}$ and $\frac{\xi''}{\xi'}$, such that the following inequality holds

$$\left\| S(t_1-t)\left[\mathcal{H},\int dx\right]S(t_2-t_1)\dots\left[\mathcal{H},\int dx\right]S(-t_n)F^0\right\|_{\xi'',\beta'} \leqslant n!\left(c\pi\xi'\left(\frac{\beta}{3}\right)^{-\frac{1}{2}}\right)^n \left\|F^0\right\|_{\xi',\beta}$$

for arbitrary $F^0 \in L_{\xi',\beta}^{\infty}$ and $n = 1,2,\dots$.

From this proposition it follows that the n–th term of the iteration series considered as the operator acting from space $L_{\xi',\beta}^{\infty}$ to space $L_{\xi'',\beta'}^{\infty}$ is bounded in norm by $\left(|t|/t_0\right)^n$, where

$$t_0 \equiv \left(\operatorname{const}\xi' Na^2\left(\frac{\beta}{3}\right)^{-\frac{1}{2}}\right)^{-1}.$$

Then the iteration series (3.2.2) converges uniformly in N, a, ξ', β for $|t| < t_0$.

In the papers of Uchiyama [U] and Spohn [Sp1] original techniques were used to produce estimates similar to (3.2.11).

In Shinbrot's paper [Sh] the local existence theorem for the BBGKY hierarchy for a system of N hard spheres in bounded domain $\Lambda \subset \mathbb{R}^3$ was proved in another way.

The space $L^\infty_{\xi',\beta}$ with a parameter $\xi' = \left(\frac{\beta}{2\pi}\right)^{-3/2} |\Lambda| \, \xi$ was considered. The auxiliary space $\mathcal{L}$ of the continuous functions $f : [0, t_0] \to L^\infty_{\xi(t_0),\beta(t_0)}$ such that

$$|f| = \max_t \|f(t)\|_{\xi(t_0),\beta(t_0)} < \infty$$

was also introduced. In [Sh] the normalization condition for $F(t)$ was such that before the expression $\left[\mathcal{H}, \int dx\right] F(t)$ in the BBGKY hierarchy there is a factor $\lambda = \{\lambda_i\}_{i \geqslant 1}$, where $\lambda_i \equiv (N - i)\, a^2$.

The following theorem holds.

Theorem 3.2.2. *Let $F^0 \in L^\infty_{\xi^0,\beta^0}$, then for sufficiently small $\lambda / |\Lambda|$ there exists a unique solution of the Cauchy problem for the BBGKY hierarchy (3.2.1), which is given by the formula*

$$F(t) =$$

$$\sum_{n=0}^{\infty} \int_0^t dt_1 \cdots \int_0^{t_{n-1}} dt_n S(-t)\lambda S(t_1)\left[\mathcal{H}, \int dx\right] S(-t_1)\ldots\lambda S(t_n)\left[\mathcal{H}, \int dx\right] S(-t_n)F^0. \tag{3.2.12}$$

This solution belongs to the space $\mathcal{L}$ and the following estimate holds

$$|F| \leqslant c \left\|F^0\right\|_{\xi^0,\beta^0},$$

where the constant c does not depend on N.

Proof. Since for a system with a fixed number N of particles the last equation of the hierarchy (3.2.1) coincides with the Liouville equation, then $F_N(t) = S^N(-t)\, F_N^0$. Using the integral form of the BBGKY hierarchy

$$F(t) = S(-t)\, F^0 + \lambda \int_0^t d\tau S(\tau - t) \left[\mathcal{H}, \int dx\right] F(\tau), \tag{3.2.13}$$

we can calculate the function $F_{N-1}(t)$ by $F_N(t)$ and so on. Thus we obtain the unique formal solution (3.2.12). Now let us show, that the operator $\lambda\int\limits_0^t d\tau S(\tau - t)$ $\times\left[\mathcal{H}, \int dx\right]$ is a mapping of the space $\mathcal{L}$ on itself. For this purpose we note that if

$$|F_s(t, x_1, \ldots, x_s)| \leqslant \xi^s(t)\frac{1}{|\Lambda|^s}\left(\frac{\beta(t)}{2\pi}\right)^{\frac{3}{2}s}\exp\left\{-\beta(t)\sum_{i=1}^{s}\frac{p_i^2}{2}\right\}\|F(t)\|_{\xi(t),\beta(t)} \tag{3.2.14}$$

then

$$\left|\left(\lambda\left[\mathcal{H}, \int dx\right]F(t)\right)_s(x_1, \ldots, x_s)\right|$$

$$\leqslant c_1\frac{\lambda_s}{|\Lambda|}\frac{d}{dt}\left[\xi^s(t)\frac{1}{|\Lambda|^s}\left(\frac{\beta(t)}{2\pi}\right)^{\frac{3}{2}s}\exp\left\{-\beta(t)\sum_{i=1}^{s}\frac{p_i^2}{2}\right\}\right]\|F(t)\|_{\xi(t),\beta(t)}, \tag{3.2.15}$$

where $\xi, \beta : [0, t_0] \to [0, \infty]$ are continuous functions such that both $\xi(t)$ and the function $\xi(t)\left(\frac{\beta(t)}{2\pi}\right)^{\frac{3}{2}}\frac{1}{|\Lambda|}\exp\left\{-\beta(t)\frac{p^2}{2}\right\}$ are increasing and $\beta(t)$ is decreasing for $t \in [0, t_0]$, $t_0 > 0$. Indeed, from the definition of the operator $\left[\mathcal{H}, \int dx\right]$ and taking into account energy conservation, we have

$$\left|\left(\lambda\left[\mathcal{H}, \int dx\right]F(t)\right)_s(x_1, \ldots, x_s)\right|$$

$$\leqslant c\lambda_s\xi^{s+1}(t)\frac{1}{|\Lambda|^s}\left(\frac{\beta(t)}{2\pi}\right)^{\frac{3}{2}s}\exp\left\{-\beta(t)\sum_{i=1}^{s}\frac{p_i^2}{2}\right\}\int\limits_{\mathbb{R}^3}dp\left(s|p| + \sum_{i=1}^{s}|p_i|\right)$$

$$\times\frac{1}{|\Lambda|}\left(\frac{\beta(t)}{2\pi}\right)^{\frac{3}{2}}\exp\left\{-\beta(t)\frac{p^2}{2}\right\}\|F(t)\|_{\xi(t),\beta(t)}$$

$$\leqslant c\frac{\lambda_s}{|\Lambda|}\xi^{s+1}(t)\frac{1}{|\Lambda|^s}\left(\frac{\beta(t)}{2\pi}\right)^{\frac{3}{2}s}\exp\left\{-\beta(t)\sum_{i=1}^{s}\frac{p_i^2}{2}\right\}$$

$$\times\left(\frac{s}{\sqrt{\beta(t)}} + \sum_{i=1}^{s}|p_i|\right)\|F(t)\|_{\xi(t),\beta(t)}$$

$$\leqslant c\frac{\lambda_s}{|\Lambda|}\xi^{s+1}(t)\frac{1}{|\Lambda|^s}\left(\frac{\beta(t)}{2\pi}\right)^{\frac{3}{2}s}\exp\left\{-\beta(t)\sum_{i=1}^{s}\frac{p_i^2}{2}\right\}\left[s\left(\frac{1}{\sqrt{\beta(t)}} + 1\right)\right.$$

$$+\sum_{i=1}^{s}|p_i|^2\Bigg]\,\|F(t)\|_{\xi(t),\beta(t)}\,.$$

We obtain the estimate (3.2.15), if the following inequality holds

$$\xi^{s+1}\exp\left\{-\beta(t)\sum_{i=1}^{s}\frac{p_i^2}{2}\right\}\frac{1}{|\Lambda|^s}\left(\frac{\beta(t)}{2\pi}\right)^{\frac{3}{2}s}\left[s\left(\frac{1}{\sqrt{\beta(t)}}+1\right)+\sum_{i=1}^{s}|p_i|^2\right]$$

$$\leqslant c_2\frac{d}{dt}\left[\xi^s(t)\frac{1}{|\Lambda|^s}\left(\frac{\beta(t)}{2\pi}\right)^{\frac{3}{2}s}\exp\left\{-\beta(t)\sum_{i=1}^{s}\frac{p_i^2}{2}\right\}\right]$$

or

$$\begin{cases}\dfrac{1}{\sqrt{\beta(t)}}+1\leqslant c_2\left(\dfrac{1}{\xi^2(t)}\dfrac{d\xi(t)}{dt}+\dfrac{3}{2\beta(t)\xi(t)}\dfrac{d\beta(t)}{dt}\right),\\[2ex] \xi(t)\leqslant -c_2\dfrac{d\beta(t)}{dt}.\end{cases}$$

The following functions satisfy these inequalities for $t\in[0,t_0]$, where $t_0>0$ is a sufficiently small number

$$\beta(t)=\beta(0)\left[1+2t_0\ln\left(1-\frac{t}{2t_0}\right)\right],$$

$$\xi(t)=\xi(0)\left[1-\frac{t}{2t_0}\right]^{-1},$$

$\xi(0)$, $\beta(0)$ and t_0 are some constants.

Now we return to the estimate of the norm of the operator $\lambda\int_0^t d\tau S(\tau-t)$ $\times\left[\mathcal{H},\int dx\right]$. Let $F^0\in\mathcal{L}$, then, because the operator $S(t)$ conserves energy, we obtain from (3.2.15)

$$\left|\left(\lambda S(\tau-t)\left[\mathcal{H},\int dx\right]F(\tau)\right)_s(x_1,\ldots,x_s)\right|$$

$$\leqslant c\frac{\lambda_s}{|\Lambda|}\frac{d}{d\tau}\left[\xi^s(t)\frac{1}{|\Lambda|^s}\left(\frac{\beta(\tau)}{2\pi}\right)^{\frac{3}{2}s}\exp\left\{-\beta(\tau)\sum_{i=1}^{s}\frac{p_i^2}{2}\right\}|F|\right]$$

or

$$\left|\lambda\int_0^t d\tau S(\tau-t)\left[\mathcal{H},\int dx\right]F(\tau)\right|\leqslant c\frac{\lambda}{|\Lambda|}|F|. \qquad (3.2.16)$$

Thus, according to (3.2.16), the BBGKY hierarchy in the integral form (3.2.13) has a unique solution $F(t) \in \mathcal{L}$ for arbitrary $S(-t)F^0 \in \mathcal{L}$, if $\lambda/|\Lambda|$ is sufficiently small. The smallness condition for $\lambda/|\Lambda|$ can be reformulated as a smallness condition for t_0, using the renormalization of parameters.

Hence, the operator $\int_0^t d\tau S(\tau - t)\left[\mathcal{H}, \int dx\right]$ is a contraction mapping of the space $\mathcal{L}$ on itself and thus the iteration series converges.

The estimate

$$|F| \leqslant c\left\|F^0\right\|_{\xi^0, \beta^0}$$

follows from the estimates for the iteration series.

3.2.4 The thermodynamic limit For a system of a finite number of particles in a bounded domain Λ and if $F_\Lambda^0 \in L_\xi^\infty \bigcap L_0^1(\Lambda)$ there exists a unique solution of the Cauchy problem for the BBGKY hierarchy $F_\Lambda(t) \in L^1(\Lambda)$ and it is represented by formula (2.3.9).

The sequences of equilibrium distribution functions $F_\Lambda \in L_\xi^\infty \bigcap L^1(\Lambda)$ (see Section 1.5) in the thermodynamic limit converge uniformly on each compact to the limit sequence of distribution functions F. Let us suppose that the sequences of initial distribution functions F_Λ^0 belong to the subspace $L_\xi^{\infty,0} \bigcap L_0^1(\Lambda)$ and converge in the same sense as equilibrium sequences to the limit sequence F^0 and show that sequences $F_\Lambda(t) = U_\Lambda(t) F_\Lambda^0$ in (3.2.2) also converge uniformly on compacts to the limit sequence of distribution functions $F(t)$, which satisfly the BBGKY hierarchy for infinite system (3.2.1).

Using Theorem 3.2.1 we will show that the thermodynamic limit exists for solutions of the initial value problem for the BBGKY hierarchy of a finite system for t in the interval (3.2.10).

Theorem 3.2.3. *If the initial data $F_{\Lambda,n}^0(x_1,\ldots,x_n)$, $n = 1,2,\ldots$ converge uniformly with respect to $x_1,\ldots,x_n$ on each compact to the limit distribution functions $F_n^0(x_1,\ldots,x_n)$, $n = 1,2,\ldots$ when $\Lambda \nearrow \mathbb{R}^\nu$, $\langle N\rangle \to \infty$, $\langle N\rangle/|\Lambda| = 1/v < \infty$ (in the thermodynamic limit), and $F_\Lambda^0 \in L_\xi^\infty \bigcap L_0^1(\Lambda)$, $F^0 \in L_\xi^{\infty,0}$, then the distribution functions $F_{\Lambda,s}(t,x_1,\ldots,x_s)$, $s = 1,2,\ldots$, $F_\Lambda(t) = U_\Lambda(t) F_\Lambda^0$ in the thermodynamic limit converge uniformly with respect to $x_1,\ldots,x_s$ on each compact and with respect to $t \in [-t_0, t_0]$ to the limit distribution functions $F_s(t,x_1,\ldots,x_s)$ defined according (3.2.2).*

Proof. Indeed, the following estimate holds

$$\begin{aligned}&\left|F_{\Lambda,s}\left(t,x_1,\ldots,x_s\right)-F_s\left(t,x_1,\ldots,x_s\right)\right|\\&\leqslant\sum_{n=0}^{n_0}a^{2n}\left|I_{\Lambda,n}\left(x_1,\ldots,x_s\right)-I_n\left(x_1,\ldots,x_s\right)\right|\\&+\left(\left\|F_\Lambda^0\right\|+\left\|F^0\right\|\right)2\left(2\xi\right)^s\exp\left\{-\frac{1}{2}\beta\sum_{i=1}^{s}\frac{p_i^2}{2}\right\}R_{n_0}.\end{aligned}\tag{3.2.17}$$

For $t\in[-t_0,t_0]$ and for arbitrary Λ ($V=|\Lambda|$) the second term, because of the constant R_{n_0}, can be made as small as desired for sufficiently large n_0.

In the expressions $I_{\Lambda,n}(x_1,\ldots,x_s)$ and $I_n(x_1,\ldots,x_s)$ let us split the domain of integration over the momentum variables into two parts–a finite sphere of radius p_0 and its complement. The part, in which the integration is carried out over values greater than p_0 for given n_0 and for arbitrary Λ can be made as small as desired by increasing p_0, because the corresponding integrals are absolutely convergent with respect to momenta.

Let us estimate the remaining part of the first term in (3.2.17). Suppose $x_1,\ldots,x_s$ belong to an arbitrarily fixed compact, then according to the definition in (2.5.6), also $x_{s+1},\ldots,x_{s+n}$ belong to compact.

From (2.5.6) it follows that one can choose the volume V of the domain Λ (radius of the sphere Λ) large enough so that all $s+n_0$ particles cannot reach the boundary $\partial\Lambda$ in the time interval $[-t_0,t_0]$. Then,

$$S_\Lambda(t)=S(t)\ .$$

The remaining part of (3.2.17) can be estimated by the expression

$$\varepsilon(\Lambda)\,2\,(2\xi)^s\exp\left\{-\frac{1}{2}\beta\sum_{i=1}^{s}\frac{p_i^2}{2}\right\},$$

where

$$\varepsilon(\Lambda)=\sup_{n\leqslant n_0}\ \sup_{x_1,\ldots,x_{n+s}}\ \xi^{-(s+n)}\exp\left\{\beta\sum_{i=1}^{s+n}\frac{p_i^2}{2}\right\}\left|\left(F_{\Lambda,s+n}^0-F_{s+n}^0\right)(x_1,\ldots,x_{s+n})\right|$$

and $x_1,\ldots,x_{s+n}$ belongs t_0 the compact. For a fixed n_0 by increasing Λ we can make this term as small as desired since $\lim_{\Lambda\to\mathbb{R}^\nu}\varepsilon(\Lambda)=0$ according to assumption.

Thus, we have

$$\lim_{\Lambda\nearrow\mathbb{R}^\nu}\left|\left(F_\Lambda(t)-F(t)\right)_s(x_1,\ldots,x_s)\right|=0.$$

uniformly with respect to $x_1,\ldots,x_s$ on each compact and with respect to $t\in[-t_0,t_0]$.

3.3 Global Existence Theorems

3.3.1 The continuation of the solution in time. In the previous section we have proved the existence of the solution for the initial value problem of the BBGKY hierarchy of a hard sphere particle system for arbitrary initial data in $L_\xi^{\infty,0}$ but only on a bounded time interval. The goal of the present subsection is to show in what way a global solution for the BBGKY hierarchy can be constructed for initial data in some subspace of $L_\xi^{\infty,0}$, natural from physical point of view.

The idea of the continuation of the local in time solution consists in the following. Let us take the solution constructed above (Theorem 3.2.1) for $t \in [0, t_0]$ as the initial data at the instant of time $t = t_0$ and attempt to determine the solution again for $t \in [t_0, 2t_0]$ and etc. For this purpose it is necessary, first of all, that the sequence of distribution functions $F(t_0)$ satisfies the same conditions as the sequence of initial data $F^0 \in L_\xi^{\infty,0}$. However the condition $F^0(t) \in L_\xi^{\infty,0}$. does not follow from the estimate (3.2.11).

Let us show that for some subsets of $L_\xi^{\infty,0}$. consisting of sequences of distribution functions close to equilibrium, the sequence of functions $F(t)$ in (3.2.2) satisfies the same estimate as the initial data F^0, i.e. $F(t) \in L_\xi^{\infty}$.

The following result holds.

Theorem 3.3.1. *If the initial data $F^0 \in L_\xi^{\infty,0}$ are sequences of locally perturbed equilibrium distribution functions (1.5.15), then the sequence $F(t)$ (3.2.2) exists, $F(t) = U(t) F^0 \in L_\xi^{\infty,0}$ for arbitrary $t \in \mathbb{R}^1$ and it is a weak solution of the Cauchy problem for BBGKY hierarchy (3.2.1). The evolution operator $U(t)$ (3.2.2) has a group property.*

Proof. The proof this theorem is divided into three steps. First, we show that above suggested procedure of continuation of the local solution can be realized.

Proof of the existence of a global solution (first step).

In the case, when $F_\Lambda^0 \in L_\Lambda^1(\Lambda) \cap L_\xi^{\infty,0}$ are locally perturbed equilibrium functions (1.5.15), from the estimate (1.5.16) it follows that the sequence $F_\Lambda(t) = U_\Lambda(t) F_\Lambda^0 \in L^1(\Lambda)$ is completely defined and for any $V \equiv |\Lambda|$, $t \in \mathbb{R}^1$ belongs to the space $L_\xi^{\infty,0}$

$$|F_{\Lambda,s}(t, x_1, \ldots, x_s)| = \left| U_\Lambda(t) \left(e^{\int\limits_\Lambda dx} \Psi_\Lambda \delta \Psi_\Lambda \right)_0^{-1} \left(e^{\int\limits_\Lambda dx} \Psi_\Lambda \delta \Psi_\Lambda \right)_s (x_1, \ldots, x_s) \right|$$

$$= \left(e^{\int\limits_\Lambda dx}\Psi_\Lambda\delta\Psi_\Lambda\right)_0^{-1}\left|\left(e^{\int\limits_\Lambda dx}S_\Lambda(-t)\,\Psi_\Lambda\delta\Psi_\Lambda\right)_s(x_1,\ldots,x_s)\right|$$

$$\leqslant cc_1(\Lambda)\left|\left(e^{\int\limits_\Lambda dx}\Psi_\Lambda\right)_0^{-1}\left(e^{\int\limits_\Lambda dx}\Psi_\Lambda\right)_s(x_1,\ldots,x_s)\right| \qquad (3.3.1)$$

Therefore it follows that the sequence of functions $F(t) = U(t)\,F^0$ in (3.2.2) for $t \in [-t_0, +t_0]$ also belongs to the space $L^{\infty,0}_{\xi}$ as the uniform component–wise limit of the sequence $F_\Lambda(t) \in L^{\infty,0}_{\xi}$.

Moreover, the sequence $F_\Lambda(t_0)$ converges to $F(t_0)$ in the same sense as initial data F^0_Λ to F^0 and satisfies the boundary conditions (1.4.4). Consequently the corresponding series which represents the solutions $F_\Lambda(t) = U_\Lambda(t-t_0)\,F_\Lambda(t_0)$ and $F(t) = U(t-t_0)\,F(t_0)$ for $t \in [t_0, 2t_0]$ with initial data $F_\Lambda(t_0)$ and $F(t_0)$ exist and these series converge. The existence of the terms of the series representing $U_\Lambda(t-t_0)\,F_\Lambda(t_0)$, $U(t-t_0)\,F(t_0)$ can be shown in the same way as in the proof of existence of $U(\Delta t)\,U(t)$ (see the end of the proof of Theorem 2.2.1; recall here that the locally perturbed equilibrium distribution functions are equal to zero by definition in some ε–neighbourhood of the forbidden configurations). We also obtain that the thermodynamic limit of the sequence $F_\Lambda(t)$ exists also for $t \in [t_0, 2t_0]$ and is equal to $F(t)$. Repeating this procedure we can construct the global solution for the Cauchy problem of the BBGKY hierarchy (3.2.1).

3.3.2 Global existence of a weak solution Let us show that the sequence of distribution functions $F(t)$ constructed above is a weak solution of the initial value problem for the BBGKY hierarchy (3.2.1) (second step of the proof of Theorem 3.3.1)

Consider the functionals

$$(\varphi_s, F_{\Lambda,s}(t)) = \int dx_1 \ldots dx_s\varphi_s(x_1,\ldots,x_s)\,F_{\Lambda,s}(t,x_1,\ldots,x_s)\,,\ \ s \geqslant 1,$$

where $\varphi = \{\varphi_s(x_1,\ldots,x_s)\}_{s\geqslant 1}$ is a finite sequence of bounded continuously differentiable functions φ_s with compact supports, equal to zero in some ε–neighbourhood of $\partial\left(\mathbb{R}^{3s}\backslash\mathbb{W}\right)$.

As the initial data are sequences of locally perturbed equilibrium distribution functions (for a system of hard spheres, considered in the bounded domain $\Lambda \subset \mathbb{R}^3$, these sequences are always finite), then according to the inequalities (2.5.16) or (3.3.1) for the solutions of the Cauchy problem for the BBGKY hierarchy the following estimate holds

$$|F_{\Lambda,s}(t,x_1,\ldots,x_s)| \leqslant c_1 c(\Lambda)\left\|F_\Lambda^0\right\|\xi^s \exp\left\{-\beta\sum_{i=1}^{s}\frac{p_i^2}{2}\right\},\quad s=1,2,\ldots . \tag{3.3.2}$$

The functions $F_{\Lambda,s}(t,x_1,\ldots,x_s)$ and $\left(\left[\mathcal{H}_\Lambda, \int_\Lambda dx\right] F_\Lambda(t)\right)_s (x_1,\ldots,x_s)$ are defined almost everywhere with respect to $x_1,\ldots,x_s$. Then from (3.3.2) it follows that the functionals $(\varphi_s, F_{\Lambda,s}(t))$, $((\mathcal{H}_\Lambda\varphi)_s, F_{\Lambda,s}(t))$ and $\left(\varphi_s, \left(\left[\mathcal{H}_\Lambda, \int_\Lambda dx\right] F_\Lambda(t)\right)_s\right)$ exist in the space of bounded functions with compact supports on the phase space of hard spheres system (see the end of the proof of Theorem 3.2.1 and Remark 1).

From Theorem 3.2.3 it follows, that the three functionals $(\varphi_s, F_{\Lambda,s}(t))$, $((\mathcal{H}_\Lambda\varphi)_s, F_{\Lambda,s}(t))$ and $\left(\varphi_s, \left(\left[\mathcal{H}_\Lambda, \int_\Lambda dx\right] F_\Lambda(t)\right)\right)_s$ in the thermodynamic limit converge to the corresponding limiting functionals $(\varphi_s, F_s(t))$, $((\mathcal{H}\varphi)_s, F_s(t))$ and $\left(\varphi_s, \left(\left[\mathcal{H}, \int dx dx\right] F(t)\right)_s\right)$. (We remark that the proof of convergence of the functionals $\left(\varphi_s, \left(\left[\mathcal{H}_\Lambda, \int_\Lambda dx\right] F_\Lambda\right)_s\right)$ is the same as in Theorem 3.2.3 for $(\varphi_s, F_{\Lambda,s}(t))$ and can be performed on all intervals $[0,t_0]$, $[t_0, 2t_0]$,...).

For $F_\Lambda^0 \in L_0^1(\Lambda)$ (where $L_0^1(\Lambda)$ was defined in Chapter I, Sect. 1.3), the sequence $F_\Lambda(t)$ of functions $(F_\Lambda(t))_s(x_1,\ldots,x_s)$ is differentiable with respect to t in the strong sense in $L^1(\Lambda)$ and satisfies the BBGKY hierarchy (3.2.1) for $t \in \mathbb{R}^1$ (Theorems 2.2.1 and 2.3.1). Multiplying the left and right–hand side of BBGKY hierarchy (3.2.1) by $\varphi_s(x_1,\ldots,x_s)$ and integrating over these variables, we obtain

$$\frac{d}{dt}(\varphi_s, F_{\Lambda,s}(t)) = ((\mathcal{H}_\Lambda\varphi)_s, F_{\Lambda,s}(t)) + \left(\varphi_s, \left(\left[\mathcal{H}_\Lambda, \int_\Lambda dx\right] F_\Lambda(t)\right)_s\right). \tag{3.3.3}$$

In this equation we can pass to the thermodynamic limit in the right–hand side. Then the functional $\frac{d}{dt}(\varphi_s, F_{\Lambda,s}(t))$ also has a limit, and the following equality holds

$$\lim_{\Lambda\nearrow\mathbb{R}^3}\frac{d}{dt}(\varphi_s, F_{\Lambda,s}(t)) = ((\mathcal{H}\varphi)_s, F_s(t)) + \left(\varphi_s, \left(\left[\mathcal{H}, \int dx\right] F(t)\right)_s\right).$$

Now, we show that the functional $(\varphi_s, F_{\Lambda.s}(t))$ is differentiable with respect to t and the following relation is valid

$$\frac{d}{dt}(\varphi_s, F_s(t)) = ((\mathcal{H}\varphi)_s, F_s(t)) + \left(\varphi_s, \left(\left[\mathcal{H}, \int dx\right] F(t)\right)_s\right). \tag{3.3.4}$$

In fact, the functional $(\varphi_s, F_{\Lambda.s}(t))$ is continuously differentiable with respect to t and in view of (3.3.3) we have

$$(\varphi_s, (F_\Lambda(t))_s) = \int_0^t d\tau \frac{d}{d\tau}(\varphi_s, F_{\Lambda,s}(\tau))$$

$$= \int_0^t d\tau \left[((\mathcal{H}_\Lambda, \varphi)_s, F_{\Lambda.s}(\tau)) + \left(\varphi_s, \left(\left[\mathcal{H}_\Lambda, \int_\Lambda dx\right] F_\Lambda(\tau)\right)_s\right)\right]. \tag{3.3.5}$$

The functions $F_{\Lambda,s}(t)$ and

$$\left(\mathcal{H}_\Lambda + \left[\mathcal{H}_\Lambda, \int_\Lambda dx\right]\right) F_\Lambda(t) = U_\Lambda(t)\left(\mathcal{H}_\Lambda + \left[\mathcal{H}_\Lambda, \int_\Lambda dx\right] F_\Lambda^0\right), \quad F_\Lambda^0 \in L_0^1(\Lambda)$$

are strongly continuous with respect to t, and hence the functionals $(\varphi_s, F_{\Lambda,s}(t))$, $((\mathcal{H}_\Lambda\varphi)_s, F_{\Lambda.s}(t))$, $\left(\varphi_s, \left(\left[\mathcal{H}_\Lambda, \int_\Lambda dx\right] F_\Lambda(t)\right)_s\right)$ are also continuous with respect to t. These functionals in the thermodynamic limit converge uniformly with respect to t on any compact to $(\varphi_s, F_s(t))$, $((\mathcal{H}\varphi)_s, F_s(t))$, $(\varphi_s, ([\mathcal{H}, \int dx] F(t))_s)$, thus the limit functionals are also continuous with respect to t and in equality (3.3.5) we can pass to the limit under the integral sign. As a result of this passage to the thermodynamic limit we obtain

$$(\varphi_s, F_s(t)) = \int_0^t d\tau \left[((\mathcal{H}\varphi)_s, F_s(\tau)) + \left(\varphi_s, \left(\left[\mathcal{H}, \int dx\right] F(\tau)\right)_s\right)\right].$$

The integrand in this equation is continuous with respect to τ, and hence the functional $(\varphi_s, F_s(t))$ is continuously differentiable with respect to t. Thus the equalities (3.3.4) are valid and this means that the sequence of limiting distribution functions (3.2.2) satisfies in a weak sense the BBGKY hierarchy (3.2.1).

We note, that in the proof the strong differentiability over t of the solution $F_\Lambda(t)$, which is represented by the formula (2.3.9), was used . On the other hand the solution in the thermodynamic limit is based on formula (3.2.2) for the iteration series. As it was shown above in Theorem 3.2.1, this expression is also a weak solution of the BBGKY hierarchy (3.2.1) and these two representations of the solutions coincide, and may be used.

To finish the proof of the Theorem 3.3.1, let us show that the evolution operator $U(t)$ in (3.2.2) for an infinite number of hard spheres has a group property (third step of the proof).

Let $x_1, \ldots, x_s$ belong to a compact from $\mathbb{R}^{3s} \times (\Lambda^s \backslash \mathbb{W}_s)$ then for arbitrary t_1 and t_2 the following inequality holds

$$\begin{aligned} &\left|\left(U(t_1+t_2)F^0 - U(t_1)U(t_2)F^0\right)_s(x_1,\ldots,x_s)\right| \\ &\quad\leqslant \left|\left(U(t_1+t_2)F^0 - U(t_1+t_2)\mathcal{X}_\Lambda F^0\right)_s(x_1,\ldots,x_s)\right| \\ &\qquad + \left|\left(U(t_1+t_2)\mathcal{X}_\Lambda F^0 - U(t_1)U(t_2)F^0\right)_s(x_1,\ldots,x_s)\right| \end{aligned}$$

where $\mathcal{X}_\Lambda$ is the characteristic function of a domain $\Lambda \subset \mathbb{R}^3$. Since $\mathcal{X}_\Lambda F^0 \in L^1(\Lambda)$, then

$$U(t_1+t_2)\mathcal{X}_\Lambda F^0 = U(t_1)U(t_2)\mathcal{X}_\Lambda F^0.$$

Using this equality and choosing the domain Λ' such that $\Lambda' \subset \Lambda$ and $x_1, \ldots, x_s$ belong to compact in $\mathbb{R}^{3s} \times \left(\Lambda'^s \backslash \mathbb{W}_s\right)$, we obtain

$$\begin{aligned} &\left|\left(U(t_1+t_2)F^0 - U(t_1)U(t_2)F^0\right)_s(x_1,\ldots,x_s)\right| \\ &\quad\leqslant \left|\left(U(t_1+t_2)F^0 - U(t_1+t_2)\mathcal{X}_\Lambda F^0\right)_s(x_1,\ldots,x_s)\right| \\ &\quad + \left|\left(U(t_1)\mathcal{X}_{\Lambda'}\left(U(t_2)\mathcal{X}_\Lambda F^0 - U(t_2)F^0\right)\right)_s(x_1,\ldots,x_s)\right| \\ &\quad + \left|\left(U(t_1)\left(1-\mathcal{X}_{\Lambda'}\right)\left(U(t_2)\mathcal{X}_\Lambda F^0 - U(t_2)F^0\right)\right)_s(x_1,\ldots,x_s)\right|. \end{aligned}$$

Taking into account that the domain Λ' is arbitrary, the first two terms tend to zero, according to Theorem 3.2.3, and the third term tends to zero because $(1-\mathcal{X}_{\Lambda'}) \to 0$ in the limit Λ, $\Lambda' \nearrow \mathbb{R}^3$, we obtain the necessary result.

3.3.3 Global solution for the BBGKY hierarchy of a one–dimensional hard sphere system We now consider in detail the initial value problem in L^∞_ξ for the BBGKY hierarchy of hard spheres in the one–dimensional case (2.4.24). We can define the solution of the Cauchy problem (2.4.24) through a series of iterations of the BBGKY hierarchy by the formula

$$F(t) = U(t)F^0 = \sum_{n=0}^{\infty} \int_0^t dt_1 \ldots \int_0^{t_n - 1} dt_n S(t_1 - t)\left[\mathcal{H}, \int dx^+ + \int dx^-\right] S(t_2 - t_1)$$
$$\ldots S(t_n - t_{n-1})\left[\mathcal{H}, \int dx^+ + \int dx^-\right] S(-t_n) F^0, \tag{3.3.6}$$

where $S(t)$ is the evolution operator of the system under consideration and the bracket $\left[\mathcal{H}, \int dx^+ + \int dx^-\right]$ is the operator appearing in the right-hand side of the hierarchy (2.4.24).

Let us rewrite the formula (3.3.6) in a component–wise form. With this purpose in mind, we introduce the following operators

$$\left[\mathcal{H}, \int dx^+ + \int dx^-\right] = \overset{*}{\mathfrak{A}}(0) - \mathfrak{A}(0),$$

$$\overset{*}{\mathfrak{A}}(0) = \overset{*}{\mathfrak{A}}{}^+(0) + \overset{*}{\mathfrak{A}}{}^-(0) \quad , \quad \mathfrak{A}(0) = \mathfrak{A}^+(0) + \mathfrak{A}^-(0),$$

where each of the operators $\mathfrak{A}^\pm(0)$, $\overset{*}{\mathfrak{A}}{}^\pm(0)$ are defined by the corresponding part of the operator $\left[\mathcal{H}, \int dx^+ + \int dx^-\right]$

$$\left(\overset{*}{\mathfrak{A}}{}^+(0)f\right)_s(x_{-s_2}, \ldots, x_{s_1}) = \int_0^\infty dV\, V f_{s+1}\Big(x_{-s_2}, \ldots, x_{s_1-1}; q_{s_1}, \overset{*}{v}_{s_1}(v_{s_1}, V);$$
$$q_{s_1} + (a_{s_1} + a_{s_2}), \overset{*}{v}_{s_1+1}(v_{s_1}, V)\Big),$$

$$\left(\mathfrak{A}^+(0)f\right)_s(x_{-s_2}, \ldots, x_{s_1}) = \int_0^\infty dV\, V f_{s+1}\Big(x_{-s_2}, \ldots, x_{s_1};$$
$$q_{s_1} + (a_{s_1} + a_{s_2}), v_{s_1} - V\Big),$$

$$\left(\overset{*}{\mathfrak{A}}{}^-(0)f\right)_s(x_{-s_2}, \ldots, x_{s_1}) = \int_0^\infty dV\, V f_{s+1}\Big(\overset{*}{v}_{-(s_2+1)}(v_{-s_2}, V), q_{-s_2} - (a_{-s_2}$$
$$+ a_{-(s_2+1)}), \overset{*}{v}_{-s_2}(v_{-s_2}, V), q_{-s_2}; x_{-(s_2-1)}, \ldots, x_{s_1}\Big),$$

$$\left(\mathfrak{A}^{-}(0)f\right)_{s}(x_{-s_2},\ldots,x_{s_1}) = \int_0^\infty dV\, V f_{s+1}\Big(v_{-s_2}+V, q_{-s_2}-(a_{-s_2}+a_{-(s_2+1)});$$
$$x_{-s_2},\ldots,x_{s_1}\Big), \qquad s = s_1+s_2. \tag{3.3.7}$$

We also define the following operators

$$\overset{*}{\mathfrak{A}}{}^{\pm}(\tau_i) = S(\tau_i)\overset{*}{\mathfrak{A}}{}^{\pm}(0),$$
$$\mathfrak{A}^{\pm}(\tau_i) = S(\tau_i)\mathfrak{A}^{\pm}(0), \qquad \text{where } \tau_i = t_i - t_{i-1}.$$

In terms of these operators the expression for the iteration series (3.3.6) reads as follows:

$$F_s(t, x_{-s_2},\ldots,x_{s_1}) = \sum_{n=0}^{\infty}\int_0^t dt_1 \ldots \int_0^{t_{n-1}} dt_n \sum_{k=0}^{n}\sum_{1\leqslant j_1<\cdots<j_k}^{n}(-1)^{n-k}$$
$$\times\Big(\mathfrak{A}(\tau_1)\ldots\mathfrak{A}(\tau_{j_1-1})\overset{*}{\mathfrak{A}}(\tau_{j_1})\mathfrak{A}(\tau_{j_1+1})\ldots\mathfrak{A}(\tau_{j_m-1})\overset{*}{\mathfrak{A}}(\tau_{j_m})\mathfrak{A}(\tau_{j_m+1})$$
$$\ldots\mathfrak{A}(\tau_k)S(-t_k)F^0\Big)_s(x_{-s_2},\ldots,x_{s_1}), \qquad s = s_1+s_2 \geqslant 1. \tag{3.3.8}$$

This formula gives a classification of all possible terms of the iteration series (3.3.6) and represents them in an explicit form, as in the multidimensional case (see Section 2.5).

We shall consider the Banach space L^∞_ξ of sequences $f = \{f_n(x_{-n_2},\ldots, x_{n_1})\}_{n=n_1+n_2\geqslant 0}$ of measurable functions $f_n(x_{-n_2},\ldots,x_{n_1})$, which are equal to zero on the set $\mathbf{W}_n \equiv \big\{(q_{-n_2},\ldots,q_{n_1}) \in \mathbb{R}^{n_1+n_2} \mid q_{i+1} < q_i + (a_i + a_{i+1}),$ at least for one pair $(i, i+1) \in [(-n_2, -n_2+1),\ldots,(-1,1),\ldots,(n_1-1,n_1)], n = n_1+n_2\big\}$, with the norm

$$\|f\| = \sup_{n\geqslant 0}\xi^{-n}\sup_{\substack{n_1,n_2\geqslant 0\\ n_1+n_2=n}}\sup_{x_{-n_2},\ldots,x_{n_1}}|f_n(x_{-n_2},\ldots,x_{n_1})|\exp\Big\{\beta\sum_{i=-n_2}^{n_1}\frac{m_i v_i^2}{2}\Big\}.$$

Now, we define the evolution operator (3.3.6) in L^∞_ξ. The following result holds:

Theorem 3.3.2. *For arbitrary $F^0 \in L^{\infty,0}_\xi$, the sequence $F(t) = U(t)F^0$ in (3.3.6) exists and the iteration series for functions $F_s(t, x_{-s_2},\ldots,x_{s_1})$, $s = s_1+s_2 \geqslant 1$*

converges uniformly with respect to $x_{-s_2},\ldots,x_{s_1}$ *on each compact for any* $t\in\mathbb{R}^1$. *The following estimate holds*

$$|F_s(t,x_{-s_2},\ldots,x_{s_1})|\leqslant\|F^0\|\,\xi^s\exp\left\{-\frac{\beta}{3}\sum_{i=-n_2}^{n_1}\frac{m_iv_i^2}{2}\right\}\sum_{n=0}^{\infty}\frac{1}{\sqrt{n!}}\left(\frac{t}{t_0}\right)^n. \quad (3.3.9)$$

where $t_0=m_{\min}\beta\left(2^3 3\sqrt{2\pi}\,\xi\right)^{-1}$. *The sequence of functions* $F_s(t)$ *is a weak unique solutions of the Cauchy problem for the BBGKY hierarchy (2.4.24).*

Proof. To prove this theorem, let us consider the typical term of n-th term of the series (3.3.8), for example,

$$\left(\mathfrak{A}^+(\tau_1)\ldots\mathfrak{A}^+(\tau_k)S(-t_k)F^0\right)_s(x_{-s_2},\ldots x_{s_1})$$

and construct an estimate for it. To this purpose we preliminary transform the integrand. Let us introduce new integration variables in it by the formula: $V_i=v_{s_1+i-1}-v_{s_1+i}$, where $i=1,\ldots,n$. Then the integrand in the expression $(\mathfrak{A}^+(\tau_1)\ldots\times\mathfrak{A}^+(\tau_k)S(-t_k)F^0)_s(x_{-s_2},\ldots x_{s_1})$ takes on the form

$$\prod_{i=1}^{n}\left(V_{s_1+i-1}(t_i-t_{i-1},\ldots)-v_{s_1+i}\right)F^0_{s+n}\left(X_{-s_2}(-t_n,\ldots),\ldots,X_{s_1+n}(-t_n,\ldots)\right), \quad (3.3.10)$$

where the functions $V_{s_1+i-1}(t_i-t_{i-1},\ldots),X_l(-t_{n_1},\ldots)$ depend on the corresponding initial data, according to the formula (3.3.8) and the definitions (3.3.7).

For arbitrary $F^0\in L^{\infty,0}_\xi$ the expression (3.3.10) is estimated using the inequality (a consequence of energy conservation):

$$\left|F^0_{s+n}(X_{-s_2}(-t_n,\ldots),\ldots,X_{s_1+n}(-t_n,\ldots))\right|\leqslant\xi^{n+s}\|F^0\|\exp\left\{-\beta\sum_{i=-s_2}^{s_1+n}\frac{m_iv_i^2}{2}\right\} \quad (3.3.11)$$

and the following lemma.

Lemma 3.3.1. *The inequality*

$$\prod_{i=1}^{n}\left(|V_{s_1+i-1}(t_i-t_{i-1},\ldots)|+|v_{s_1+i}|\right)\exp\left\{-\alpha\sum_{j=-s_2}^{s_1+n}v_j^2\right\}\leqslant(2\alpha^{-1})^{\frac{n}{2}}\sqrt{n!} \quad (3.3.12)$$

is valid.

Proof. Indeed, using energy conservation, we have

$$\prod_{i=1}^{n}\Big(\left|V_{s_1+i-1}\left(t_i-t_{i-1},\dots\right)\right|+\left|v_{s_1+i}\right|\Big)$$
$$=\prod_{i=1}^{n}\left(\sqrt{\left(V_{s_1+i-1}(t_i-t_{i-1},\dots)\right)^2}+\left|v_{s_1+i}\right|\right)$$
$$\leqslant\prod_{i=1}^{n}\left(\left(\sum_{j=-s_2}^{s_1+i-1}V_j^2\left(t_i-t_{i-1},\dots\right)\right)^{\frac{1}{2}}+\left|v_{s_1+i}\right|\right)$$
$$=\prod_{i=1}^{n}\left(\left(\sum_{j=-s_2}^{s_1+i-1}v_j^2\right)^{\frac{1}{2}}+\left|v_{s_1+i}\right|\right)$$
$$\leqslant 2^n\left(\sum_{j=-s_2}^{s_1+n}v_j^2\right)^{\frac{n}{2}}.$$

Calculating the value of $\sup\limits_{\{v_i\}}\left(\sum\limits_{j=-s_2}^{s_1+n}v_j^2\right)^{\frac{n}{2}}\exp\left\{-\alpha\sum\limits_{j=-s_2}^{s_1+n}v_j^2\right\}$ we obtain the estimate (3.3.12).

Estimates similar to (3.3.12) are valid for the terms of the n–th–term of the series (3.3.8), which are defined by the operators $\left\{\overset{*}{\mathfrak{A}}{}^{\pm}(\tau_i)\right\}$, as well. This follows immediately from its explicit form and the explicit form of expressions (1.4.25′) $\{v_l^*\left(v_l,V\right)\}$. Analogously one can estimate the terms defined via the operators $\mathfrak{A}^+\left(\tau_i\right)$, $\mathfrak{A}^-\left(\tau_i\right)$, $\overset{*}{\mathfrak{A}}\left(\tau_i\right)$ and $\mathfrak{A}\left(\tau_i\right)$.

Let us chose in (3.3.12) $\alpha=\frac{1}{6}\beta m_{\min}$, where $m_{\min}=\min\{m_i\}_{i\in\mathbb{Z}^1\setminus\{0\}}$. As a result, for $F^0\in L_\xi^{\infty.0}$ every integral over the velocity in the series (3.3.8) is estimated by the value $\sqrt{\pi\alpha^{-1}}$,and we obtain the inequality (3.3.9).

The proposition, that the functions $F_s(t)$ in (3.3.8) are a weak solution of the BBGKY hierarchy (2.4.24) can be proved in a way similar to the multidimensional case. Hence we have obtained the statement of the Theorem 3.3.2. Now we give some comments about a system of hard rods in the interval Λ. Let us denote by $F_\Lambda(t)=U_\Lambda(t)F_\Lambda^0$ the solution of the BBGKY hierarchy for a finite system of hard rods. The thermodynamic limit for these functions exists in the following sense.

Theorem 3.3.3. *The distribution functions* $F_{\Lambda,s}\left(t,x_{-s_2},\dots,x_{s_1}\right)$; $F_\Lambda(t)=U_\Lambda(t)F_\Lambda^0$, *where* $F_\Lambda^0\in L_\xi^{\infty.0}\cap L_\alpha^1(\Lambda)$, *converge uniformly on each compact as*

$\Lambda \nearrow \mathbb{R}^1$, $\langle N\rangle \to \infty$, $\langle N\rangle / |\Lambda| = \frac{1}{v}$ *to the functions* $F_s(t, x_{-s_2}, \dots, x_{s_1})$ *in (3.3.6) for any* $t \in \mathbb{R}^1$, *if the same convergence holds for the initial data* F_Λ^0 *and* F^0.

3.4 Method of the Interaction Region

3.4.1. Formulation of the problem In the previous two sections of this Chapter we considered the methods of constructing the solutions of the BBGKY hierarchy in L_ξ^∞ using the representation of the solutions by the iteration series (3.2.2),(3.3.6). The objective of the present section is to develop a method of constructing the solutions in L_ξ^∞, based on the functional series (2.3.3).

To describe the problem it is convenient to introduce the following notations:

$$(x_1, \dots, x_s) = X, (x_{s+1}, \dots, x_{s+n}) = Y$$

$$(x_{i_1}, \dots, x_{i_{n-k}}) = Z \subset Y, (i_1, \dots, i_{n-k}) \subset (s+1, \dots, s+n)$$

$$dx_{s+1} \dots dx_{s+n} = dY.$$

Then

$$F_s(t, x_1, \dots, x_s) \equiv F(t, X),$$

$$F_{s+n}(t, x_1, \dots, x_s, x_{s+1}, \dots, x_{s+n}) \equiv F(t, X \cup Y)$$

and

$$S^s(-t, x_1, \dots, x_s) = S(-t, X),$$

$$S^{s+n-k}(-t, x_1, \dots, x_s, x_{i_1}, \dots, x_{i_{n-k}}) = S(-t, X \cup Z).$$

Using these notations the formula (2.3.3) takes on the form

$$F(t, X) = S(-t, X) F^0(X) + \sum_{n=1}^{\infty} \frac{1}{N(Y)!} \int dY \sum_{Z \subset Y} (-1)^{N(Y \setminus Z)} S(-t, X \cup Z) F^0(X \cup Y), \quad (3.4.1)$$

where $N(Y \setminus Z)$ is the number of elements of the set $Y \setminus Z$.

As was shown in Chapter 2, the series (3.4.1) is meaningful in L^1 and $F(t) \in L^1$ if $F^0 \in L^1$. Our goal now is to give a mathematical meaning to the series (3.4.1) for initial data F^0 belonging to the space L_ξ^∞ or to some of its subspaces. At first sight this seems to be impossible. Indeed, if one substitutes in (3.4.1) initial data belonging to L_ξ^∞, then, since the functions $F^0(X \cup Y)$ are bounded with respect to the configuration variables, each term of the series (3.4.1) with $n \geqslant 1$ diverges. But, using a more detailed information about the particle dynamics, we shall prove that each integrand in the expression (3.4.1) (or (2.4.17)) with an arbitrarily fixed $N(Y)$ differs from zero only in a bounded region of the configuration space which will be called the interaction region. Therefore each term in (3.4.1)with a fixed $N(Y)$ is meaningful and the corresponding integrals will be absolutely convergent. In order to prove the existence of $F(t,X)$, it is then only required to show that the series (3.4.1) converges.

For the sake of simplicity we shall restrict ourselves to considering one-dimensional systems of particles. Before presenting the idea of the method, let us impose some technical conditions on the interaction potential Φ of the systems of particles considered here:

$$\text{a)}\ \Phi \in C^2[a,R], \quad 0 < a < R < \infty,$$

$$\text{b)}\ \Phi(|q|) = \begin{cases} +\infty, & |q| \in [0,a[\\ 0, & |q| \in]R,\infty[. \end{cases}$$

$$\text{c)}\ \Phi'(a+0) = 0. \qquad (3,4.2)$$

From (3.4.2) it follows that the conditions of stability (1.5.4) and regularity (1.5.5) are fulfilled, and for $q \in \mathbb{R}^s \backslash \mathbb{W}_s$ the following relations hold

$$\left| \sum_{i<j=1}^{s} \Phi(q_i - q_j) \right| \leqslant Bs, \qquad B = \sup_{q\in[a,R]} |\Phi(q)| \left(\left[\frac{R}{a} \right] + 1 \right), \qquad (3.4.3.a)$$

$$D \equiv \sup_{1\leqslant i\leqslant s} \mathcal{F}_i = \sup_{1\leqslant i\leqslant s} \left| \frac{\partial}{\partial q_i} \sum_{j=1}^{\infty} \Phi(q_i - q_j) \right| < \infty \quad \text{for any } s \in N, \qquad (3.4.3.b)$$

where $\mathcal{F}_i$ is a total force acting upon the i-th particle and $[x]$ is the integer part of $x > 0$.

Let us show that for the system of particles with short-range potential of interaction (3.4.2) the domain of integration over the configuration variables in each term of series (3.4.1) (interaction region) is bounded.

3.4.2 The interaction region We start with the definition of the concept of the interaction region.

Let us consider a system of $s+n$ particles. We fix the initial momenta and positions X of the 1-st,...,s-th particle, the initial data Y of the other $s+1$-th,...,$s+n$-th particles being arbitrary inside the set of admissible configurations. For the sake of simplicity we shall denote the particles with initial data X as the s-particles and those with initial data Y —as the n-particles.

If any of the n–particles is located sufficiently far from the others, then it is not able to interact with them during the time interval $[0, 2t]$, because the potential $\Phi(q)$ is short–range (3.4.2). Obviously, the same is also true if during the time interval $[0, t]$ the particles evolve "forward" (operator $S(-t, X \cup Z)$) or "backward" (for example, via the operator $S(t, Z)$).

Let us associate with every i–th particle at instant of the time $2t$ an interval with the length $2R+2\int_0^{2t}|p_i(\tau)|\,d\tau$. More exactly, one lays off on the left- and right-hand side of the center of i–th particle two intervals of length $R+\int_0^{2t}|p_i(\tau)|\,d\tau$. The momenta $p_i(\tau)$, $i \in (1, \dots s+n)$ are defined from the Hamilton equations and under collisions particles change momenta, *i.e.* the functions $p_i(\tau)$ can depend on the initial data of all the $s+n$ particles. Denote by $\ell_s(0)$ an interval containing all s–particles at time $t = 0$ (the n–particles may also be situated on it). Denote by $\ell_{s+n}(2t)$ an interval with length

$$|\ell_{s+n}(2t)| = |\ell_s(0)| + 4(s+n)R + 4\sum_{i=1}^{s+n}\int_0^{2t}|p_i(\tau)|\,d\tau$$

(the lengths of the intervals $\ell_{s+n}(2t)$, $\ell_s(0)$ are denoted by $|\ell_{s+n}(2t)|$ and $|\ell_s(0)|$). From the geometrical point of view the interval $\ell_{s+n}(2t)$ is obtained if one lays off on the left- and right- hand side of the interval $\ell_s(0)$ the intervals of the length $2(s+n)R + 2\sum_{i=1}^{s+n}\int_0^{2t}|p_i(\tau)|\,d\tau$.

As it was mentioned above the momenta $p_i(\tau)$ can depend on positions and momenta of all $s+n$ particles. In what follows we need an estimate of the length of the interval $\ell_{s+n}(2t)$ independent of the initial positions of the n–particles. In order to do this let us estimate from above the length $|\ell_{s+n}(2t)|$. Taking into account that at the collisions instants, the particles change momenta and between collisions their momenta and positions are determined by the Hamilton equations, one obtains the following estimate

$$|\ell_{s+n}(2t)| \leqslant |\ell_s(0)| + 4(s+n)R + 8\sum_{i=1}^{s+n}\left(|p_i(0)|\,t + Dt^2\right) \qquad (3.4.4)$$

We see that the estimate (3.4.4) does not depend on the initial positions of the n–particles.

One says that two particles interact with each other during the time interval $[0, 2t]$, if the intervals associated with them intersect. Then the estimate (3.4.4) defines the maximum length of such interval $\ell_{s+n}(2t)$ that the n–particles located within it at $t = 0$ can interact with each other and with the s–particles during the time interval $[0, 2t]$.

One can give the following interpretation of the estimate (3.4.4). Let us define an imaginary dynamics of particles in which every particle is associated with an interval of length $2R + 4\,|\,p_i(0)\,|\,t + 4Dt^2$ at istant of time $2t$ for arbitrary initial conditions. If the i–th and j–th particles have interacted, then they are associated with a single interval with length

$$4R + 4(|\,p_i(o)\,| + |\,p_j(o)\,|)t + 8Dt^2$$

which is equal to the sum of the lengths of the intervals associated with these particles.

If more than two particles have interacted, then they are associated again with a single interval whose length is equal to the sum of the lengths of the intervals associated with all these particles and so on.

Let us divide the infinite interval of configurational variables of each of the n–particles into intervals of length $|\,\ell_{s+n}(2t)\,|$, defined in (3.4.4), and let us lay them on the left- and right- hand side of the interval $\ell_{s+n}(2t)$ which contains all the s–particles.

Let us denote by $L_{s+n}(2t)$ the interval consisting of the interval $\ell_{s+n}(2t)$ and two intervals with length $|\ell_{s+n}(2t)\,|$ on the left- and right- hand side of $\ell_{s+n}(2t)$. The region $\Omega_n(t)$ in the configuration space of the subsystem of the n–particles consisting of the intervals $L_{s+n}(2t)$ with respect to each of all n configuration variables is called the interaction region (recall that the momenta of the n–particles are arbitrary but fixed)

$$\Omega_n(t) = (\otimes L_{s+n}(2t))^n\,.$$

Assume that at the initial time $t = 0$ one of the n–particles, let us say with number j, is located outside the interval $L_{s+n}(2t)$, for example to the right. Since $\ell_{s+n}(2t)$ is the maximum length interval such that the n–particles located within it can interact with each other and with the s–particles, the whole group of particles divides into at least two groups which do not interact with each other during the time interval $[0, 2t]$ and one of the groups also contains the s–particles. It follows from the fact that it is impossible to cover an interval between the j–th particle and the last one at the right of the s–particles which are located on $\ell_{s+n}(2t)$. using the intervals associated with particles.

Denote by Y'' the phase points of a subset of the n–particles which do not interact with the s–particles and with the rest of the n–particles with phase points Y'. Represent the sets $X \cup Y$, $X \cup Z$ as follows

$$X \cup Y = X \cup Y' \cup Y'', \quad X \cup Z = X \cup Z' \cup Z'',$$

where $Z' \subset Y'$, $Z'' \subset Y''$. Recall that two groups with initial phase points $X \cup Y'$ and Y'' do not interact if the intervals between two neighbouring particles from $X \cup Y'$ and Y'' do not intersect.

The particles with initial phase points Z'' also do not interact with particles with phase points Z', because the subsets Z'' and Z' are obtained from the subsets Y'' and Y' respectively by omitting in them some points. Therefore if in the subsets $X \cup Y'$ and Y'' one has not enough intervals to cover the interval between $X \cup Y'$ and Y'', then it follows that one has not enough intervals to cover the interval between $X \cup Z'$ and Z''.

The above statements are expressed analitically by the formulae

$$\begin{aligned} S(-t, X \cup Y) &= S(-t, X \cup Y')\, S(-t, Y''), \\ S(-t, X \cup Z) &= S(-t, X \cup Z')\, S(-t, Z''), \\ S(t, Z) &= S(t, Z')\, S(t, Z''). \end{aligned} \tag{3.4.5}$$

From the definition of the interaction region $\Omega_n(t)$ and the estimate (3.4.4) it follows

Proposition 3.4.1. If the interaction potential $\Phi(q)$ satisfies the condition (3.4.2) then, for fixed initial momenta of all the n–particles, the following estimate of the volume $V_n(2t)$ of the region $\Omega_n(t)$ holds

$$V_n(2t) \leqslant 3^n \Big[\, |\ell_s(0)| + 4(s+n)R + 8 \sum_{i=1}^{s+n} \big(|p_i(0)|\,t + Dt^2\big) \Big]^n .$$

Using the notation

$$|\ell_s(t)| = |\ell_s(0)| + 4sR + 8 \sum_{i=1}^{s} \big(|p_i(0)|\,t + Dt^2\big)$$

one obtain the following estimate of $V_n(t)$

$$V_n(t) \leqslant 3^n \Big(|\ell_s(t)| + 4nR + 8 \sum_{i=s+1}^{s+n} (|p_i(0)|t + Dt^2) \Big)^n \tag{3.4.6}$$

We remark that in the case of a system of particles interacting with their nearest neighbours (1.4.12) the analogous estimate holds

$$V_{n_1+n_2}(t) \leqslant \frac{1}{n_1!\,n_2!} \bigg(|\ell_{s_1+s_2}(t)| + 4n_1R + 8 \sum_{i=s_1+1}^{s_1+n_1} \Big(|p_i(0)|t + Dt^2 \Big) \bigg)^{n_1}$$
$$\times \bigg(\ell_{s_1+s_2}(t) + 4n_2R + 8 \sum_{i=s_2+1}^{s_2+n_2} \Big(|p_{-i}(0)|t + Dt^2 \Big) \bigg)^{n_2}.$$

Let us now prove the following statement about properties of the interaction region.

Proposition 3.4.2. If the interaction potential satisfies conditions (3.4.2), then the integrand in the expression (3.4.1) is non-zero only in the interaction region $\Omega_n(t)$.

Proof. Using the Liouville theorem (Section 1.2) and the symmetry of the distribution functions we can identically transform the expression (3.4.1) as follows

$$F(t,X) = S(-t,X)F^0(X) + \sum_{n=1}^{\infty} \frac{1}{N(Y)!} \int dY$$
$$\times \sum_{Z \subset Y} (-1)^{N(Y \setminus Z)} S(t,Z) S(-t, X \cup Z) F^0(X \cup Y). \tag{3.4.1*}$$

If some of the configuration variables of a subsystem of the n–particles are initially outside the interaction region $\Omega_n(t)$, then, as shown above, the whole system of $s+n$ particles separates into two or more groups of particles (one of which contains the s–particles) noninteracting over the time interval $[0, 2t]$ and as a result its evolution operator $S(-t, X \cup Z)$ (and also $S(t, Z)$) factorizes into evolution operators of noninteracting subsystems according to (3.4.5). Using the equalities (3.4.5) we obtain

$$\sum_{Z\subset Y}(-1)^{N(Z)}S(t,Z)S(-t,X\cup Z)F^0(X\cup Y)$$
$$=\sum_{Z'\subset Y'}\sum_{Z''\subset Y''}(-1)^{N(Z'\cup Z'')}S(t,Z')S(t,Z'')S(-t,X\cup Z')$$
$$\times S(-t,Z'')F^0(X\cup Y)$$
$$=\sum_{Z'\subset Y'}(-1)^{N(Z')}\sum_{Z''\subset Y''}(-1)^{N(Z'')}S(t,Z')$$
$$\times S(-t,X\cup Z')F^0(X\cup Y)=0.$$

Thus the proposition is proved.

Note, that the proposition 3.4.2 is valid also for the functional series (2.4.17) of nonsymmetrical system of particles.

3.4.3. Basic estimate We will now prove that for any initial data $F^0\in L_\xi^{\infty,0}$ the series (3.4.1) (or (3.4.1*)) converges for bounded time interval uniformly with respect to X on compact sets.

Theorem 3.4.1. *If the interaction potential satisfies the conditions (3.4.2) and the initial data F^0 are in $L_\xi^{\infty,0}$, then the series (3.4.1) converges uniformly in X on any compact if*

$$t\in[-t_0,+t_0],$$
$$t_0=\sqrt{\frac{d}{2D}\left(\frac{\sqrt{\beta}}{2^7\xi de^{4(\beta B+1)}}-1\right)},\tag{3.4.7}$$

where $d=\max\left(D,8(\beta D)^{-1},1\right)$.

Proof. Since $F^0\in L_\xi^{\infty,0}$ the following inequality holds

$$|F(t,X)|\leqslant\|F^0\|\sum_{n=0}^{\infty}\frac{\xi^{s+n}}{n!}\int_{\mathbb{R}^n\times\Omega_n(t)}dY$$
$$\times\sum_{Z\subset Y}S(t,Z)S(-t,X\cup Z)\exp\left\{-\beta\sum_{i=1}^{s+n}\frac{p_i^2}{2}\right\},\tag{3.4.8}$$

where $\Omega_n(t)$ is the interaction region.

Taking into account the invariance of the Gibbs distributions (the Maxwellian distribution can be extended to become a Gibbs distribution) with respect to the action of evolution operators $S(\pm t)$ and using the conditions (3.4.3), we have

$$S^{n-k}(t, x_{s+1}, \dots, x_{s+n-k}) \exp\left\{-\beta \sum_{i=s+1}^{s+n-k} \frac{p_i^2}{2}\right\} \leqslant e^{2\beta B n} \exp\left\{-\beta \sum_{i=s+1}^{s+n-k} \frac{p_i^2}{2}\right\},$$

$$S^{s+n-k}(-t, x_1, \dots, x_{s+n-k}) \exp\left\{-\beta \sum_{i=1}^{s+n-k} \frac{p_i^2}{2}\right\} \leqslant e^{2\beta B(s+n)} \exp\left\{-\beta \sum_{i=1}^{s+n-k} \frac{p_i^2}{2}\right\}.$$

According to these inequalities and to the estimate (3.4.6), we obtain the majorant for (3.4.8)

$$\begin{aligned} |F(t, X)| \leqslant \| F^0 \| \exp\left\{-\beta \sum_{i=1}^{s} \frac{p_i^2}{2}\right\} \sum_{n=0}^{\infty} \frac{\xi^{s+n}}{n!} e^{2\beta B(s+2n)} 2^{2n} \\ \times \int_{-\infty}^{\infty} dp_{s+1} \dots dp_{s+n} \exp\left\{-\beta \sum_{i=s+1}^{s+n} \frac{p_i^2}{2}\right\} 3^n \\ \times \left[|\ell_s(t)| + 8t \sum_{i=s+1}^{s+n} |p_i| + 4n\,(R + 2t^2 D) \right]^n. \qquad (3.4.9) \end{aligned}$$

Using the following equality for the integrand in (3.4.9)

$$\begin{aligned} \left[|\ell_s(t)| + 8t \sum_{i=s+1}^{s+n} |p_i| + 4n\,(R + 2t^2 D) \right]^n \\ = \sum_{k=0}^{n} \frac{n!}{k!} |\ell_s(t)|^k \sum_{r=0}^{n-k} \frac{1}{r!} \left[4n\,(R + 2t^2 D)\right]^r (8t)^{n-k-r} \\ \times \sum_{\substack{m_1, \dots, m_n = 0 \\ m_1 + \dots + m_n = n-k-r}}^{n-k-r} \frac{1}{m_1! \dots m_m!} |p_{s+1}|^{m_1} \dots |p_{s+n}|^{m_n} \end{aligned}$$

we can calculate in the majorant (3.4.9) the integrals over the momentum variables. As a result we obtain

$$|F(t,X)| \leqslant \| F^0 \| \exp\left\{-\beta \sum_{i=1}^{s} \frac{p_i^2}{2}\right\} \sum_{n=0}^{\infty} \xi^{s+n} e^{2\beta B(s+2n)} 3^n 2^{2n}$$
$$\times \sum_{k=0}^{n} \frac{1}{k!} |\ell_s(t)|^k \sum_{r=0}^{n-k} \frac{1}{r!} \left[4n\left(R+2t^2 D\right)\right]^r (8t)^{n-k-r}$$
$$\times \left(\frac{2}{\beta}\right)^{n-\frac{1}{2}(k+r)} \sum_{\substack{m_1,\dots,m_n=0 \\ m_1+\dots+m_n=n-k-r}}^{n-k-r} \frac{\Gamma\left(\frac{m_1+1}{2}\right)}{m_1!} \cdots \frac{\Gamma\left(\frac{m_n+1}{2}\right)}{m_n!}, \tag{3.4.10}$$

where $\Gamma(x)$ is the Γ–function.

Taking into account that

$$\frac{1}{m_i!}\Gamma\left(\frac{m_i+1}{2}\right) \leqslant 1,$$

$$\sum_{\substack{m_1,\dots,m_n=0 \\ m_1+\dots+m_n=n-k-r}}^{n-k-r} 1 = C_{2n-k-r-1}^{n-k-r} \leqslant 2^{2n},$$

the estimate (3.4.10) is transformed in the following way:

$$|F(t,X)| \leqslant \| F^0 \| \exp\left\{-\beta \sum_{i=1}^{s} \frac{p_i^2}{2}\right\} \xi^s e^{2\beta B s} \sum_{n=0}^{\infty} \xi^n e^{4\beta B n} 3^n 2^{4n} \left(\frac{2}{\beta}\right)^n$$
$$\times \sum_{k=0}^{n} \frac{1}{k!} |\ell_s(t)|^k \sum_{r=0}^{n-k} \frac{1}{r!} \left[4n\left(R+2t^2 D\right)\right]^r (8t)^{n-k-r} \left(\sqrt{\beta}\right)^{k+r}. \tag{3.4.11}$$

We will strengthen the estimate (3.4.11) by replacing the constant R by $d = \max\left\{R, 8(\beta D)^{-1}, 1\right\}$. Then, for any t, the inequalities

$$\left(d+2t^2 D\right)\frac{\sqrt{\beta}}{8t} \geqslant 1$$

and

$$\left(d+2t^2 D\right) \geqslant 1$$

hold, and hence we have

$$(R+2t^2D)^r\,(8t)^{n-k-r}(\sqrt{\beta}\,)^{r+k}\leqslant(\sqrt{\beta}\,)^n\,(d+2t^2D)^n.$$

Finally, using the inequalities

$$\sum_{k=0}^{n}\frac{1}{k!}\,|\,\ell_s(t)\,|^k\leqslant\exp\{\,|\,\ell_s(t)\,|\,\},$$

$$\sum_{r=0}^{n-k}\frac{1}{r!}\,(4n)^r\leqslant e^{4n},$$

we obtain the following majorant for (3.4.11) or (3.4.8)

$$|\,F(t,X)\,|\leqslant\|\,F^0\,\|\exp\left\{-\beta\sum_{i=1}^{s}\frac{p_i^2}{2}+|\,\ell_s(t)\,|\right\}(\xi e^{2\beta B})^s$$
$$\times\sum_{n=0}^{\infty}\xi^n\,3^n\,2^{5n}\,e^{4n(\beta B+1)}\left(\sqrt{\beta}\,\right)^{-n}(d+2Dt^2)^n. \quad (3.4.12)$$

The geometric series in (3.4.12) converges if $t\in[\,-t_0,t_0\,]$, where

$$t_0\equiv\sqrt{\frac{d}{2D}\left(\frac{\sqrt{\beta}}{\xi\,3\cdot2^7\,d\,e^{4(\beta B+1)}}-1\right)}. \quad (3.4.13)$$

It is obvious that the condition

$$\xi<\frac{\sqrt{\beta}}{3\cdot2^7\,d\,e^{4(\beta B+1)}} \quad (3.4.14)$$

must be satisfied. From a physical point of view this means that we deal with low density systems.

Thus, for $t\in[\,-t_0,t_0\,]$ (3.4.13) we have determined the evolution operator $U(t)=e^{\int dx}S(-t)\,e^{-\int dx}$ on the sequences $F^0\in L_\xi^{\infty,0}$ in the sense that the series (3.4.1) converges uniformly with respect to $x_1,\dots,x_s$ on each compact and the following estimate holds

$$|\,F(t,X)\,|\leqslant\|\,F^0\,\|\,\xi^s\,e^{(2\beta B+1)s}\exp\left\{-\beta\sum_{i=1}^{s}\frac{p_i^2}{2}+|\,\ell_s(t)\,|\right\}. \quad (3.4.15)$$

The same holds for the solutions of the BBGKY hierarchy

$$F_\Lambda(t) = e^{\int\limits_\Lambda dx} S_\Lambda(-t) e^{-\int\limits_\Lambda dx} F_\Lambda^0$$

in a bounded domain Λ.

Remark 1. In fact, we made no effort in the proof to optimize the time interval of convergence with respect to the parameters. However, it is transparent from (3.4.13) that every time interval of convergence will only be a fraction of the quantity which is proportional to the inverse of the mean free time between collisions.

3.4.4. Existence theorems and remarks Using the results of the previous subsection we can formulate the local existence of solutions of the BBGKY hierarchy. We postpone the proof of the fact that the solution so constructed is a weak one to Theorem 3.4.4.

Theorem 3.4.2. *If $F^0 \in L_\xi^{\infty,0}$ and the interaction potential satisfies the conditions (3.4.2), then the sequence $F(t)$ (3.4.1) exists and the functional series (3.4.1) converges uniformly on each compact of the set $\mathbb{R}^{\nu s} \times (\mathbb{R}^{\nu s} \backslash \mathbb{W}_s)$ for $t \in [-t_0, t_0]$, where t_0 is a finite number (3.4.7). The sequence $F(t)$ is a weak unique solution of the Cauchy problem for the one-dimensional BBGKY hierarchy.*

Remark 2. Analogous results were also obtained for two- and three-dimensional systems of particles but for a more restricted class of initial data. We refer the readers to the monograph [PGM2] and the surveys [PGM1][MP].

In order to continue the solution constructed above in time, we will use the method of the thermodynamic limit, developed earlier in this Chapter. basing ourselves on the following theorem:

Theorem 3.4.3. *The solutions $F_{\Lambda,s}(t,X)$ converge as $\Lambda \nearrow \mathbb{R}^3$ to the limit $F_s(t,X)$ uniformly with respect to X on compact sets for any $s \geqslant 0$ and $t_0 \in [-t_0, t_0]$ if $F_{\Lambda,s}^0$ converge to F_s^0 as $\Lambda \nearrow \mathbb{R}^3$ in the sense of uniform convergence on compact sets and $F_\Lambda^0 \in L_\xi^{\infty,0}$, $F_0 \in L_\xi^{\infty,0}$.*

Proof. We have the estimate

$$|F_\Lambda(t,X) - F(t,X)| \leqslant \sum_{N(Y)=0}^{n_0} \frac{1}{N(Y)!} \int dY \times \sum_{Z\subset Y} \Big| S_\Lambda(t,Z)\, S_\Lambda(-t, X\cup Z)\, F^0(X\cup Y) - S(t,Z)\, S(-t,X\cup Z)\, F^0(X\cup Y)\Big| + \Big(\| F^0 \| + \| F^0_\Lambda \| \Big) R_{n_0}(t), \tag{3.4.16}$$

where $R_{n_0}(t)$ is a remainder of the convergent series of the majorant (3.4.12).

If $t \in [-t_0, t_0]$, where t_0 is defined in (3.4.7), for sufficiently large n_0, according to the definition of $R_{n_0}(t)$, the second term can be made arbitrarily small for any Λ. In the first term of (3.4.16) we subdivide the domain of integration over the momentum variables into two parts — a finite ball of radius p_0 and its complement. The part of the first term, where the integration over at least one momentum variable goes over values larger than p_0 can, for a given n_0, be made arbitrarily small by increasing p_0 for any Λ. We estimate the remaining part of the first term, where the integration over the momentum variables is carried out over the ball of given radius p_0.

Suppose that X belongs to a fixed compact, and Y belongs to the interaction region $\Omega_n(t)$. We choose the domain Λ sufficiently large so that the particles that were in $X \cup Y$ at initial time could not reach the boundary $\partial\Lambda$ during $t \in [0, t_0]$. Then $S_\Lambda(-t, X\cup Z) = S(-t, X\cup Z)$, $S_\Lambda(t,Z) = S(t,Z)$, and the considered part of the first term in (3.4.16) can be estimated according to inequality (3.4.15) as

$$\sum_{N(Y)=0}^{n_0} \frac{1}{N(Y)!} \int dY \sum_{Z\subset Y} S(t,Z)\, S(-t,X\cup Z)\, \big| F^0_\Lambda(X\cup Y) - F^0(X\cup Y)\big| \leqslant \varepsilon(\Lambda)\, \xi^s\, e^{(2\beta B+1)s} \exp\Big\{ -\beta \sum_{i=1}^{s} \frac{p_i^2}{2} + |\ell_s(t)| \Big\}, \tag{3.4.17}$$

where

$$\varepsilon(\Lambda) = \sup_{n\leqslant n_0} \sup_{\substack{X\in\mathcal{K}^s\times\Re^s_{p_0} \\ Y\in\Omega_n(t)\times\Re^s_{p_0} \\ Z\subset Y}} \xi^{s+n} \exp\Big\{ -\beta \sum_{i=1}^{s+n} \frac{p_i^2}{2} \Big\} \times S(t,Z)\, S(-t, X\cup Z)\, \big| F^0_\Lambda(X\cup Y) - F^0(X\cup Y)\big|. \tag{3.4.18}$$

and $\mathfrak{R}^s_{p_0}$ is a s–dimensional sphere of radius p_0. $\mathcal{K}^s$ is the domain in configuration space of the s—particles consisting of points located on the segment $\ell_s(0)$.

It follows from inequality (3.4.17) that for given number n_0 by increasing the volume of the domain Λ we can make first term in (3.4.16) arbitrarily small since according to the definition (3.4.18) we have $\lim\limits_{\Lambda\nearrow\mathbb{R}^1} \varepsilon(\Lambda) = 0$. Consequently, it holds

$$\lim_{\Lambda\nearrow\mathbb{R}^1} |F_\Lambda(t,X) - F(t,X)| = 0$$

on each compact, for $t \in [-t_0, t_0]$ (3.4.7).

Remark 3. If we choose, as initial data F^0_Λ, finite sequences of distribution functions (with the N–th component different from zero) with compact support in the momentum variables as well (for each momentum $|p_i| < p_0$), then in this case we can carry out the limiting transition $\Lambda \to \mathbb{R}^\nu$ independently of the limiting transitions $N \to \infty$, $p_0 \to \infty$. We denote such initial data by $F^{(N),0}_\Lambda$.

We are now able to formulate the global existence result for the BBGKY hierarchy.

Theorem 3.4.4. *If $F^0 \subset L^{\infty,0}_\xi$ is a sequence of locally perturbed equilibrium distribution functions, then the sequence $F(t)$ in (3.4.1) exists, is given by $F(t) = U(t)F^0 \in L^\infty_{\xi,0}$ for an arbitrary $t \in \mathbb{R}^1$ and is a weak solution of the Cauchy problem for the BBGKY hierarchy of a one-dimensional system.*

Proof. To complete the proof of this theorem, it is necessary to show that the local solution (provided by Theorem 3.4.2) can be continued for all $t \in \mathbb{R}^1$ and that the expression so constructed satisfies the BBGKY hierarchy.

In order to understand the method used to continue the solution in time, we remark that, first of all, we need an estimate different from (3.4.15), holding for local solutions. In connection with this, we remind the reader that for the solution $F_\Lambda(t)$ (3.4.16) of the BBGKY hierarchy with locally perturbed equilibrium distribution functions as initial data F^0_Λ, the following estimate holds for any $\Lambda \subset \mathbb{R}^1$ and $t \in \mathbb{R}^1$

$$|F_\Lambda(t,X)| \leqslant c\, c_1(\Lambda)\, \xi^s \, \|F_{eq}\| \exp\left\{-\beta \sum_{i=1}^{s} \frac{p_i^2}{2}\right\}. \tag{3.4.19}$$

Here F_{eq} is a sequence of equilibrium distribution functions. In other words the sequence of functions $F_\Lambda(t) = e^{\int\limits_\Lambda dx} S_\Lambda(-t)\, e^{-\int\limits_\Lambda dx} F^0_\Lambda$ belongs to $L^{\infty,0}_\xi$ for all $\Lambda \subset \mathbb{R}^1$

and $t \in \mathbb{R}^1$. Then from Theorem 3.4.3 about the existence of the thermodynamic limit it follows that the sequence of functions $F(t) = e^{\int dx} S(-t) e^{-\int dx} F^0$ belongs to the space $L_\xi^{\infty,0}$ for $t \in [-t_0, t_0]$ and satisfies the boundary condition together with $F_\Lambda(t)$ as a componentwise uniform limit of the sequence of functions $F_\Lambda(t) = e^{\int_\Lambda dx} S_\Lambda(-t) e^{-\int_\Lambda dx} F_\Lambda^0$.

Now let us consider the solution $F_\Lambda(t)$ for $t \in [t_0, 2t_0]$. It can be written in the form

$$F_\Lambda(t) = e^{\int_\Lambda dx} S_\Lambda(-t + t_0) e^{-\int_\Lambda dx} F_\Lambda^0(t_0). \tag{3.4.20}$$

By analogy the sequence $F(t)$ can be defined for $t \in [t_0, 2t_0]$ according to the formula

$$F(t) = e^{\int dx} S(-t + t_0) e^{-\int dx} F^0(t_0). \tag{3.4.21}$$

Indeed, from the above consideration it follows that $F_\Lambda(t_0)$, $F(t_0) \in L_\xi^{\infty,0}$ and we can state the validity of the Theorems 3.4.1-2 for the sequences (3.4.20) and (3.4.21). Then for the sequences (3.4.20-21) the existence of the thermodynamic limit follows (by Theorem 3.4.3) and, as result, we find that the sum of series (3.4.21) belongs to $L_\xi^{\infty,0}$ as a limit of the series (3.4.20), componentwise uniform on each compact set. (We remind the reader that $F_\Lambda(t_0)$ converges to $F(t_0)$ in the same sense as F_Λ^0 to F^0). The last statement means that we can continue the described procedure as many times as we like.

Let us show that the solution constructed in this way satisfies the BBGKY hierarchy. With this aim we introduce the following functionals ($s = 1, 2, \dots$)

$$\Big(\varphi, F_\Lambda(t)\Big)_s = \int dx_1 \dots dx_s \varphi_s(x_1, \dots, x_s) F_{\Lambda,s}(t, x_1, \dots, x_s) \tag{3.4.22}$$

$$\begin{aligned}\Big(\varphi, \Big[\mathcal{H}_{\mathrm{hc}}, \int dx\Big] F_\Lambda(t)\Big)_s = \int dx_1 \dots dx_s \varphi_s(x_1, \dots, x_s) \\ \times \sum_{i=1}^{s} \int_0^\infty dP\, P \Big\{ F_{\Lambda,s+1}(t, x_1, \dots, x_s; q_i + a, p_i - P) \\ - F_{\Lambda,s+1}(t, x_1, \dots, q_i, p_i - P, \dots, x_s, q_i - a, p_i) \\ + F_{\Lambda,s+1}(t, x_1, \dots, x_s, q_i - a, p_i + P) \\ - F_{\Lambda,s+1}(t, x_1, \dots, q_i, p_i + P, \dots, x_s q_i + a, p_i) \Big\}.\end{aligned} \tag{3.4.23}$$

where $\varphi = (\varphi_1(x_1)), \ldots, \varphi_s(x_1, \ldots, x_s), \ldots$ is a finite sequence of continuously differentiable bounded functions φ_s with compact supports, vanishing in a certain ε–neighbourhood of the set W_s of forbidden configurations. $\mathcal{H}_{hc}$ means the Poisson bracket for the system of particles with hard core interaction. According to the inequality (3.4.19), the functions $F_{\Lambda,s}(t, x_1, \ldots, x_s)$ are defined almost everywhere with respect to $x_1, \ldots, x_s$ and hence the functionals (3.4.22–23) exist. We also take into account that $F_{\Lambda,s+1}(t, (x)_{s+1})$ is well defined on the hypersurface $|q_i - q_{s+1}| = a$, according to Theorem 2.5.1 and the remark in Chapter II, following the end of the prooof of Theorem 2.5.1. Like the functions φ_s, so the expressions $(\mathcal{H}_\Lambda, \varphi)_s = \{\varphi_s, H_s^\Lambda\}$, where $\{.,.\}$ is a Poisson bracket, are bounded and have compact supports. Moreover, for sufficiently large $\Lambda \subset \mathbb{R}^1$ we have $\{\varphi_s, H_s^\Lambda\} = \{\varphi_s, H_s\}$ since the supports of the functions φ_s are subsets in Λ. Then according to the inequality (3.4.19), the functionals $(s = 1, 2, \ldots)$

$$(\mathcal{H}_\Lambda \varphi, F_\Lambda(t))_s = \int dx_1 \ldots dx_s \{\varphi_s, \mathcal{H}_s^\Lambda\} F_{\Lambda,s}(t, x_1, \ldots, x_s) \tag{3.4.24}$$

exist.

Using the notations from Chapter 2, let us also introduce the functionals $(s = 1, 2, \ldots)$

$$\begin{aligned}\left(\left[\mathcal{H}_\Phi, \int dx^*\right]_\varphi, F_\Lambda(t)\right)_{s+1} = &-\int dx_1 \ldots dx_{s+1} \\ &\times \sum_{i=1}^{s} \frac{\partial}{\partial q_i} \Phi(q_i - q_{s+1}) \frac{\partial}{\partial p_i} \varphi(x_1, \ldots, x_s) \\ &\times F_{\Lambda,s+1}(t, x_1, \ldots, x_{s+1}).\end{aligned} \tag{3.4.25}$$

The function $\sum_{i=1}^{s} \frac{\partial}{\partial q_i} \Phi(q_i - q_{s+1}) \frac{\partial}{\partial p_i} \varphi_s(x_1, \ldots, x_s)$ appearing in the expression (3.4.25) is finite with respect to the variables $x_1, \ldots, x_s$ and q_{s+1}, since the potential Φ and its derivatives are finite, and does not depend upon the variable p_{s+1}. Since, according to the inequality (3.4.19) the function $F_{\Lambda,s+1}(t)$ is exponentially decreasing with respect to p_{s+1}, the functionals (3.4.25) exist.

As $\Lambda \nearrow \mathbb{R}^1$, the functions $F_{\Lambda,s}(t, x_1, \ldots, x_s)$ converge uniformly with respect to $x_1, \ldots, x_s$ on compact sets to their limit $F_s(t, x_1, \ldots, x_s)$; therefore all the functionals (3.4.22–24) converge as $\Lambda \nearrow \mathbb{R}^1$ to their limits

$$(\varphi, F(t))_s, \qquad \left(\varphi, \left[\mathcal{H}_{\mathrm{hc}}, \int dx\right] F(t)\right)_s,$$
$$(\mathcal{H}\varphi, F(t))_s, \qquad \left(\left[\mathcal{H}_\Phi, \int dx^*\right] \varphi, F(t)\right)_{s+1}.$$

For $F_\Lambda^0 \subset L_\xi^{\infty,0} \cap L^1(\Lambda)$ the sequence $F_\Lambda(t)$ is strongly differentiable $L^1(\Lambda)$ with respect to t and satisfies the BBGKY hierarchy. Multiplying both sides of the BBGKY hierarchy by φ_s and integrating over all variables, we obtain the functional equalities

$$\frac{d}{dt}(\varphi, F_\Lambda(t))_s = (\mathcal{H}^\Lambda \varphi, F_\Lambda(t))_s + \left(\left[\mathcal{H}_\Phi, \int dx^*\right] \varphi, F_\Lambda(t)\right)_{s+1}$$
$$+ \left(\varphi, \left[\mathcal{H}_{\mathrm{hc}}, \int dx\right] F_\Lambda(t)\right)_s. \tag{3.4.26}$$

Let us take the thermodynamic limit in these equalities. All functionals on the right–hand side of (3.4.26) converge to their limits as $\Lambda \nearrow \mathbb{R}^1$; hence the functionals $\frac{d}{dt}(\varphi, F_\Lambda(t))_s$ also have limits and this yields

$$\lim_{\Lambda \nearrow \mathbb{R}^1} \frac{d}{dt}(\varphi, F_\Lambda(t))_s = (\mathcal{H}\varphi, F(t))_s + \left(\left[\mathcal{H}_\Phi, \int dx^*\right] \varphi, F(t)\right)_{s+1}$$
$$+ \left(\varphi, \left[\mathcal{H}_{\mathrm{hc}}, \int dx\right] F(t)\right)_s.$$

Therefore the functionals $(\varphi, F(t))_s$, $s = 1, 2, \ldots$ are differentiable with respect to t and the following equality holds

$$\frac{d}{dt}(\varphi, F(t))_s = (\mathcal{H}\varphi, F(t))_s + \left(\left[\mathcal{H}_\Phi, \int dx^*\right] \varphi, F(t)\right)_{s+1}$$
$$+ \left(\varphi, \left[\mathcal{H}_{\mathrm{hc}}, \int dx\right] F(t)\right)_s.$$

In fact, the functionals $(\varphi, F_\Lambda(t))_s$, $s = 1, 2, \ldots$ are continuously differentiable with respect to t and thus

$$(\varphi, F_\Lambda(t))_s = \int_0^t d\tau \frac{d}{d\tau}(\varphi, F_\Lambda(\tau))_s$$
$$= \int_0^t d\tau \Big\{ (\mathcal{H}\varphi, F_\Lambda(\tau))_s + \Big(\Big[\mathcal{H}_\Phi, \int dx^*\Big] \varphi, F_\Lambda(\tau) \Big)_{s+1}$$
$$+ \Big(\varphi, \Big[\mathcal{H}_{\text{hc}}, \int dx\Big] F_\Lambda(\tau) \Big)_s \Big\}. \qquad (3.4.27)$$

The functions $F_{\Lambda,s}(t)$ are strongly continuous with respect to t in L^1 and therefore the functionals $(\varphi, F_\Lambda(t))_s$ are continuous with respect to time as well. Moreover, the functions $F_s(t)$ are also continuous with respect to the variable t and defined almost everywhere with respect to $x_1, \ldots, x_s$ on a compact set, since they are uniform limits as $\Lambda \nearrow \mathbb{R}^1$ (with respect to $x_1, \ldots, x_s$ and t on compact sets) of the functions $F_{\Lambda,s}(t)$.

The functionals $(\varphi, F_\Lambda(t))_s$, $(\mathcal{H}\varphi, F_\Lambda(t))_s$, $\Big(\Big[\mathcal{H}_\Phi, \int dx^*\Big]\varphi, F_\Lambda(t)\Big)_{s+1}$, $\Big(\varphi, \Big[\mathcal{H}_{\text{hc}}, \int dx\Big] F_\Lambda(t)\Big)_s$, are continuous functions of $t \in \mathbb{R}^1$ and converge as $\Lambda \nearrow \mathbb{R}^1$ uniformly with respect to t to the corresponding limits $(\varphi, F(t))_s$, $(\mathcal{H}\varphi, F(t))_s$, $\Big(\Big[\mathcal{H}_\Phi, \int dx^*\Big]\varphi, F(t)\Big)_{s+1}$, $\Big(\varphi, \Big[\mathcal{H}_{\text{hc}}, \int dx\Big] F(t)\Big)_s$. Therefore the limiting functionals are also continuous functions of $t \in \mathbb{R}^1$, and we can pass to the limit under the integral sign in (3.4.27). As a result, we obtain the equality

$$(\varphi, F(t))_s = \int_0^t d\tau \Big\{ (\mathcal{H}\varphi, F(\tau))_s + \Big(\Big[\mathcal{H}_\Phi, \int dx^*\Big] \varphi, F(\tau) \Big)_{s+1}$$
$$+ \Big(\varphi, \Big[\mathcal{H}_{\text{hc}}, \int dx\Big] F(\tau) \Big)_s \Big\}. \qquad (3.4.28)$$

The integrand in (3.4.27) is continuous with respect to t; therefore the function $(\varphi, F(t))_s$ is a continuously differentiable function of the variable t and this yields $(s = 1, 2, \ldots)$

$$\frac{d}{dt}(\varphi, F(t))_s = (\mathcal{H}\varphi, F(t))_s + \Big(\Big[\mathcal{H}_\Phi, \int dx^*\Big] \varphi, F(t) \Big)_{s+1}$$
$$+ \Big(\varphi, \Big[\mathcal{H}_{\text{hc}}, \int dx\Big] F(t) \Big)_s .$$

These equalities mean that the solution $F(t) = e^{\int dx} S(-t) \times e^{-\int dx} F^0$, as constructed above. satisfies the BBGKY hierarchy in a weak sense.

Remark 4. The results of the Theorem 3.4.4 hold for solutions which are represented by the formula (2.4.17) of the initial value problem of the BBGKY hierarchy for nonsymmetrical systems.

CHAPTER IV

The Derivation of the Boltzmann Equation

4.1. Introduction : On the Boltzmann–Grad limit

In the present chapter we shall consider an example of a rigorous derivation of the kinetic equation, namely, the Boltzmann equation for a hard sphere particle system. The problem of the derivation of kinetic equations is treated in this chapter as the construction of the relevant asymptotics of the solution to the initial value problem of the BBGKY hierarchy.

Before attacking the problem of a rigorous justification, let us make some comments on the physical argument that (consciously or unconsciously) Boltzmann must have had in mind when he derived his equation.

In a rarefied gas the number of particles N is a very large number and the molecular diameter a (expressed in common units, such as, *e. g.*, centimeters) is very small; to fix the ideas. let us say that we have a box whose volume is 1 cm^3 at room temperature and atmospheric pressure. Then $N \cong 10^{20}$ and $a = 10^{-8} cm$. Then $(N - 1)a^2 \cong Na^2 = 10^4 \ cm^2 = 1 \ m^2$ is a sizable quantity, while we can neglect the difference between q_1 and $q_1 + a\eta$. This means that the equation to be written can be rigorously valid only in the so called *Boltzmann-Grad limit*, when $N \to \infty$, $a \to 0$ with Na^2 finite.

Equivalently we can say that we consider a low density gas in a situation where the range of the interaction potential of the particles is sufficiently small in comparison with the average length of a free path between two subsequent collisions. In terms of the density $1/v$ or the activity z, the Boltzmann–Grad limit for a system of infinitely many hard spheres can be formulated as follows: the diameter a of the particles tends to zero and the activity z (or the density $1/v$) tends to infinity so that the quantity $a^2 z = \alpha < \infty$ ($a^2/v = \alpha < \infty$), which is inversely proportional to the length of a free path of particles, remains fixed. We note, that in this limit the dimensionless density a^3/v tends to zero.

In fact. since the volume occupied by the particles is about $Na^3 \cong 10^{-4} cm^3$, the collisions between two preselected particles is a rather rare event. Thus two spheres that happen to collide can be thought to be two randomly chosen particles and it makes sense to assume that the probability density of finding the first molecule at q_1 with momentum p_1 and the second at q_2 with momentum p_2 is the product of the probability density of finding the first molecule at q_1 with momentum p_1 times the probability density of finding the second molecule at q_2 with momentum p_2. If we accept this we can write (assumption of *molecular chaos*):

$$F_2(t, q_1, p_1, q_2, p_2) = F_1(t, q_1, p_1)F_1(t, q_2, p_2) \tag{4.1.1}$$

for two particles that are about to collide, or:

$$F_2(t, q_1, p_1, q_1 + a\eta, p_2) = F_1(t, q_1, p_1)F_1(t, q_1 + a\eta, p_2)$$
$$\text{for } \langle (p_2 - p_1), \eta \rangle < 0. \tag{4.1.2}$$

Thus we can apply this *recipe* to the loss term in the first equation of the hierarchy (3.2.1) but the same relation does not hold for two particles that have just collided (because the collision has created a correlation). It is possible, however, to apply Eq. (4.1.2) (with an appropriate modification of the arguments) to the form of the gain term given in the previous chapters, because the transformation (1.2.5) maps the hemisphere $\mathbb{S}^2_-$ onto the hemisphere $\mathbb{S}^2_+$.

If we accept all the simplifying assumptions made by Boltzmann, we obtain the following form for the gain and loss terms in the evolution equations for F_1:

$$G = Na^2 \int_{\mathbb{R}^3} \int_{\mathbb{S}_+} F_1(t, q_1, p_1^*)F_1(t, q_1, p_2^*)\langle (p_2 - p_1), \eta \rangle dp_2 d\eta, \tag{4.1.3}$$

$$L = Na^2 \int_{\mathbb{R}^3} \int_{\mathbb{S}_+} F_1(t, q_1, p_1)F_1(t, q_1, p_2)\langle (p_2 - p_1), \eta \rangle dp_2 d\eta. \tag{4.1.4}$$

By inserting these expressions in the first equation of the BBGKY hierarchy, we can write the *Boltzmann equation* in the following form:

$$\frac{\partial F_1(t, q_1, p_1)}{\partial t} + \langle p_1, \frac{\partial F_1(t, q_1, p_1)}{\partial q_1} \rangle = \tag{4.1.5}$$

$$= Na^2 \int_{\mathbb{R}^3} \int_{\mathbb{S}_+} [F_1(t, q_1, p_1^*)F_1(t, q_1, p_2^*) - F_1(t, q_1, p_1)F_1(t, q_1, p_2)]$$

$$\langle (p_2 - p_1), \eta \rangle dp_2 d\eta$$

The Boltzmann equation is an evolution equation for F_1, without any reference to F_2. This is its main advantage. It has been obtained, however, at the price of several assumptions; the chaos assumption present in Eqs. (4.1.3) and (4.1.4) is particularly strong and will be presently discussed.

The molecular chaos, assumed in Eqs. (4.1.3)-(4.1.4), is clearly a property of randomness. Intuitively, one feels that collisions exert a randomizing influence, but it would be completely wrong to argue that the statistical independence described by Eq. (4.1.2) is a consequence of the dynamics. It is quite clear that we cannot expect every choice of the initial value for distribution function at the level of the Liouville equation to give a F_1 which agrees with the solution of the Boltzmann equation in the Boltzmann-Grad limit. In other words, molecular chaos must be present initially and we can only ask whether it is preserved by the time evolution of the system of hard spheres.

It is evident that the chaos property (4.1.1-2), if initially present, is almost immediately destroyed, if we insist that it should be valid everywhere. In fact, if it were strictly valid everywhere, the gain and loss terms, in the Boltzmann-Grad limit, would be exactly equal and there would be no effect of the collisions on the time evolution of F_1. The essential point is that we need the chaos property only for particles which are about to collide, *i. e.* in the precise form stated in Eq. (4.1.3). It is clear, then, that even if F_1, as predicted by the Liouville equation, converges nicely to a solution of the Boltzmann equation, F_2 may converge to a product, as stated in Eq. (4.1.2), only in a way, which is, in a certain sense, very singular. In fact, it is not enough to show that the convergence is almost everywhere, because we need to use the chaos property in a zero measure set. On the other hand we cannot try to show convergence everywhere, because this would be false; in fact, we have just remarked that Eq. (4.1.2) is simply not true for molecules which have just collided.

How can we approach the question of justifying the Boltzmann equation without invoking the molecular chaos assumption as an *a priori* hypothesis? Obviously, since F_2 appears in the evolution equation for F_1, we must investigate the time evolution for F_2; now, as the BBGKY hierarchy tells us, the evolution equation for F_2 contains F_3 and, in general, the evolution equation for F_s contains the next function F_{s+1}, till we reach $s = N$; in fact F_N satisfies the Liouville equation. It is clear thus that we cannot proceed unless we handle all the F_s at the same time and attempt to prove a generalized form of molecular chaos, *i.e.*

$$F_s(t, q_1, p_1, , q_2, p_2, \dots , q_s, p_s) = \prod_{j=1}^{s} F_1(t, q_j, p_j) \tag{4.1.6}$$

The task then becomes to show that, if true at $t = 0$, this property remains preserved (for any fixed s) in the Boltzmann-Grad limit. This will is the main task of this chapter.

There remains the problem of justifying the *initial chaos assumption*, according to which Eq. (4.1.1) is satisfied at $t = 0$. One can give two justifications, one of them being physical in nature and the second mathematical; essentially, they say the same thing, i.e. that it is hard to prepare an initial state for which Eq. (4.1.1) does not hold. The physical reason for this is that, in general, we cannot handle the single molecules, but rather act on the gas as a whole at a macroscopic level, usually starting from an equilibrium state (for which Eq. (4.1.1) holds). The mathematical argument indicates that, if we choose the initial data for the molecules at random there is an overwhelming probability [Ce1, Ce2] that Eq. (4.1.6) is satisfied for $t = 0$.

A word should be said about boundary conditions. When proving that chaos is preserved in the limit, it is absolutely necessary to have a boundary condition compatible (at least in the limit) with Eq. (4.1.1). If the boundary conditions are those of periodicity or specular reflection, no problems arise. More in general, it is sufficient that the particles are scattered without losses from the boundary in a way that does not depend on the state of the other particles of the gas.

Let us consider the Boltzmann-Grad limit ($N \to \infty$ and $a \to 0$ in such a way that Na^2 remains finite). Then, we obtain (for each fixed s) that, if each F_s tends to a limit (which we denote by the same symbol) and this limit is sufficiently smooth, the finite hierarchy becomes in the limit:

$$\frac{\partial F_s(t, x_1, \dots , x_s)}{\partial t} + \sum_{i=1}^{s} p_i, \frac{\partial F_s(t, x_1, \dots , x_s)}{\partial q_i} =$$

$$Na^2 \sum_{i=1}^{s} \int_{\mathbb{R}^3} \int_{\mathbb{S}^2_+} (F_{s+1}(t. x_1, \dots , q_i, p_i^*, \dots , x_s, q_i, p_{s+1}^*)$$

$$-F_{s+1}(t, x_1, \dots , q_i, p_i, \dots , x_s, q_i, p_{s+1}))\langle (p_i - p_{s+1}). \eta \rangle d\eta dp_{s+1} \tag{4.1.7}$$

where the arguments of F_{s+1} are the same as above, except for the fact that $q_{s+1} = q_i$ in agreement with the BBGKY hierarchy for $a \to 0$. Eqs. (4.1.7) give a complete description of the time evolution of a Boltzmann gas (*i. e.* the ideal gas which is obtained in the Boltzmann-Grad limit), provided the initial value problem is well posed for this infinite system of equations, which appears to have been first written in [Ce3] and is usually called *the Boltzmann hierarchy.*

As we already know, Eq. (4.1.7) is not equivalent to the Boltzmann equation, unless a special assumption on the initial data is made. Indeed, as discussed by Spohn [Sp1], the solutions of the Boltzmann hierarchy describe the evolution of a Boltzmann gas, when the chaos assumption is not satisfied by the initial data at $t = 0$. T. Here we shall assume that the data satisfy Eq. (4.1.6), which we rewrite here for $t = 0$:

$$F_s(0, q_1, p_1, q_2, p_2, \dots, q_s, p_s) = \prod_{j=1}^{s} F_1(0, q_j, p_j), \tag{4.1.8}$$

It is now a simple remark, made in [Ce3], that if Eq. (4.18) is satisfied and the Boltzmann equation, given by Eq. (4.1.5) or, shortly by:

$$\frac{\partial F_1(t, q_1, p_1)}{\partial t} + \langle p_1, \frac{\partial F_1(t, q_1, p_1)}{\partial q_1} \rangle$$
$$= Na^2 \int_{\mathbb{R}^3} \int_{\mathbb{S}^2_+} (F_1(t, q_1, p_1^*) F_1(t, q_1, p_2^*) - F_1(t, q_1, p_1) F_1(t, q_1, p_2)) \langle p_1 - p_2, \eta \rangle dp_2 d\eta \tag{4.1.9}$$

admits a solution $F_1(t, q_s, p_s)$ for given initial data $F_1(0, q_s, p_s)$, then the Boltzmann hierarchy, Eq. (4.1.7) has at least a solution, given by

$$F_s(t, q_1, p_1, q_2, p_2, \dots, q_s, p_s) = \prod_{j=1}^{s} F_1(t, q_j, p_j) \tag{4.1.10}$$

Therefore the chaos assumption, Eq. (4.1.10), is not inconsistent with the dynamics of rigid spheres in the Boltzmann-Grad limit; actually, if the Boltzmann hierarchy has a unique solution for data which satisfy Eq. (4.1.8), then Eq. (4.1.10) necessarily holds at any time if holds at $t = 0$. Then the Boltzmann equation is justified.

We stress, however, the fact that we made several assumptions (existence of limits, their smoothness, an existence theorem for the Boltzmann equation, a uniqueness theorem for the Boltzmann hierarchy) that might not be satisfied.

As said before, in this chapter it will be established that the asymptotics of the solutions of the BBGKY hierarchy of hard spheres in the specific kinetic limit (Boltzmann–Grad limit)is described by the so–called Boltzmann hierarchy and, as a consequence, for factorized initial data we shall show that the equation determining the evolution of an initial state is a closed equation for a one–particle distribution function, i.e., Boltzmann's kinetic equation.

The Boltzmann–Grad limit was first studied in Grad's papers [Gra1,Gra2]. Then the existence of this limit for the case of hard spheres was discussed in detail from the mathematical point of view by Cercignani [Ce3] and later by Lanford [Lan3] and recently it has been given a rigorous mathematical justification in a series of papers [PG3,GP2] by Petrina and Gerasimenko.

The detailed analysis of the problem of the construction of this limit for solutions of the BBGKY hierarchy shows that the basic difficulty consists in proving the term by term convergence of the iteration series for solutions of BBGKY hierarchy towards the corresponding limit, i.e., to the solution of the Boltzmann hierarchy. This difficulty is related to the fact that the integrands in each term of the iteration series do not converge to the limit uniformly on the whole domain of integration. Namely, for fixed initial positions of the particles there exist cones in momentum space such that the particles can interact and the evolution of hard spheres differs from that of free particles. The volumes of these cones are proportional to the square of the ratio of the diameter to the distance between particles and, thus, they tend to infinity if the distances tend to zero. But in the integrands of the various terms of the iteration series, there exist initial data such that the particles touch each other, *i.e.* their distances are zero, or they are located in such a way that the ratio is different from zero for all values of a. It is necessary to estimate the contribution of these "dangerous" sets and to show that it tends to zero in the Boltzmann-Grad limit when a tends to zero (and N to infinity, with Na^2 bounded). We note that in early works on the justification of the Boltzmann–Grad limit attention was not properly paid to this property and also a precise mathematical meaning was not given to the individual terms of the series giving the solutions of the BBGKY hierarchy. In this chapter we present a complete discussion of these problems, following the papers [PG3,GP2].

We shall now formulate the problems arising in the rigorous justification of the Boltzmann kinetic equation. It is necessary to prove that:

(i) There exists a unique solution to the Cauchy problem of the Boltzmann hierarchy (Section 4.3),

(ii) There exists, in a certain sense, the term by term convergence of the iteration series of the BBGKY hierarchy towards the corresponding series of the Boltzmann hierarchy (Section 4.4),

(iii) There exists the Boltzmann–Grad limit for the solution to the initial value problem of the BBGKY hierarchy (Section 4.4); in particular this applies to the equilibrium states (Section 4.2).

We remark that the Boltzmann–Grad limit problem for special models of the BBGKY hierarchy (the soft sphere and the Uchiyama model) is also discussed in some papers and a recent book [Ce4,U1,CIP].

4.2 The Boltzmann–Grad Limit for Equilibrium States

4.2.1 The infinite system of hard spheres. Before we begin to study the asymptotic properties of the solutions of the initial value problem for the BBGKY hierarchy (3.2.1) in the Boltzmann–Grad limit, it is useful to consider this topic in the case of the Gibbs distribution functions (1.5.7), which will provide us with an example of an explicit solution.

We shall first examine an infinite 3–dimensional hard sphere particles system. For such system let us reformulate the definition of the Gibbs distribution functions in a form suitable to our task.

Let us consider the following sequence of distribution functions $\rho^{(a)} = \{ a^{2s} \times \rho_s(q_1, \dots, q_s) \}_{s \geqslant 1}$, where the functions $\rho_s(q_1, \dots, q_s)$, $s \geqslant 1$ are the solution of the Kirkwood–Saltsburg equation (1.5.8). The sequence $\rho^{(a)}$ satisfies the equation

$$\rho^{(a)} = \alpha \rho_0 + \alpha \left(K_1^{(a)} + K_2^{(a)} \right) \rho^{(a)}, \tag{4.2.1}$$

where $\alpha = a^2 z$, $\rho_0 = \{1, 0, 0, \dots\}$ and

$$K_1^{(a)} = K_1,$$

$$\left(K_2^{(a)} \rho^{(a)} \right)_s (q_1, \dots, q_s) = \mathcal{X}_{W_s^1} \sum_{n=1}^{\infty} \frac{a^{-2n}}{n!} \int_{\mathbb{R}^{3n}} dy_1 \dots dy_n \prod_{i=1}^{n} \varphi_{q_1}(y_i) \times \rho_{s+n-1}^{(a)}(q_2, \dots, q_s, y_1, \dots, y_n),$$

where K_1 and φ_{q_1} are defined as above (Section 1.5), and $\mathcal{X}_{W_s^1}$ is the characteristic function of the set $|q_1 - q_i| > a$, $i = 2, \ldots, s$.

In the space L_ξ^∞ of sequences $f = \{f_n(q_1, \ldots, q_n)\}_{n \geqslant 0}$ of functions $f_n(q_1, \ldots, q_n)$ equal to zero on the set $\mathbb{W}_n$ of forbidden configurations with the norm

$$\|f\| = \sup_{n \geqslant 0} \sup_{q_1, \ldots, q_n} \xi^{-n} |f_n(q_1, \ldots, q_n)|$$

the following result holds

Lemma 4.2.1. *The equation (4.2.1) has in L_ξ^∞ a unique solution, represented by the following formula*

$$\rho^{(a)} = \left(1 - \alpha\left(K_1^{(a)} + K_2^{(a)}\right)\right)^{-1} \alpha\rho_0, \tag{4.2.2}$$

if the conditions: $\xi = k\alpha$ and $a < a_0 (k\alpha)^{-1} \ln 2$, where $a_0 = 3/4\pi$ and $k > 2$, are satisfied.

The second condition in Lemma 4.2.1 means that the size a of the particles is smaller than the average length of the free path, which is proportional to α^{-1}. The statement of Lemma 4.2.1 follows from the following estimates for the operators $K_1^{(a)}$ and $K_2^{(a)}$:

$$\begin{aligned} \left\|\alpha K_1^{(a)}\right\| &\leqslant \alpha\xi^{-1}, \\ \left\|\alpha K_2^{(a)}\right\| &\leqslant \alpha\xi^{-1}\left(e^{\xi(a/a_0)} - 1\right). \end{aligned} \tag{4.2.3}$$

Therefore $\left\|\alpha K_1^{(a)}\right\| + \left\|\alpha K_2^{(a)}\right\| < 1$, and consequently the formula (4.2.2) describes the unique solution of the equation (4.2.1).

The sequence (4.2.2) defines on the phase space the sequence $F^{(a)} = \{F_s^{(a)}(x_1, \ldots, x_s)\}_{s \geqslant 0}$ of the Gibbs distribution functions

$$F_s^{(a)}(x_1, \ldots, x_s) = \left(\frac{2\pi}{\beta}\right)^{-\frac{3}{2}s} \exp\left\{-\beta \sum_{i=1}^{s} \frac{p_i^2}{2}\right\} \rho_s^{(a)}(q_1, \ldots, q_s), \tag{4.2.4}$$

which belongs to the space $L_{\xi,\beta}^\infty$ defined by (1.5.6).

Let us consider also in L_ξ^∞ the other equation

$$\rho^{(0)} = \alpha\rho_0 + \alpha K^{(0)} \rho^{(0)}, \tag{4.2.5}$$

where

$$\left(K^{(0)}\rho^{(0)}\right)_s (q_1, \dots, q_s) = (1 - \delta_{s,1})\,\rho^{(0)}_{s-1}(q_2, \dots, q_s).$$

This equation follows from the equation (4.2.1) as a result of the formal limit transition: $a \to 0$, $z \to \infty$, $a^2 z = \alpha < \infty$. For the equation (4.2.5) the following statement holds:

Lemma 4.2.2. *In L^∞_ξ the equation (4.2.5) has the unique solution*

$$\rho^{(0)} = \left(1 - \alpha K^{(0)}\right)^{-1} \alpha \rho_0 = \left(\alpha, \alpha^2, \dots, \alpha^n, \dots\right) \in L^\infty_\xi. \tag{4.2.6}$$

Sequence (4.2.6) defines on the phase space the sequence $F^{(0)} = \left\{ F^{(0)}_s(x_1, \dots, x_s) \right\}_{s \geqslant 0}$ of the Maxwell distribution functions by the formula

$$F^{(0)}_s (x_1, \dots, x_s) = \alpha^s \left(\frac{2\pi}{\beta}\right)^{-\frac{3}{2}s} \exp\left\{-\beta \sum_{i=1}^{s} \frac{p_i^2}{2}\right\}. \tag{4.2.7}$$

The asymptotic properties of the Gibbs distribution functions (4.2.4) in the Boltzmann–Grad limit are described by the following Theorem.

Theorem 4.2.1. *In the limit $a \to 0$, $z \to \infty$, $a^2 z = \alpha < \infty$ the functions $F^{(a)}_s (x_1, \dots, x_s)$, $s \geqslant 1$ (4.2.4) converge uniformly on any compact set in $\mathbb{R}^{3s} \times (\mathbb{R}^{3s}\backslash \mathbb{W}_s)$ to the functions $F^{(0)}_s (x_1, \dots, x_s)$, $s \geqslant 1$ defined by (4.2.7), if the conditions $\xi = k\alpha$, and $a < a_0 (k\alpha)^{-1} \ln 2$, where $k > 2$ and $a_0 = \frac{3}{4\pi}$, are satisfied.*

Proof. Let us denote by $\mathcal{X}_{\mathbb{R}^{3s}\backslash \mathbb{W}_s}$ the characteristic function of the set of configurations $\mathbb{R}^{3s}\backslash \mathbb{W}_s$ for some fixed value of the parameter a. Since the expressions $\mathcal{X}_{\mathbb{R}^{3s}\backslash \mathbb{W}_s}\left(K^{(a)}_1 f\right)_s (q_1, \dots, q_s)$ and $\mathcal{X}_{\mathbb{R}^{3s}\backslash \mathbb{W}_s}\left(K^{(0)} f\right)_s (q_1, \dots, q_s)$ coincide for arbitrary $f \in L^\infty_\xi$, then taking into account the estimates (4.2.3) and the equality

$$(1 - x)^{-2} = \sum_{n=0}^{\infty} (n + 1)\, x^n,$$

we deduce the following estimate

$$\left|\mathcal{X}_{\mathbb{R}^{3s}\setminus W_s}\left(\rho_s^{(a)}(q_1,\ldots,q_s)-\rho_s^{(0)}(q_1,\ldots,q_s)\right)\right|$$

$$\leqslant \|\alpha K_2^{(a)}\|\xi^s\left(1-\alpha(\|K_1^{(a)}\|+\|K_2^{(a)}\|)\right)^{-2}\|\alpha\rho_0\|$$

$$\leqslant \left(\alpha\xi^{-1}\right)^2\left(e^{\xi\frac{a}{a_0}}-1\right)\xi^s\left(1-\alpha(\|K_1^{(a)}\|+\|K_2^{(a)}\|)\right)^{-2}$$

$$\leqslant \operatorname{const}(\alpha,\xi,s)\,a+o(a)\,.$$

The last inequality proves the statement of Theorem 4.2.1.

4.2.2 Extension: the smooth potential. Theorem 4.2.1 holds also for the system of particles interacting through a smooth potential of finite range:

$$\phi(q)=\begin{cases}\varphi(|q|), & \|q\|\leqslant R,\\ 0, & \|q\|>R,\end{cases} \tag{4.2.8}$$

which satisfy the stability condition (1.5.4). In this case the Boltzmann–Grad limit means: $R\to 0$, $z\to\infty$, $R^{(\nu-1)}z=\alpha<\infty$.

To prove this fact let us consider the following sequence of distribution functions $\rho^{(R)}=\left\{R^{(\nu-1)s}\rho_s(q_1,\ldots,q_s)\right\}_{s\geqslant 1}$, where the functions $\rho_s(q_1,\ldots,q_s)$, $s\geqslant 1$ are the solutions of the Kirkwood–Saltsburg equation (1.5.8). The sequence $\rho^{(R)}$ satisfies the equation

$$\rho^{(R)}=\alpha\rho_0+\alpha\left(K_1^{(R)}+K_2^{(R)}\right)\rho^{(R)}, \tag{4.2.9}$$

where $\alpha=R^{(\nu-1)}z$, $\rho_0=(1,0,\ldots)$ and

$$K_1^{(R)}=K_1,$$

$$\left(K_2^{(R)}\rho^{(R)}\right)_s(q_1,\ldots,q_s)=\exp\left\{-\beta\sum_{i=2}^{s}\phi(q_1-q_i)\right\}\sum_{n=1}^{\infty}\frac{R^{-(\nu-1)n}}{n!}\int\limits_{\mathbb{R}^{\nu n}}dy_1\ldots dy_n$$

$$\times\prod_{i=1}^{n}\varphi_{q_1}(y_i)\,\rho_{s+n-1}^{(R)}(q_2,\ldots,q_s,y_1,\ldots,y_n)\,.$$

Here

$$\varphi_{q_1}(y_i)=\begin{cases}\exp\{-\beta\varphi(|q_1-y_i|)\}-1, & \text{if } |q_1-y_i|\leqslant R,\\ 0, & \text{if } |q_1-y_i|>R,\end{cases}$$

and according to the stability condition (1.5.4), for this function the following estimate holds

$$\int dy\, |\varphi_q(y)| \leqslant R^{\nu} \left(e^{2\beta B} + 1\right) v_1, \tag{4.2.10}$$

where v_1 is a volume of a v–dimensional solid sphere of the unit radius

$$v_1 = \begin{cases} \pi \dfrac{(\nu - 1)}{\nu\left(\frac{\nu}{2} - 1\right)\left(\frac{\nu}{2} - 2\right)\cdots\frac{1}{2}}, & \text{if } \nu \text{ is odd,} \\ \dfrac{2\pi^{\nu/2}}{\nu\left(\frac{\nu}{2} - 1\right)!}, & \text{if } \nu \text{ is even.} \end{cases}$$

In the space L^{∞}_{ξ} of sequences $f = \{ f_n(q_1, \ldots, q_n) \}_{n \geqslant 0}$ of functions $f_n(q_1, \ldots, q_n)$ with the norm

$$\| f \| = \sup_{n \geqslant 0} \xi^{-n} \sup_{q_1, \ldots, q_n} | f_n(q_1, \ldots, q_n) |$$

there exists a unique solution of equation (4.2.9)

$$\rho^{(R)} = \left(1 - \alpha\left(K_1^{(R)} + K_2^{(R)}\right)\right)^{-1} \alpha \rho_0$$

if

$$\xi = k\alpha e^{2\beta B} \tag{4.2.11}$$

and

$$R < \frac{\ln 2}{k\alpha} \left(e^{2\beta B}\left(e^{2\beta B} + 1\right) v_1\right)^{-1},$$

where $k > 2$. This statement follows directly from the estimates for the operators $K_1^{(R)}$ and $K_2^{(R)}$, and the inequality (4.2.10)

$$\left\| \alpha K_1^{(R)} \right\| \leqslant \frac{\alpha}{\xi} e^{2\beta B},$$

$$\left\| \alpha K_2^{(R)} \right\| \leqslant \frac{\alpha}{\xi} e^{2\beta B} \left(\exp\{R\xi\left(e^{2\beta B} + 1\right) v_1\} - 1\right).$$

Strictly speaking, the last inequalities hold for the corresponding symmetrized operators $K^{(R)}$, $K_1^{(R)}$, $K_2^{(R)}$, respectively. We refer the reader to the monograph [PGM2].

The asymptotic properties of the Gibbs distribution functions (1.5.7) in the Boltzmann–Grad limit are described by

Theorem 4.2.2. *If the finite range potential (4.2.8) satisfies the stability condition (1.5.4) then in the limit* $R \to 0$, $z \to \infty$, $R^{(\nu-1)}z = \alpha < \infty$ *the functions*

$$F_s^{(R)}(x_1,\ldots,x_s) = \left(\frac{2\pi}{\beta}\right)^{-\frac{\nu}{2}s} \exp\left\{-\beta\sum_{i=1}^{s}\frac{p_i^2}{2}\right\} \rho_s^{(R)}(q_1,\ldots,q_s), \; s \geqslant 1$$

converge uniformly on any compact set to the Maxwell distribution functions

$$F_s^{(0)}(x_1,\ldots,x_s) = \alpha^s \left(\frac{2\pi}{\beta}\right)^{-\frac{\nu}{2}s} \exp\left\{-\beta\sum_{i=1}^{s}\frac{p_i^2}{2}\right\}, \; s \geqslant 1,$$

if the conditions (4.2.11) hold.

The details of the proof are similar to those of the proof of Theorem 4.2.1.

4.3 The Boltzmann Hierarchy and the Boltzmann Equation

4.3.1 On the Boltzmann hierarchy. In order to show in what way the Boltzmann equation can be derived from the BBGKY hierarchy, as a preliminary, we need to study a hierarchy of kinetic equations, namely the so–called Boltzmann hierarchy.

To define the Boltzmann hierarchy, let us assume that there exists in a certain sense the Boltzmann–Grad limit for solution $F(t) = \{F_s(t, x_1, \ldots, x_s)\}_{s\geqslant 1}$ of the BBGKY hierarchy

$$\lim_{\substack{a\to 0, z\to\infty \\ a^2 z = \alpha < \infty}} a^{2s} F_s(t, x_1, \ldots, x_s) = f_s(t, x_1, \ldots, x_s).$$

Then, as a result of the transition to the limit: $a \to 0$, $z \to \infty$, $a^2 z = \alpha < \infty$ in the BBGKY hierarchy (3.2.1), we formally obtain the following chain of equations for the limiting sequences $f(t) = \{f_s(t, x_1, \ldots, x_s)\}_{s\geqslant 1}$ (Boltzmann hierarchy)

$$\frac{d}{dt} f(t) = -\mathcal{H}_0 f(t) + Q f(t), \tag{4.3.1}$$

where

$$(Qf(t))_s(x_1,\ldots,x_s) = \operatorname{sign}(t)\sum_{i=1}^{s}\int_{\Re^3\times\mathbb{S}^2_+} d\eta\, dp_{s+1}\,\langle\eta,(p_i-p_{s+1})\rangle$$
$$\times\Big(f_{s+1}\big(t,x_1;\ldots;q_i,p_i-\eta\langle\eta,(p_i-p_{s+1})\rangle;$$
$$\ldots;x_s;q_i,p_{s+1}+\eta\langle\eta,(p_i-p_{s+1})\rangle\big)$$
$$-f_{s+1}(t,x_1,\ldots,x_s;q_i,p_{s+1})\Big)$$

and $(-\mathcal{H}_0)$ is the infinitesimal generator of the group $S_0(-t)$ of the evolution operators of non–interacting point particles system

$$\big(\mathcal{H}_0 f(t)\big)_s(x_1,\ldots,x_s) = \sum_{i=1}^{s}\Big\langle p_i,\frac{\partial}{\partial q_i}\Big\rangle f_s(t,x_1,\ldots,x_s).$$

The meaning of $\mathbb{S}^2_+$ is the same as for the BBGKY hierarchy.

Let us construct the solution to the Cauchy problem of the Boltzmann hierarchy (4.3.1) in the Banach space L^∞_ξ of the sequences of continuous functions $f=\{f_n(x_1,\ldots,x_n)\}_{n\geqslant 0}$ with the norm

$$\|f\| = \sup_{n\geqslant 0}\xi^{-n}\sup_{x_1,\ldots,x_n}|f_n(x_1,\ldots,x_n)|\exp\Big\{\beta\sum_{i=1}^{n}\frac{p_i^2}{2}\Big\}.$$

We note that above the space L^∞_ξ of sequences of functions $f_n(x_1,\ldots,x_n)$ that are equal to zero on the set $\mathbb{W}_n$ of forbidden configurations, was used. For the Boltzmann hierachy the sets $\mathbb{W}_n$ are empty.

A solution to the Cauchy problem of the hierarchy (4.3.1) for $f^0\in L^\infty_\xi$ is defined in the form of a series of iterations of this hierarchy

$$f(t) = U_0(t)f^0$$
$$=\sum_{n=0}^{\infty}\int_0^t dt_1\cdots\int_0^{t_n} dt_n\, S_0(-t)S_0(t_1)QS_0(-t_1)\ldots S_0(t_n)QS_0(-t_n)f^0. \tag{4.3.2}$$

Each term of this series exists and its explicit form will be considered later.

The following result holds

Theorem 4.3.1. *If $f^0 \in L^\infty_\xi$, then the sequence $f(t) = U_0(t) f^0$ (4.3.2) exists and the series for a function $f_s(t, x_1, \ldots, x_s)$ converges uniformly on each compact if $t \in [-t_0', t_0']$, where*

$$t_0' = (2^6 \pi^2 \xi \beta)^{-1} \tag{4.3.3}$$

is a finite number and the following estimate holds

$$|f_s(t, x_1, \ldots, x_s)| \leqslant \| f^0 \| (2\xi)^s \exp\left\{ -\frac{1}{2} \beta \sum_{i=1}^{s} \frac{p_i^2}{2} \right\}. \tag{4.3.4}$$

The functions $\{ f_s(t, x_1, \ldots, x_s) \}_{s \geqslant 1}$ are a weak unique solution to the Cauchy problem for the Boltzmann hierarchy (4.3.1).

The details of the proof are analogous to the details of proof of Theorem 3.2.1.

We shall now indicate an characteristic property of the solutions of the Boltzmann hierarchy (4.3.1), which is interpreted from a physical point of view as the propagation of chaos.

Let

$$f_n^0(x_1, \ldots, x_n) = f_1^0(x_1) \ldots f_1^0(x_n), \ n \geqslant 1$$

be initial data for the Boltzmann hierarchy (4.3.1); in other words the initial state of the particles is such that they are statistically independent (there are no correlations in the system). Then this property is preserved for any time, i.e., there is a solution of the Boltzmann hierarchy having the following form

$$f_s(t, x_1, \ldots, x_s) = f_1(t, x_1) \ldots f_1(t, x_s), \ s \geqslant 1,$$

where $f_1(t, x)$ is a solution of the Boltzmann equation

$$\frac{\partial}{\partial t} f_1(t, x) = -\left\langle p, \frac{\partial}{\partial q} \right\rangle f_1(t, x) + Q(f, f)(t, x),$$

$$Q(f, f)(t, x) = \text{sign}(t) \int_{\mathbb{R}^3 \times \mathbb{S}^2_+} d\eta \, dp_0 \, \langle \eta, (p - p_0) \rangle$$

$$\times \Big(f_1(t, q, p - \eta \langle \eta, (p - p_0) \rangle) f_1(t, q, p_0 + \eta \langle \eta, (p - p_0) \rangle)$$

$$- f_1(t, x) f_1(t, q, p_0) \Big)$$

with the initial data $f_1^0(x)$. This property of the solution (4.3.2) follows directly from the fact that the Boltzmann hierarchy admits separation of variables.

4.3.2 The main result. It is clear that the problem of justifying the derivation of the kinetic Boltzmann equation consists in determing in what sense the solutions of the BBGKY hierarchy converge in the Boltzmann–Grad limit to the solutions of the Boltzmann hierarchy.

Let $\Re^s_{p_0}$ be a ball of radius $p_0 < \infty$ in the momentum subspace of the phase space of a s–particles system, and $K^s_0 \in \mathbb{R}^{3s} \backslash \mathbb{W}_s$ is an any compact that consists of such configurations of hard spheres for which $|q_i - q_j| \geqslant a_0(a) + a$, $i \neq j \in (1, \dots, s)$, where $a_0(a)$ is a fixed value such that $a_0(a) \to 0$ and $a/a_0(a) \to 0$ as $a \to 0$. Then the following Boltzmann–Grad limit theorem holds.

Theorem 4.3.2. *If the initial data* $a^{2n} F^0_n(x_1, \dots, x_n)$, $n \geqslant 1$ *converge in the limit* $a \to 0$, $z \to \infty$, $a^2 z = \alpha < \infty$ *uniformly on every compact set in* $\mathbb{R}^{3n} \times (\mathbb{R}^{3n} \backslash \mathbb{W}_n)$ *to the functions* $f^0_n(x_1, \dots, x_n)$, $n \geqslant 1$ *and* $F^0, f^0 \in L^{\infty,0}_\xi$, *then for* $t \in [-T, T]$, *the solutions* $a^{2s} F_s(t, x_1, \dots, x_s)$, $s \geqslant 1$, *(3.2.2) of the BBGKY hierarchy converge in the limit* $a \to 0$, $z \to \infty$, $a^2 z = \alpha < \infty$ *uniformly with respect to* $(q_1, \dots, q_s) \in K^0_s$ *and in a weak sense with respect to* $(p_1, \dots, p_s)$ *to the limit* $f_s(t, x_1, \dots, x_s)$, $s \geqslant 1$, *which is the solution (4.3.2) of the Boltzmann hierarchy.*

In this theorem the time interval $[-T, T]$ means $T = \min(t_0, t'_0)$, where t_0 and t'_0 are defined by the corresponding formulas (3.2.10) and (4.3.3), and in (3.2.10) the factor a^2 is omitted, because of the different normalization $F^{(a)}_s(t) = a^{2s} F_s(t)$.

The details of the proof of Theorem 4.3.2. form the subject of the next section.

We note that this result holds for a system of hard spheres of any dimension, except $\nu = 1$. The Boltzmann–Grad limit for a one–dimensional hard spheres system means: $a \to 0$ for a fixed value of the activity z. Therefore in this limit the BBGKY hierarchy is reduced to a set of uncoupled Liouville–type equations for non–interacting point particles, and, thus, to a trivial kinetic equation. Such result has a clear physical sense, because the collisions in this case are trivial.

4.3.3 On the H–function and irreversibility concept. The collision term of the Boltzmann equation, Q(f,f), defined at the end of the subsection 4.3.1 has a particular property. If we multiply both sides of this equation by $\log f_1$ and integrate with respect to p, we obtain:

$$\frac{\partial \mathcal{H}}{\partial t} + \langle \frac{\partial}{\partial q}, \mathcal{J} \rangle = \mathcal{S} \tag{4.3.5}$$

where

$$\mathcal{H} = \int\limits_{\mathbb{R}^3} f_1 \log f_1 dp \tag{4.3.6}$$

$$\mathcal{J} = \int\limits_{\mathbb{R}^3} p f_1 \log f_1 dp \tag{4.3.7}$$

$$\mathcal{S} = a^2 \int\limits_{\mathbb{R}^3} \log f_1 Q(f,f) dp \tag{4.3.8}$$

Now an inequality, due to Boltzmann, easy to prove [Ce1,Ce2,CIP], shows that

$$\mathcal{S} \leq 0 \quad \text{and} \quad \mathcal{S} = 0 \quad \text{iff} \quad f \quad \text{is a Maxwellian} \tag{4.3.9}$$

Because of this inequality, Eq. (4.3.5) plays an important role in the theory of the Boltzmann equation. We illustrate the role of Eq. (4.3.9) in the case of space homogeneous solutions. In this case the various quantities do not depend on q and Eq. (4.3.5) reduces to

$$\frac{\partial \mathcal{H}}{\partial t} = \mathcal{S} \leq 0 \tag{4.3.10}$$

This implies the so-called H-theorem (for the space homogeneous case): $\mathcal{H}$ is a decreasing quantity, unless f is a Maxwellian (in which case the time derivative of $\mathcal{H}$ is zero). In this case the densities of mass, momentum and energy are constant in time: we can thus build a Maxwellian M which has, at any time, the same conserved densities as any solution f corresponding to given initial data. Since $\mathcal{H}$ decreases unless f is a Maxwellian (*i. e.* $f = M$), it is tempting to conclude that f tends to M when $t \to \infty$. The temptation is strengthened when we realize that $\mathcal{H}$ is bounded from below by the value taken by the functional $\mathcal{H}$ when $f = M$, *i.e* $\mathcal{H}_M$ (thanks to the inequality $z\log z - z\log y + y - z \geq 0$, valid for nonnegative y and z and the fact that $\int\limits_{\mathbb{R}^3} \log M(f - M) dp = 0$). In fact $\mathcal{H}$ is decreasing, its derivative is nonpositive unless it takes the value $\mathcal{H}_M$; one feels that $\mathcal{H}$ tends to $\mathcal{H}_M$! This conclusion is, however, unwarranted, without a more detailed consideration of the source term $\mathcal{S}$ in Eq. (4.3.10) see, *e.g* [CIP]. Here we only remark that if $\mathcal{H}$ tends to $\mathcal{H}_M$, then it is easy to conclude that f tends to M, thanks to the inequality:

$$f \log f - f \, logM + M - f \geq cg(\frac{|f-M|}{M})|f-M|\,, \tag{4.3.11}$$

where c is a constant (independent of f) and

$$g(z) = \begin{cases} z & \text{if} \quad 0 \leq z \leq 1 \\ 1 & \text{if} \quad z \geq 1\,. \end{cases} \tag{4.3.12}$$

Integrating both sides of eq. (4.3.11) gives

$$\mathcal{H} - \mathcal{H}_M \geq c[\int_{L_t} |f-M|dp + \int_{S_t} |f-M|^2 M^{-1} dp]\,, \tag{4.3.13}$$

where L_t and S_t denote the sets (depending on t) where $| \, f - M \, |$ is larger (resp. smaller) than M. Since $\mathcal{H}$ is assumed to tend to $\mathcal{H}_M$, it follows that both integrals tend to zero when $t \to \infty$. The fact that the second integral tends to zero implies, by Schwarz's inequality, that

$$\int_{S_t} |f-M|dp \leq [\int_{S_t} |f-M|^2 M^{-1} dp]^{1/2} [\int_{S_t} M dp]^{1/2} \to 0 \tag{4.3.14}$$

Then

$$\int_{\mathbb{R}^3} |f-M|dp = \int_{L_t} |f-M|dp + \int_{S_t} |f-M|dp \tag{4.3.15}$$

also tends to zero and f tends strongly to M in L^1.

If the state of the gas is not space homogeneous, the situation becomes more complicated. For a discussion on this point we refer to [CIP].

Boltzmann's H-theorem is of basic importance because it shows that his equation has a basic feature of irreversibility: the quantities $\mathcal{H}$ (in the space homogeneous case) and H (in other cases where the gas does not exchange mass and energy with a solid boundary) always decrease in time. This result seems to be in conflict with the fact that the molecules constituting the gas follow the laws of classical mechanics which are time reversible. Accordingly, given a motion at $t = t_0$ with momenta $p_1, p_2, \ldots, p_N$, we can always consider the motion with momenta $-p_1, -p_2, \ldots, -p_N$ (and the same positions as before) at $t = t_0$; the backward evolution of the latter state will be equal to the forward evolution of the original one.

Therefore if $dH/dt < 0$ in the first case, we shall have $dH/d(-t) < 0$ or $dH/dt > 0$ in the second case, which contradicts Boltzmann's H-theorem.

This paradox is mentioned by Thomson in a short paper [Tho], which is seldom quoted . This paper appeared in 1874 and contains a substantial part of the physical aspects of the modern interpretation of irreversibility not only for gases, but also for more general systems made up of molecules. The impossibility to observe macroscopic phenomena which run backwards with respect to those which are actually observed is, in the last analysis, due to the large number of molecules present even in macroscopically small volumes.

Josef Loschmidt, to whom the paradox is usually attributed, mentioned it briefly in [Lo]. In spite of the obscure arguments of Loschmidt, Boltzmann quickly got the point and gave a thorough discussion of the paradox [Bol1], ending up with a conclusion similar to that of Thomson.

Nowadays we are more prepared to discuss this kind of questions. In fact, we remark that, when giving a justification of the Boltzmann equation, as we have done heuristically in the first section of this chapter, and we shall do rigorously later, we use the laws of elastic collisions and the continuity of the probability density at the impact to express the distribution functions corresponding to an after-collision state in terms of the distribution functions corresponding to the state before the collision, rather than the latter in terms of the former. It is obvious that the first way is the right one to follow if the equations are to be used to predict the future from the past and not *vice versa*; it is clear, however, that this choice introduces a connection with the everyday concepts of past and future which are extraneous to molecular dynamics and are based on our macroscopic experience. When we take the Boltzmann-Grad limit we obtaine equations which describe the statistical behavior of the gas molecules: a striking consequence of our choice is that the Boltzmann equation describes motions for which the quantity H (or $\mathcal{H}$) has a tendency to increase, while the opposite choice would lead to an equation having a negative sign in front of the collision term, and hence describing only motions with increasing H. We must remark that, in order to derive the Boltzmann equation, we take special (although highly probable) initial data; thus certain special data were excluded. As the discussion in the first section shows, these excluded data correspond to a state in which the molecular momenta of the molecules that are about to collide show an unusual correlation. This situation can be simulated by studying the dynamics of many interacting particles on a computer and leads to an evolution in which there is an increasing H, as expected, while "randomly" chosen initial data invariably lead to an evolution with decreasing H [Ce1,Ce2,TM]. In other words, the fact that H decreases is not an intrinsic property of the dynamical system but a property of the level of description.

It is not the place here to discuss the relation of the H-theorem with the notions of past and future [Ce1,Ce2,Ce5].

We remark that the term "time-reversible" has different meanings in the mathematical literature; a detailed discussion of the various definitions is given in a paper by R. Illner and H. Neunzert [IN].

It is not perhaps out of place to comment on a statement which is frequently made, to the effect that no kind of irreversibility can follow by correct mathematics from the analytical dynamics of a conservative system and hence some assumption of kinetic theory must contradict analytical dynamics. It should be clear that it is not a new assumption that is introduced, but the fact that we study asymptotic properties of a conservative system in the Boltzmann-Grad limit, under the assumption that the initial probability distributions are factorized in the way indicated in Sect. 1.

There is another objection which can be raised against the H-theorem when presented as a rigorous consequence of the laws of dynamics. The starting point is a theorem of Poincaré [Poi1, Arn1] (the so-called "recurrence theorem"), which says that any conservative system, whose possible states form a compact set in phase space, will return arbitrarily close to its initial state, for almost any choice of the latter, provided we wait long enough. This applies to a gas of hard sphere molecules, enclosed in a specularly reflecting box, because the set of the possible states S with a given energy is compact and has a finite measure $\mu(S)$ (induced by the Lebesgue measure). If A is a subset of S with measure $\mu(A)$, which evolves into a set A_t at time t (according to the dynamics of the system), then $\mu(A_t) = \mu(A)$. To prove Poincaré's theorem, let us assume that there is a subset A whose points will never come back to A. We choose A small enough and τ large enough so that A_τ and A do not overlap (if this is impossible, the theorem is trivially true); then none of the sets $A_{2\tau}$, $A_{3\tau}, \ldots$ overlap, because, if $A_{n\tau}$ and $A_{(n+k)\tau}$ had points in common, then, by tracing the motion backwards and using the uniqueness of the motion through any given phase space point, it would follow that A and $A_{k\tau}$ must have common points and this would contradict the definition of A. If A, A_τ, $A_{2\tau}, \ldots$ do not overlap, then, since $\mu(A) = \mu(A_\tau) = \mu(A_{2\tau})=..$, the total measure of the union of these disjoint sets would be infinite (which is impossible because $\mu(S) < \infty$), unless $\mu(A) = 0$, and Poincaré's recurrence theorem is proved.

This theorem implies that our molecules can have, after a "recurrence time", positions and momenta so close to the initial ones that the one particle distribution function f would be practically the same; therefore H should also be practically the same, and if it decreased initially, then must have increased at some later time. This paradox goes under the name of Zermelo, who stated it in 1896 [Zer], but was actually mentioned before in a short paper by Poincaré [Poi2]. The traditional answer to Zermelo's paradox was given by Boltzmann himself [Bol2]: the recurrence time is so large that, practically speaking, one would never observe a significant portion of the recurrence cycle. In fact, according to an estimate made by Boltzmann [Bol2], the recurrence time for a typical amount of gas is a huge number even if the estimated age of the Universe is taken as time unit.

In view of the fact that we claim validity for the Boltzmann equation in the Boltzmann-Grad limit only, we do not have to worry about the recurrence paradox; in fact, the set S is no longer compact when $N \to \infty$ and the recurrence time is expected to go to infinity with N (at a much faster rate).

4.4 The Boltzmann–Grad Limit for the Solutions of Initial Value Problem for the BBGKY Hierarchy

4.4.1 Auxiliary lemmas. Before proving the main convergence result (Theorem 4.3.2), we determine in what sense each term of the iteration series (3.2.2) for the BBGKY hierarchy converges to the corresponding term of the iteration series (4.3.2) for the Boltzmann hierarchy.

As above, let $\Re^s_{p_0}$ be a ball of radius $p_0 < \infty$ in the momentum subspace of the phase space of a system of s particles, and $K^s_0 \in \mathbb{R}^{3s}\backslash \mathbb{W}_s$ be an arbitrary compact set of the admissible configurations of a system of s particles that consists of those configurations of hard spheres for which $|\, q_i - q_j \,| \geqslant a_0\,(a) + a$, $i \neq j \in (1, \ldots, s)$, where $a_0\,(a)$ is a fixed value such that $a_0\,(a) \to 0$ and $a/a_0\,(a) \to 0$ as $a \to 0$.

Let us denote by $\mathfrak{M}_s \subset \mathcal{K}^s_0 \times \Re^s_{p_0}$ the set whose complement $\left(\mathcal{K}^s_0 \times \Re^s_{p_0}\right) \backslash \mathfrak{M}_s$ is the set of all possible initial data of a system of s hard spheres such that the particles do not interact during the evolution for $t \in \mathbb{R}^1$, i.e.

$$\left(\mathcal{K}^s_0 \times \Re^s_{p_0}\right) \backslash \mathfrak{M}_s = \Big\{ \left(x_1, \ldots, x_s\right) \in \mathcal{K}^s_0 \times \Re^s_{p_0} : X^0_i\left(t, x_1, \ldots, x_s\right) \\ = X^0_i\left(t, x_i\right) = \left(q_i + p_i t, p_i\right), i \in (1, \ldots, s), t \in \mathbb{R}^1 \Big\}.$$

This means that the s-particles can interact only if their initial data $(x_1, \ldots, x_s)$ belong to $\mathfrak{M}_s$.

We now estimate the volume of the set $\mathfrak{M}_s$.

Lemma 4.4.1. *(on "sweeping up"). The following estimate holds for the volume mes $\mathfrak{M}_s$ of the set $\mathfrak{M}_s \in \mathcal{K}_0^s \times \Re_{p_0}^s$*

$$\text{mes } \mathfrak{M}_s \leqslant \text{mes } \mathcal{K}_0^s \text{ const}\,(s, p_0) \left(\frac{a}{a_0\,(a)} \right)^2 . \tag{4.4.1}$$

Proof. According to the definition of the set $\mathfrak{M}_s$ we have: $|\, q_i - q_j \,| \geqslant a_0\,(a) + a$, $i \neq j \in (1, \ldots, s)$ and $\sum_{i=1}^{s} p_i^2 < p_0^2$. Let us fix the initial data for the i–th and j–th $\big(i \neq j \in (1, \ldots, s)\big)$ particles.

We will determine for what values of their momenta a collision (interaction) between the i–th and j–th particles can occur.

Let $\eta = (q_i - q_j)/\,|\, q_i - q_j \,|$, then the particles can collide if $\langle (p_i - p_j)\,, \eta \rangle \leqslant 0$. The instant of collision

$$t_* = \frac{|\, q_i - q_j \,| - a}{|\, \langle (p_i - p_j)\,, \eta \rangle \,|}$$

can be estimated as follows:

$$t_* \geqslant \frac{a_0\,(a)}{|\, \langle (p_i - p_j)\,, \eta \rangle \,|} \geqslant \frac{a_0\,(a)}{|\, p_i - p_j \,|} \geqslant \frac{a_0\,(a)}{2p_0}.$$

Let $p_i^{\perp} = p_i - \eta \langle \eta, p_i \rangle$ is the transversal component of the momentum p_i, then the collision of the particles is possible if the transversal components of the momentum of the i–th and j–th particles satisfy the inequality

$$|\, p_i^{\perp} - p_j^{\perp} \,| \leqslant \frac{2a}{t_*} \leqslant \frac{2a}{a_0\,(a)}\, |\, \langle (p_i - p_j)\,, \eta \rangle \,| \leqslant 4p_0 \frac{a}{a_0\,(a)}.$$

Thus the collision of the i–th and j–th particles is only possible in the case when the relative momentum of these particles in a suitable system of coordinate (where the j–th particle rests) belongs to the domain that is a cone with base radius $4p_0 \frac{a}{a_0\,(a)}$ and height $2p_0$. Consequently, the volume of the set in the 6-dimensional momentum space of the i–th and j–th particles for which the collisions are possible can be estimated by the quantity

$$V_{ij} \leqslant \frac{32}{3}\pi p_0^3 \left(\frac{a}{a_0(a)}\right)^2 \frac{4}{3}\pi p_0^3.$$

In order to estimate the volume of the set in the momentum space of a system of s particles we are interested in, we have to take into account only pair collisions, since the set $\mathcal{M}_s^0$ of initial data for which multiple collisions of particles occur in the system has the Lebesque measure zero. From s particles we can form $\frac{1}{2}s(s-1)$pairs. Thus the volume of the corresponding set can be estimated by

$$V_s \leqslant \frac{1}{2}s(s-1)V_{ij} \equiv \text{const}(s, p_0)\left(\frac{a}{a_0(a)}\right)^2.$$

Hence the statement of the lemma follows directly.

From the definition of the set $\mathfrak{M}_s$ we obtain the following result:

Lemma 4.4.2. *The evolution operator $S^s(t, x_1, \ldots, x_s)$ of a sytem of hard spheres coincides with the evolution operator $S_0^s(t, x_1, \ldots, x_s)$ of non–interacting particle system on the set $\mathcal{K}_0^s \times \mathfrak{R}_{p_0}^s \backslash \mathfrak{M}_s$, i.e. $(x_1, \ldots, x_s) \in \mathcal{K}_0^s \times \mathfrak{R}_{p_0}^s \backslash \mathfrak{M}_s$.*

Thus Lemmas 4.4.1–2 define in what sense the group of evolution operators $S(t)$ of a system of hard spheres coincide with the group of operators $S_0(t)$ of a system of non–interacting point particles.

Let us now consider the problem of the term by term convergence in the general case. For this we recall some notations and introduce new ones.

We will use the representation (2.5.6) for the n–th term I_n of the series (2.5.2) for the BBGKY hierarchy (see Chapter II) with only one difference. We will omit the value $a^{n(\nu-1)}$ before the expression for the n–th term and remain the same notation I_n for it.

We also rewrite the series (4.3.2) for the Boltzmann hierarchy in a componentwise form as we did in (2.5.6)

$$I_n^{(0)} = \int_0^t dt_1 \cdots \int_0^{t_n} dt_n\, S_0(t_1 - t)\, Q S_0(t_2 - t_1) \ldots S_0(t_n - t_{n-1})\, Q S_0(-t_n)\, f^0,$$

where the following operators are used:

$$\left(S_0\left(t_k-t_{k-1}\right)Qf\right)_{s+k-1}\left(x_1,\dots,x_{s+k-1}\right) = \sum_{i_k=1}^{s+k-1}\left(\left(\overset{*}{a}{}_{i_k}^{(0)}-a_{i_k}^{(0)}\right)f\right)_{s+k-1}\left(x_1,\dots,x_{s+k-1}\right),$$

$$\left(a_{i_k}^{(0)}f\right)_{s+k-1}\left(x_1,\dots,x_{s+k-1}\right) = \int\limits_{\mathbb{R}^\nu\times\mathbb{S}_+^{\nu-1}} d\eta_k\, dp_{s+k}\left\langle\eta_k,\left(p_{i_k}-p_{s+k}\right)\right\rangle \times f_{s+k}\Big(X_1^{(0)}\left(t_k-t_{k-1}\right),\dots,X_{s+k-1}^{(0)}\left(t_k-t_{k-1}\right); Q_{i_k}^{(0)}\left(t_k-t_{k-1}\right),p_{s+k}\Big),$$

$$\left(\overset{*}{a}{}_{i_k}^{(0)}f\right)_{s+k-1}\left(x_1,\dots,x_{s+k-1}\right) = \int\limits_{\mathbb{R}^\nu\times\mathbb{S}_+^{\nu-1}} d\eta_k\, dp_{s+k}\left\langle\eta_k,\left(p_{i_k}-p_{s+k}\right)\right\rangle \times f_{s+k}\Big(X_1^{(0)}\left(t_k-t_{k-1}\right);\dots;Q_{i_k}^{(0)}\left(t_k-t_{k-1}\right), \overset{*}{p}_{i_k},\dots X_{s+k-1}^{(0)}\left(t_k-t_{k-1}\right); Q_{i_k}^{(0)}\left(t_k-t_{k-1}\right),\overset{*}{p}_{s+k}\Big) \tag{4.4.2}$$

and $X_i^{(0)}\left(t_k-t_{k-1}\right)$ denotes the phase trajectory in the free motion of the i–th particle. Thus, we have

$$\left(I_n^{(0)}\right)_s\left(x_1,\dots,x_s\right) = \int_0^t dt_1\cdots\int_0^{t_{n-1}} dt_n \sum_{i_1=1}^{s}\cdots\sum_{i_n=1}^{s+n-1} \times\left(\prod_{k=1}^{n}\left(\overset{*}{a}{}_{i_k}^{(0)}-a_{i_k}^{(0)}\right)S_0\left(-t_n\right)f^0\right)_s\left(x_1,\dots,x_s\right)$$
$$= \sum_{m=0}^{n}(-1)^{n-m}\sum_{i_1=1}^{s}\cdots\sum_{i_n=1}^{s+n-1}\sum_{1\leqslant j_1<\dots<j_m}^{n} \times I_{\{i_1,\dots,i_n\}}^{(0),m}\left(x_1,\dots,x_s\right), \tag{4.4.3}$$

where

$$I_{\{i_1,\dots,i_n\}}^{(0),m}\left(x_1,\dots,x_s\right) = \int_0^t dt_1\cdots\int_0^{t_{n-1}} dt_n\Big(a_{i_1}^{(0)}\dots a_{i_{j_1-1}}^{(0)}\overset{*}{a}{}_{i_{j_1}}^{(0)}a_{i_{j_1+1}}^{(0)}\dots a_{i_{j_m-1}}^{(0)}\overset{*}{a}{}_{i_{j_m}}^{(0)} \times a_{i_{j_m+1}}^{(0)}\dots a_{i_n}^{(0)}S_0\left(-t_n\right)f^0\Big)_s\left(x_1,\dots,x_s\right).$$

This formula allows us to classify all possible elements of the n–th term of the series (4.3.2) and write them in an explicit form.

4.4.2 The approximating functionals. We introduce the functionals

$$(\varphi_s, I_{n,s}) = \int dp_1 \ldots dp_s I_{n,s}(x_1,\ldots,x_s)\,\varphi_s(p_1,\ldots,p_s), \quad s \geqslant 1$$

and

$$\left(\varphi_s, I_{n,s}^{(0)}\right) = \int dp_1 \ldots dp_s I_{n,s}^{(0)}(x_1,\ldots,x_s)\,\varphi_s(p_1,\ldots,p_s),$$

where $\varphi_s = \varphi_s(p_1,\ldots,p_s)$ is a continuous function with compact support. We emphasize that the integration in these functionals is carried out with respect to $s+n$ momentum variables $p_1,\ldots,p_s,p_{s+1},\ldots,p_{s+n}$. According to Theorem 2.5.1 and estimates (3.2.11), (4.3.4) the functionals $\left(\varphi_s, I_{n,s}\right)$ and $\left(\varphi_s, I_{n,s}^{(0)}\right)$ exist.

We denote by $\left(\overline{\varphi_s, I_{n,s}}\right)$ and $\left(\overline{\varphi_s, I_{n,s}^{(0)}}\right)$ the functionals that approximate respectively the expressions $\left(\varphi_s, I_{n,s}\right)$ and $\left(\varphi_s, I_{n,s}^{(0)}\right)$, respectively, as $a \to 0$. in addition, $\left(\overline{\varphi_s, I_{n,s}}\right)$ tends to $\left(\varphi_s, I_{n,s}^{(0)}\right)$, as $a \to 0$.. These functionals can be constructed in a few steps.

We first consider the functional $\left(\varphi_s, I_{n,s}\right)$.

First step. Let $\delta > 0$ be a sufficiently small number independent of the parameter a. δ will be fixed in the sequel. We exclude from the functional $\left(\varphi_s, I_{n,s}\right)$ the following time intervals

$$|\, t_{i+1} - t_i \,| < \delta, \qquad i \in (0,1,\ldots,n-1)\,.$$

Thus we obtain an expression, denoted by $(\varphi_s, I_{n,s}^{\delta})$, which approximates $\left(\varphi_s, I_{n,s}\right)$ with accuracy of order δ, since for $F^0 \in L_{\xi}^{\infty,0}$ the integrals with respect to the remaining variables converge absolutely. With accuracy of order $\varepsilon(p_0)$ (where $\lim_{p_0\to\infty} \varepsilon(p_0) = 0$), we can for any a restrict the integration with respect to the momentum variables to a ball of finite radius $\sum_{i=1}^{s+n} p_i^2 < p_0^2$, since the functional $(\varphi_s, I_{n,s}^{\delta})$ exists and the integral with respect to all the $s+n$ momentum variables converges absolutely. We denote this functional by $(\varphi_s, I_{n,s}^{\delta})_1$.

Second step. Further, in the functional $(\varphi_s, I_{n,s}^{\delta})_1$ we exclude from the domain of integration with respect to the variables $(p_1,\ldots,p_s)$ the set $\mathfrak{M}_s$ (for a fixed value of $(q_1, \ldots, q_s) \in \mathcal{K} \subset \mathcal{K}_0^s)$. We denote this functional by $\left(\varphi_s, I_{n,s}^{\delta}\right)_2$. Moreover, according to Lemma 4.4.1 (on "sweeping up") the functional formed $\left(\varphi_s, I_{n,s}^{\delta}\right)_2$ approximates $\left(\varphi_s, I_{n,s}^{\delta}\right)_1$ with accuracy of order $\left(\frac{a}{a_0(a)}\right)^2$.

In the sequel we restrict our consideration to a term $\left(\varphi_s, I_{\{s,s+1,\ldots,s+n-1\}}^{0,\delta}\right)_2$ of the functional $\left(\varphi_s, I_{n,s}^{\delta}\right)_2$ as a typical example. This means that the $s+1$-th particle is shifted by $a\eta$ with respect to the s-th particle, the $s+2$-th particle is shifted by $a\eta$ with respect to the $s+1$-th particle, *etc.* With the help of the explicit form (2.5.7) for $I_{\{s,s+1,\ldots,s+n-1\}}^0$, we consecutively consider the evolution of every subsystem consisting of the $s+1$-th, $\ldots, s+n$-th particles.

Third step. Consider the subsystem of $s+1$ particles and recall that during the evolution of the subsystem of s particles the $s+1$–th particle moves with momentum p_s together with the s–th particle. Each subsystem of $s+1$ particles is in a state $\left(X_1^0(t_1-t, x_1), \ldots, X_s^0(t_1-t, x_s), Q_s^0(t_1-t, x_s) - a\eta_1, p_{s+1}\right)$. We take this state as the initial one for the evolution of the subsystem under consideration (with evolution operator $S^{s+1}\left(t_2-t_1, X_1^0(t_1-t, x_1), \ldots, X_s^0(t_1-t, x_s), Q_s^0(t_1-t, x_s) - a\eta_1, p_{s+1}\right)$ and $X_i^0(t_1-t, x_i) = (q_i + p_i(t_1-t), p_i)$, $i = 1, \ldots, s$). In the subsystem of $s+1$ particles we exclude those initial configurations in which the $(s+1)$–th particle is at a distance smaller than $a_0(a) + a$ from the 1–st,...,$(s-1)$–th particles. This can be achieved if in the functional $\left(\varphi_s, I_{\{s,s+1,\ldots,s+n-1\}}^{0,\delta}\right)_2$ we exclude from the domain of integration with respect to the momentum variables certain domains that are balls of radius proportional to

$$\frac{(a + a_0(a))}{|t_1 - t|} \leqslant (a + a_0(a))\,\delta^{-1}.$$

Thus we make an error of order $(a + a_0(a))^3 \delta^{-3}$. In order to prove this statement, we distinguish two cases:

1) The distances between the $s+1$-th and the 1-st,...$s-1$-th particles are less than $a + a_0(a)$ at the state $x_1, \ldots, x_s, q_s - a\eta_1, p_{s+1}$, *i.e.*

$$|q_1 - q_s + a\eta_1| \leq a + a_0(a), \ldots, |q_{s+1} - q_s + a\eta_1| \leq a + a_0(a).$$

2) Some of these distances are greater than $a + a_0(a)$, for example $|q_j - q_s + a\eta_1| > a + a_0(a)$.

In the first case, the $(s+1)$–th and the j-th particles $(j \in (1,\ldots,s-1))$ particles occupy, after the action of the operator $S^s(t_1-t,x_1,\ldots,x_s)$, the configurations mentioned above if the relative momentum of the j–th and the s–th particles satisfies the inequality

$$|p_j - p_s||t_1 - t| \leqslant (a + a_0(a)).$$

From the last inequalities it follows that the distances between the $s+1$-th and and the 1-st,...,$s-1$-th particles are less than $a+a_0(a)$ after the evolution of s particles only if the relative momenta belong to the spheres $|p_j - p_s| < (a+a_0(a))\,\delta^{-1}$ $(j=1,\ldots,s)$ with volumes less than $(4/3)\pi\,(a+a_0(a))^3\,\delta^{-3}$.

Let us proceed now to the second case, when, for example, $|q_j - q_s + a\eta_1| > a + a_0(a)$. In this case the j-th and the $s+1$-th particles do not interact during the evolution of the subsystem of s particles when the *s+1*-th particle moves with momentum p_s together with the s-th particle. Buth their distance may became less than $a+a_0(a)$ (obviously greater than a). Denote by $(p_j-p_s)^*$ a relative momentum for which the distance between the j-th and $s+1$-th particles are equal to $a+a_0(a)$ at a fixed time instant $(t_1-t)^*$ during the evolution of s particles. then the distance between them at that instant can be less than $a+a_0(a)$ only if $|(p_j-p_{s+1})-(p_j-p_{s+1})^*||(t_1-t)^*| < a+a_0(a)$, *i.e.*if the relative momenta belong to a sphere with a center $(p_j-p_s)^*$ and its volume is less than $(4/3)\pi\,(a+a_0(a))^3\,\delta^{-3}$. The 1-st,...,$j-1$-th, $j+1$-th, ..., $s-1$-th particles are at distances less than $a+a_0(a)$ from the $s+1$-th particle and can be treated as in the first case.

The case when more than one particle is at a distance greater than $a+a_0(a)$ from the $s+1$-th one can be treated in the same way.

In both cases we exclude all the corresponding spheres with volumes less than $(4/3)\pi\,(a+a_0(a))^3\,\delta^{-3}$ and thus make an error of order $(a+a_0(a))^3\,\delta^{-3}$. The third step is thus completed.

Fourth step. In the expression obtained at the initial instant t_1-t of the evolution of a subsystem of $s+1$ particles, the $(s+1)$–th particle is in the state of collision with s–th one and at a distance greater than $a_0(a)+a$ from the 1–st,...,$(s-1)$–th particles with which it can interact in the evolution process. We recall again that the point $(X_1^0(t_1-t,x_1),\ldots,X_s^0(t_1-t,x_s),Q_s^0(t_1-t,x_s)-a\eta_1,p_{s+1})$ is an initial state for the evolution of the system of $s+1$ particles (operator $S^{s+1}(t_2-t_1,X_1^0(t_1-t,x_1),\ldots,X_s^0(t_1-t,x_s),Q_s^0(t_1-t,x_s)-a\eta_1,p_{s+1}))$. We exclude ("sweep up") from the functional $\left(\varphi_s, I^{0,\delta}_{\{s,s+1,\ldots,s+n-1\}}\right)_2$, by Lemma 4.4.1, the corresponding domain $\mathfrak{M}_{s+1}$ with respect to the momentum variables of the $(s+1)$–th particle, whose volume is of order $\left(a/a_0\,(a)\right)^2$. In this case the $(s+1)$–th particle will not interact in the evolution process with the remaining s particles which do not interact with each other. Obviously the $s+1$-th particle will not interact with the s-th one, because both of them are in a collision state at the initial time $t_1 - t$ and can interact only indirectly via the interactions with the 1–st,...,$(s-1)$–th particles. But these particles cannot interact because their positions belong to $\mathcal{K}_0$ and the set $\mathfrak{M}_s$ was excluded. After the evolution according to the $s+1$ particle dynamics they will all be in the state

$$\left(X_1^0(t_1-t,x_1),\ldots,X_s^0(t_2-t,x_s),Q_{s+1}^0(t_2-t_1,Q_(t_1-t,x_s)-a\eta_1,p_{s+1}\right)\,.$$

In a similar way we exclude consecutively the corresponding integration domains for $(s+2),\ldots,(s+n)$–particle subsystems.

Fifth step. Having proved the result in the previous step, if we consider other expressions $\left(\varphi_s, I^{k,\delta}_{\{i_1,\ldots,i_n\}}\right)_2$, $k \in (1,\ldots n)$, then the construction remains valid. Indeed, these expressions differ from those previously considered because they contain the momentum arguments of F^0_{s+n} that are connected with integration variables by the transformation (1.2.5) (momenta after a collision, see Chapter I). After the corresponding changes of the momentum variables in the functionals $\left(\varphi_s, I^{k,\delta}_{\{i_1,\ldots,i_n\}}\right)_2$, the volumes of the integration domains that are excluded can be estimated as before, since the Jacobian of this transformation is unity.

Summing up, we obtain for the functional $\left(\varphi_s, I_{n,s}\right)$ the approximate expression $\left(\overline{\varphi_s, I_{n,s}}\right)$ that is defined as the result of consecutive applications of the procedure described above to $\left(\varphi_s, I_{n,s}\right)$. We note that according to Lemma 4.4.2 the operators $S_0\,(t)$ of non–interacting particles appear in the functionals $\left(\overline{\varphi_s, I_{n,s}}\right)$.

We repeat these steps for the functionals $\left(\varphi_s, I^{(0)}_{n,s}\right)$ of the Boltzmann hierarchy. That is, we exclude from $\left(\varphi_s, I^{(0)}_{n,s}\right)$ the same sets as from $\left(\varphi_s, I_{n,s}\right)$.

For the approximate functionals $\left(\overline{\varphi_s, I_{n,s}}\right)$ and $\left(\overline{\varphi_s, I^{(0)}_{n,s}}\right)$ the following statement holds for continuous initial data defined on forbidden configurations as well:

Lemma 4.4.3. *In the case of continuous initial data, for any $\varepsilon > 0$ there is a value a such that*

$$\left| \left(\overline{\varphi_s, I_{n.s}}\right) - \left(\overline{\varphi_s, I_{n,s}^{(0)}}\right) \right| < \varepsilon\,(a)\,\mathrm{const}\,(p_0, s, n).$$

Proof. This inequality is a consequence of the fact that the operators $S_0\,(t)$ of the free evolution appear in the functionals $\left(\overline{\varphi_s, I_{n,s}}\right)$ and $\left(\overline{\varphi_s, I_{n,s}^{(0)}}\right)$ and the operator $a^{-2}\left[\mathcal{H}, \int dx\right]$ in (3.2.1) differs from the operator Q in (4.3.1) by the shifts $\{\, a\eta_i \,\}_{i=1}^{n}$ with respect to the corresponding variables. Therefore the above estimate holds for continuous initial distribution functions.

From Lemma 4.4.3 and the definitions of the functionals $\left(\overline{\varphi_s, I_{n,s}}\right)$ and $\left(\overline{\varphi_s, I_{n,s}^{(0)}}\right)$ there follows a result on the convergence in the Boltzmann–Grad limit of the n–th term of the series for the BBGKY hierarchy to its limit.

Lemma 4.4.4. *Let $F^0 \in L_{\xi}^{\infty,0}$ and $(q_1, \ldots, q_s) \in \mathcal{K}_0^s$, then the expression $I_{n,s}$in (2.5.2) converges uniformly with respect to $(q_1, \ldots, q_s)$ and in a weak sense with respect to $(p_1, \ldots, p_s)$ to the expression $I_{n,s}^{(0)}$ in (4.4.3) as $a \to 0$.*

Remark. Note that the subsystems of s particles (with suffixes from 1 to s) and n partiucles (with suffixes from $s+1$ to $s+n$) should be treated in a different way. One can fix the positions of the s particles and then let the diameter a go to zero. At some stage the distances $a_0(a)$ between them become such that the ratio $a/a_0(a)$ tends to zero and the volumes of the corresponding cones in momentum space also tend to zero. The trajectories of the s spheres tend to the corresponding trajectories of free particles in this limit. But the situation of the subsystem made of n particles is quite different. They are attached?? to a particle of either one subsystem or the other. In the case when they are attached?? to a particle of the same subsystem and are not in a state of collision, there can exist positions, with certain vectors η of full measure on $\mathbb{S}^2$, such that the distances $a_0(a)$ between them tend to zero but $a/a_0(a)$ is different from zero for all values of a. Then the volumes of the corresponding cones in momentum space are also different from zero for all a. The trajectories of $s+n$ hard spheres with these initial data (of full measure with respect to the vectors η and the momenta) will not converge to the trajectories of free particles.

To prove that the contributions of these "dangerous" set in the terms of the iteration series tend to zero, it is necessary to use the "sweep up" procedure descxribed above after eliminating the corresponding time intervals.

4.4.3 The Boltzmann–Grad limit for non–equilibrium states. Let $(q_1, \dots, q_s) \in \mathcal{K}_0^s$ and using the formulas (3.2.2) and (4.3.2) we construct the following functionals $\left(\varphi_s\,,\, a^{2s}\left(U(t)\,F^0\right)_s\right)$ and $\left(\varphi_s\,,\,\left(U_0(t)\,f^0\right)_s\right)$, where $\varphi_s = \varphi_s(p_1, \dots, p_s)$ is a continuous function with a compact support.

The following estimate holds

$$\left|\left(\varphi_s\,,\, a^{2s}\left(U(t)\,F^0\right)_s\right) - \left(\varphi_s\,,\,\left(U_0(t)\,f^0\right)_s\right)\right|$$
$$\leqslant \sum_{n=0}^{n_0}\left|\left(\varphi_s\,,\, a^{2(s+n)} I_{n,s}\left(F^0\right)\right) - \left(\varphi_s\,,\, I_{n,s}^{(0)}\left(f^0\right)\right)\right| + R_{n_0}(t)\,, \quad (4.4.4)$$

where for the n–th terms $I_{n,s}$ and $I_{n,s}^{(0)}$ of the corresponding series we introduced the notations $I_{n,s}\left(F^0\right)$ and $I_{n,s}^{(0)}\left(f^0\right)$ that indicate explicitly on what initial data these quantities depend and by $R_{n_0}(t)$ we denote the remainder of the sum of the numerical series (3.2.11) and (4.3.2).

If $t \in [-T, T]$, where T is the smaller between the numbers $t_0 a^2$ and t_0' defined according to (3.2.10) and (4.3.3), then the remainder $R_{n_0}(t)$ in (4.4.4) can be made arbitrarily small for a sufficiently large n_0 independently of a. For each integer n the first term can be estimated as follows

$$\left|\left(\varphi_s\,,\, a^{2(s+n)} I_{n,s}\left(F^0\right)\right) - \left(\varphi_s\,,\, I_{n,s}^{(0)}\left(f^0\right)\right)\right|$$
$$\leqslant \left|\left(\varphi_s\,,\, a^{2(s+n)} I_{n,s}\left(F^0\right)\right) - \left(\overline{\varphi_s\,,\, a^{2(s+n)} I_{n,s}\left(F^0\right)}\right)\right|$$
$$+ \left|\left(\varphi_s\,,\, I_{n.s}^{(0)}\left(f^0\right)\right) - \left(\overline{\varphi_s\,,\, I_{n,s}^{(0)}\left(f^0\right)}\right)\right|$$
$$+ \left|\left(\overline{\varphi_s\,,\, I_{n,s}\left(f^0\right)}\right) - \left(\overline{\varphi_s\,,\, I_{n,s}^{(0)}\left(f^0\right)}\right)\right|$$
$$+ \left|\left(\overline{\varphi_s\,,\, a^{2(s+n)} I_{n,s}\left(F^0\right)}\right) - \left(\overline{\varphi_s\,,\, I_{n,s}\left(f^0\right)}\right)\right|. \quad (4.4.5)$$

For a fixed n_0 we fix a δ sufficiently small and a p_0 sufficiently large; then we let the diameter a go to zero. According to Lemma 4.4.4 the first three terms in the estimate (4.4.5) converge in the Boltzmann–Grad limit to zero. The difference $a^{2(s+n)}F^0_{s+n} - f^0_{s+n}$ in the last term in (4.4.5) appears in the approximate functional and therefore it can only be considered on compact sets in $\mathbb{R}^{3(s+n)} \times (\mathbb{R}^{3(s+n)} \setminus \mathbb{W}_{s+n})$from $\mathcal{K}_0^{s+n} \times \Re_{p_0}^{s+n}$, where according to the assumption of Theorem 4.3.2 it converges to zero in the Boltzmann–Grad limit.

Thus we have proved the existence of the Boltzmann–Grad limit for solutions of the BBGKY hierarchy (3.2.1) if $t \in [-T, T]$ (Theorem 4.3.2).

Remarks.

a) Using the result of Proposition 2.5.1 for initial data in $X_{\xi,\beta} \subset L^1$ (see subsection 2.5.4) it is clear that Theorem 4.3.2 on the existence of the Boltzmann–Grad limit holds for any $t \in \mathbb{R}^1$.

b) In a series of papers [CDPP,U] the problem of deriving the Boltzmann–Grad limit of discrete velocity models of the Boltzmann equation was considered.

4.5 The Boltzmann–Grad Limit for Equilibrium States of Systems of Hard Spheres in the Framework of the Canonical Ensemble

In this section, we consider the problem of the existence of the Boltzmann–Grad limit for the equilibrium distribution functions of a system of hard spheres in the framework of the canonical ensemble. (Note that the same problem was solved In Sect. 4.2 in the framework of grand canonical ensemble.) Here we prove that the equilibrium distribution functions normalized on unity tend to zero if one simultaneously passes to the Boltzmann–Grad limit and to the thermodynamic limit. The nonzero limit distribution functions with this normalization can be obtained only for the systems of particles located in a bounded domain.

The system of particles located in the whole space was also investigated for a case of the distribution functions normalized to the number of particles. In this case the limit distribution functions are nonzero only if the diameter of spheres tends to zero and their density tends to infinity so that their product is constant.

4.5.1 The existence of the Boltzmann–Grad limit for normalized equilibrium distribution functions of hard spheres in the framework of the canonical ensemble. Statement of the problem

Let us consider N hard spheres in a domain $\Lambda \subset \mathbb{R}^3$ with the volume $V(\Lambda) = V$. Denote by $D^{(N)}(t, x_1, \dots, x_N)$ a distribution function in the phase space of N hard spheres, normalized to unity, $x = (p, q)$ is a point of phase space, p is momentum, q is position of the center of hard sphere. Let us denote by $F^{(N)}(t)$ the sequence of reduced distribution functions

$$F^{(N)}(t) = \Big(F_1^{(N)}(t, x_1), F_2^{(N)}(t, x_1, x_2), \ldots, \\ F_s^{(N)}(t, x_1, \ldots, x_s), \ldots, F_N^{(N)}(t, x_1, \ldots, x_N), 0, \ldots \Big) \quad (4.5.1)$$

which are defined trough the function $D^{(N)}(t, x_1, \ldots, x_N)$ by the following formula

$$F_s^{(N)}(t, x_1, \ldots, x_s) = \int_{\Lambda^{N-s}} dx_{s+1} \ldots dx_N D^{(N)}(t, x_1, \ldots, x_s, x_{s+1}, \ldots, x_N), \\ 1 \leqslant s \leqslant N, \quad (4.5.2)$$

where

$$\int_{\Lambda} dx = \int_{\Lambda} dq \int_{\mathbb{R}^3} dp.$$

The distribution functions defined according to (4.5.2) are called normalized (normalized to unity).

The functions $F_s^{(N)}(t, x_1, \ldots, x_s)$ satisfy the following equations, known as the BBGKY hierarchy (for $t > 0$)

$$\frac{\partial F_s^{(N)}(t, x_1, \ldots, x_s)}{\partial t} = -\sum_{i=1}^{s} \langle p_i, \frac{\partial}{\partial q_i} \rangle F_s^{(N)}(t, x_1, \ldots, x_s) \\ + (N - s) a^2 \sum_{i=1}^{s} \int_{\mathbb{R}^3 \times \mathbb{S}_+^2} d\eta \, dp_{s+1} \langle \eta, (p_i - p_{s+1}) \rangle \\ \times \Big[F_{s+1}^{(N)} \Big(t, x_1, \ldots, q_i, p_i - \eta \langle \eta, (p_i - p_{s+1}) \rangle, \ldots, x_s, q_i - a\eta, \\ p_{s+1} + \eta \langle \eta, (p_i - p_{s+1}) \rangle \Big) \\ - F_{s+1}^{(N)} \Big(t, x_1, \ldots, x_s, q_i + a\eta, p_{s+1} \Big) \Big] \quad (4.5.3)$$

where a is the diameter of a sphere, η a unit vector, $|\eta| = 1$.

The functions $F_s^{(N)}(t, x_1, \dots, x_s)$ are symmetrical (invariant under permutations), equal to zero on forbidden configurations

$$\mathbb{W}_s\Big(q_1, \dots, q_s \,\Big|\, |\, q_i - q_j \,| < a,\ (i,j) \in (1, \dots, s)\Big)$$

and satisfy the boundary conditions

$$\begin{aligned} &F_s^{(N)}(t, x_1, \dots, x_i, \dots, x_j, \dots, x_s) \\ &\qquad = F_s^{(N)}(t, x_1, \dots, x_i^*, \dots, x_j^*, \dots, x_s), \qquad \text{if } |\, q_i - q_j \,| = a. \end{aligned}$$

Here $x_i^* = \big(q_i, p_i - \eta\langle\eta(p_i - p_j)\rangle\big)$, $x_j^* = \big(q_j, p_j + \eta\langle\eta(p_i - p_j)\rangle\big)$.

They are also equal to zero outside of Λ, $F_s^{(N)}(t, x_1, \dots, x_s) = 0$, if $q_i \notin \Lambda$, $i = 1, 2, \dots, s$.

Let us stress that the multiplier $(N - s)$ appeared in the second term of right–hand side of equations (4.5.3) due to definition (4.5.2). If one employs an usual definition of reduced distribution functions

$$\begin{aligned} F_s^{(N)}(t, x_1, \dots, x_s) &= N(N-1)\dots(N-s+1) \\ &\times \int\limits_{\Lambda^{N-s}} dx_{s+1} \dots dx_N D^{(N)}(t, x_1, \dots, x_s, x_{s+1}, \dots, x_N), \end{aligned} \qquad (4.5.4)$$

then the multiplier $(N - s)$ will be absent in the second term of right–hand side of equations (4.5.3) and the equations for the distribution functions (4.5.4) will have the following form

$$\begin{aligned} \frac{\partial F_s^{(N)}(t, x_1, \dots, x_s)}{\partial t} &= -\sum_{i=1}^{s} p_i \frac{\partial}{\partial q_i} F_s^{(N)}(t, x_1, \dots, x_s) \\ &+ a^2 \sum_{i=1}^{s} \int\limits_{\mathbb{R}^3 \times S_+^2} d\eta\, dp_{s+1} \langle\eta(p_i - p_{s+1})\rangle \\ &\times \Big[F_{s+1}^{(N)}\Big(t, \dots, q_i, p_i - \eta\eta\,(p_i - p_{s+1}), \dots, x_s, \\ &\qquad q_i - a\eta, p_{s+1} + \eta\eta\,(p_i - p_{s+1})\Big) \\ &\quad - F_{s+1}^{(N)}\Big(t, x_1, \dots, x_s, q_i + a\eta, p_{s+1}\Big)\Big] \end{aligned} \qquad (4.5.5)$$

If we define the reduced distribution functions according to

$$F_s^{(N)}\left(t, x_1, \ldots, x_s\right) = V^s \int\limits_{\Lambda^{N-s}} dx_{s+1} \ldots dx_N$$
$$\times D^{(N)}\left(t, x_1, \ldots, x_s, x_{s+1}, \ldots, x_N\right), \quad (4.5.6)$$

then the multiplier $(N-s)/V$ will be present in the second term on the right–hand side of (4.5.3) or (4.5.5). This multiplier is equal to density in thermodynamic limit. (We use the same notation $F_s^{(N)}\left(t, x_1, \ldots, x_s\right)$ for distribution functions under all normalizations)

The equilibrium distribution functions defined according to (4.5.4) or (4.5.6) are well investigated in the thermodynamic limit when $\Lambda \nearrow \mathbb{R}^3$, $V(\Lambda) \to \infty$, $N \to \infty$, $N/V(\Lambda) = 1/v = \text{const}$. For details, we refer the reader to the article of Bogolubov, Petrina and Khatset [BPK] or to the monograph [PGM1]. In the present section, we carry out analogous investigations for the equilibrium distribution functions defined according to (4.5.2) [LPe].

Definitions (4.5.2) and (4.5.4), (4.5.6) are equivalent for finite N and $V(\Lambda)$, because we can pass from one to the other using different normalizations. But this is impossible in the thermodynamic limit when N and $V(\Lambda)$ tend to infinity. Therefore, the case (4.5.2) requires an independent investigation. It is obvious that the definitions (4.5.4) and (4.5.6) are also equivalent in the thermodynamic limit because

$$\lim_{\substack{N \to \infty,\, V(\Lambda) \to \infty \\ \frac{N}{V(\Lambda)} = \frac{1}{v}}} \frac{V^s(\Lambda)}{N(N-1)\ldots(N-s+1)} = \frac{1}{v^s}.$$

4.5.2 The Kirkwood–Salsburg equations. It is well known that in order to investigate the equilibrium states it suffices to restrict oneself to equilibrium distribution functions in configuration space. The equilibrium distribution function of N particles distributed in a domain Λ is defined as follows

$$D^{(N)}\left(q_1, \ldots, q_N\right) = \frac{1}{Q(N, \Lambda)} \exp\left\{-\beta \sum_{i<j=1}^{N} \Phi\left(q_i - q_j\right)\right\},$$

$$Q(N, \Lambda) = \int\limits_{\Lambda^N} \exp\left\{-\beta \sum_{i<j=1}^{N} \Phi\left(q_i - q_j\right)\right\} dq_1 \ldots dq_N, \quad (4.5.7)$$

where Φ–is the interaction potential of hard spheres

$$\Phi(q) = \begin{cases} \infty, & |q| < a, \\ 0, & |q| > a. \end{cases}$$

The s–particle reduced distribution functions are defined according to (4.5.2)

$$F_s^{(N)}(q_1, \ldots, q_s) = \int\limits_{\Lambda^{N-s}} D^{(N)}(q_1, \ldots, q_s, q_{s+1}, \ldots, q_N)\, dq_{s+1} \ldots dq_N, \qquad 1 \leqslant s \leqslant N,$$

$$F^{(N)} = \left(F_1^{(N)}(q_1), \ldots, F_s^{(N)}(q_1, \ldots, q_s), \ldots, F_N^{(N)}(q_1, \ldots, q_N), 0, \ldots\right).$$

They satisfy the following Kirkwood–Salsburg relations

$$\begin{aligned} F_s^{(N)}(q_1, \ldots, q_s) = {} & \frac{Q(N-1, \Lambda)}{Q(N, \Lambda)} \exp\left\{-\beta \sum_{i=2}^{s} \Phi(q_1 - q_i)\right\} \times \Bigg[F_{s-1}^{(N-1)}(q_2, \ldots, q_s) \\ & + \sum_{k=1}^{N-s} \frac{1}{k!} (N-s)(N-s-1) \ldots (N-s-k+1) \\ & \times \int\limits_{\Lambda^k} \prod_{i=1}^{k} \varphi_{q_1}(y_i) \quad F_{s-1+k}^{(N-1)}(q_2, \ldots, q_s, y_1, \ldots, y_k)\, dy_1 \ldots dy_k\Bigg], \\ & (N > s > 1) \end{aligned}$$

$$\begin{aligned} F_1^{(N)}(q_1) = {} & \frac{Q(N-1, \Lambda)}{Q(N, \Lambda)} \Bigg[1 + \sum_{k=1}^{N-1} \frac{1}{k!} (N-1)(N-2) \ldots (N-k) \\ & \times \int\limits_{\Lambda^k} \prod_{i=1}^{k} \varphi_{q_1}(y_i)\, F_k^{(N-1)}(y_1, \ldots, y_k)\, dy_1 \ldots dy_k\Bigg], \end{aligned}$$

$$F_N^{(N)}(q_1, \ldots, q_N) = \frac{Q(N-1, \Lambda)}{Q(N, \Lambda)} \exp\left\{-\beta \sum_{i=2}^{N} \Phi(q_1 - q_i)\right\} F_{N-1}^{(N-1)}(q_2, \ldots, q_N),$$

$$\varphi_{q_1}(y) = \exp\{-\beta\Phi(q_1 - y)\} - 1. \tag{4.5.8}$$

For details of derivation, we refer to the monograph [PGM1]. Note that, for the functions $F_s^{(N)}(q_1, \ldots, q_s)$, defined according to (4.5.4) the multipliers $(N-s)$ $(N-s-1) \ldots (N-s-k+1)$ are absent in relations (4.5.8) connecting $F^{(N)}$ with $F^{(N-1)}$ (see below).

The Boltzmann-Grad limit means that number of particles tends to infinity $N \to \infty$, diameter of particles tends to zero $a \to 0$, but the product $Na^2 = \alpha$ is fixed. Our objective is to investigate the sequence of distribution functions $F^{(N)}$ in the Boltzmann-Grad limit. For this aim the Kirkwood-Salsburg relations (4.5.8) will be used.

Define the numbers $\alpha = \alpha_0 = Na^2, \alpha_1 = (N-1)\,a^2, \ldots, \alpha_i = (N-i)\,a^2, \ldots$, $i < N$. According to definition of the Boltzmann–Grad limit, the numbers α_i are bounded and $\alpha_i < \alpha$, $i > 1$. It is easy to transform relations (4.5.8) to the following form

$$F_s^{(N)}(q_1, \ldots, q_s) = a(N, \Lambda) \exp\left\{-\beta \sum_{i=2}^{s} \Phi(q_1 - q_i)\right\} \times \left[F_{s-1}^{(N-1)}(q_2, \ldots, q_s)\right.$$
$$+ \sum_{k=1}^{N-s} \frac{1}{k!} \prod_{j=0}^{k-1} \frac{\alpha_{s+j}}{a^2} \int_{\Lambda^k} \prod_{i=1}^{k} \varphi_{q_1}(y_i)$$
$$\left.\times F_{s-1+k}^{(N-1)}(q_2, \ldots, q_s, y_1, \ldots, y_k)\, dy_1 \ldots dy_k\right],$$
$$F_1^{(N)}(q_1) = a(N, \Lambda) \left[1 + \sum_{k=1}^{N-1} \frac{1}{k!} \prod_{j=0}^{k-1} \frac{\alpha_{1+j}}{a^2} \int_{\Lambda^k} \prod_{i=1}^{k} \varphi_{q_1}(y_i)\right.$$
$$\left.\times F_k^{(N-1)}(y_1, \ldots, y_k)\, dy_1 \ldots dy_k\right],$$
$$F_N^{(N)}(q_1, \ldots, q_N) = a(N, \Lambda) \exp\left\{-\beta \sum_{i=2}^{N} \Phi(q_1 - q_i)\right\} F_{N-1}^{(N-1)}(q_2, \ldots, q_N), \tag{4.5.9}$$

$$a(N, \Lambda) = \frac{Q(N-1, \Lambda)}{Q(N, \Lambda)}.$$

Relations (4.5.9) hold for $N > 2$. For $N = 2$, one has

$$F_1^{(2)}(q_1) = \frac{\int_{\Lambda} \exp\{-\beta\Phi(q_1 - q_2)\} dq_2}{\int_{\Lambda^2} \exp\{-\beta\Phi(q_1 - q_2)\} dq_1\, dq_2},$$

$$F_2^{(2)}(q_1, q_2) = \frac{\exp\{-\beta\Phi(q_1 - q_2)\}}{\int_{\Lambda^2} \exp\{-\beta\Phi(q_1 - q_2)\} dq_1\, dq_2}.$$

Consider a Banach space E_ξ of sequences $f = \big(f_1(q_1), \ldots, f_s((q)_s), \ldots\big)$ of bounded functions $f_s\,(q_1, \ldots, q_s) = f_s\big((q)_s\big)$, $(q_1, \ldots, q_s = (q)_s)$ with norm

$$\| f \| = \sup_{s\geqslant 1} \frac{1}{\xi^s} \sup_{(q)_s} \big| f_s\big((q)_s\big) \big|, \qquad \xi > 0.$$

Define in E_ξ an operator $K^{(N)}$ in the following way

$$\begin{aligned}
\left(K^{(N)} f\right)_s ((q)_s) &= \mathcal{X}_\Lambda\big((q)_s\big) \exp\left\{ -\beta \sum_{i=2}^{s} \Phi\,(q_1 - q_i) \right\} \Bigg[f_{s-1}\big((q)_s^1\big) \\
&\quad + \sum_{k=1}^{N-s} \frac{1}{k!} \prod_{j=0}^{k-1} \frac{\alpha_{s+j}}{a^2} \int\limits_{\Lambda^k} \prod_{i=1}^{k} \varphi_{q_1}\,(y_i) \\
&\quad \times f_{s-1+k}\big((q)_s^1, (y)_k\big)\, d\,(y)_k \Bigg], \; N > s > 1 \\
\left(K^{(N)} f\right)_1 ((q)_1) &= \mathcal{X}_\Lambda\big((q)_1\big) \sum_{k=1}^{N-1} \frac{1}{k!} \prod_{j=0}^{k-1} \frac{\alpha_{1+j}}{a^2} \\
&\quad \times \int\limits_{\Lambda^k} \prod_{i=1}^{k} \varphi_{q_1}\,(y_i)\, f_k\big((y)_k\big)\, d(y)_k \\
\left(K^{(N)} f\right)_N ((q)_N) &= \mathcal{X}_\Lambda\big((q)_N\big) \exp\left\{ -\beta \sum_{i=2}^{N} \Phi\,(q_1 - q_i) \right\} f_{N-1}\big((q)_N^1\big) \\
\left(K^{(N)} f\right)_s ((q)_s) &= 0, \quad s > N
\end{aligned} \tag{4.5.10}$$

$$(q)_s^1 = (q_2, \ldots, q_s), \quad d(y)_k = dy_1 \ldots dy_k,$$

where $\mathcal{X}_\Lambda\big((q)_s\big)$ is the characteristic function of domain Λ^s.

Taking into account that

$$\varphi_q\,(y) = \begin{cases} -1, & |\,q - y\,| < a, \\ \;\;\,0, & |\,q - y\,| > a, \end{cases} \tag{4.5.11}$$

we have the following estimate for the norm of the operator $K^{(N)}$

$$\left\| K^{(N)} \right\| \leqslant \sup_{s\geqslant 1} \xi^{s-1} \sum_{k=0}^{N-s} \left(\alpha \frac{4}{3} \pi\, a \xi \right)^k \frac{1}{k!} \leqslant \xi^{-1} e^{\frac{4}{3}\pi\, a\xi\, \alpha}. \tag{4.5.12}$$

It follows from (4.5.12) that the operator $K^{(N)}$ is uniformly bounded with respect to N and a if the parameter a lies in the finite interval $0 \leqslant a \leqslant a_0$. In what follows we assume that this condition is satisfied.

By using the operator $K^{(N)}$, relations (4.5.9) can be represented as single operator relation, namely

$$F^{(N)} = a(N,\Lambda)\left(K^{(N)}F^{(N-1)} + F_0^{(N-1)}\right), \tag{4.5.13}$$

where $F_0^{(N-1)} = (1,0,\dots) = F_0$.

Our aim is to prove the existence of the thermodynamic and Boltzmann-Grad limits for $F^{(N)}$. To do this, we first need to estimate the value $a(N,\Lambda)$.

4.5.3 Estimation of the value $a(N,\Lambda)$**.** To estimate the ratio between $Q(N-1,\Lambda)$ and $Q(N,\Lambda)$, we use the following trick. Represent $Q(N,\Lambda)$ in the following way

$$Q(N,\Lambda) = \int\limits_{\Lambda^N} e^{-\beta \sum\limits_{i<j=1}^{N} \Phi(q_i - q_j)} dq_1\, d(q)_N^1$$

$$= \int\limits_{\Lambda^{N-1}} \left\{ e^{-\beta \sum\limits_{i<j=2}^{N} \Phi(q_i - q_j)} \int\limits_{\Lambda} \prod_{i=2}^{N} \left(e^{-\beta\Phi(q_1 - q_i)} - 1 + 1\right) dq_1 \right\} d(q)_N^1 . \tag{4.5.14}$$

The value $a_i = e^{-\beta\Phi(q_1 - q_i)} - 1$ is nonpositive and $0 \geqslant a_i \geqslant -1$. Thus, $(a_i + 1) \geqslant 0$. Hence, the following inequality holds

$$\prod_{i=2}^{N} (1 + a_i) \geqslant 1 - \sum_{i=2}^{N} |\, a_i \,|$$

or, which is the same,

$$\prod_{i=2}^{N} \left(e^{-\beta\Phi(q_1 - q_i)} - 1 + 1\right) \geqslant 1 - \sum_{i=2}^{N} \left| e^{-\beta\Phi(q_1 - q_i)} - 1 \right| . \tag{4.5.15}$$

Substituting inequality (4.5.15) in (4.5.14), one gets

$$Q(N,\Lambda) \geqslant \int\limits_{\Lambda^{N-1}} \left\{ e^{-\beta \sum\limits_{i<j=2}^{N} \Phi(q_i - q_j)} \int\limits_{\Lambda} \left(1 - \sum_{i=2}^{N} \left| e^{-\beta\Phi(q_1 - q_i)} - 1 \right| \right) dq_1 \right\} d(q)_N^1$$

$$\geqslant \int\limits_{\Lambda^{N-1}} e^{-\beta \sum\limits_{i<j=2}^{N} \Phi(q_i - q_j)} d(q)_N^1 \, V(\Lambda)$$

$$- \int\limits_{\Lambda^{N-1}} e^{-\beta \sum\limits_{i<j=2}^{N} \Phi(q_i - q_j)} d(q)_N^1 \, (N-1) \int\limits_{\mathbb{R}^3} \left| e^{-\beta\Phi(q)} - 1 \right| dq$$

$$= Q(N-1,\Lambda) \left[V(\Lambda) - (N-1)\frac{4}{3}\pi a^3 \right].$$

It is obvious that $V(\Lambda) - (N-1)\frac{4}{3}\pi a^3 > 0$, because N spheres are situated in the volume $V(\Lambda)$. Thus, one obtains the desired estimate

$$a(N,\Lambda) = \frac{Q(N-1,\Lambda)}{Q(N,\Lambda)} \leqslant \frac{Q(N-1,\Lambda)}{Q(N-1,\Lambda)\left(V(\Lambda) - (N-1)\frac{4}{3}\pi a^3\right)} = \frac{1}{V(\Lambda) - (N-1)\frac{4}{3}\pi a^3}. \tag{4.5.16}$$

Let us suppose that the thermodynamic and Boltzmann-Grad limits are performed simultaneously. The value $Na^2 = \alpha$ is fixed and, therefore, $Na^3 \to 0$ in the Boltzmann–Grad limit. It follows from estimate (4.5.16) that the following lemma is true.

Lemma 4.5.1. *The ratio* $a(N,\Lambda) = \dfrac{Q(N-1,\Lambda)}{Q(N,\Lambda)}$ *tends to zero if the following parameters tend to their limits simultaneously*

$$N \to \infty, \;\; a \to 0, \;\; Na^2 = \alpha = const, \;\; \Lambda \nearrow \mathbb{R}^3, \;\; V(\Lambda) \to \infty, \;\; \frac{N}{V(\Lambda)} = \frac{1}{v} = const, \tag{4.5.17}$$

i.e. if the thermodynamic and Boltzmann-Grad limits are performed simultaneously.

It is easy to see that the ratios $a\,(N-i,\Lambda)=\frac{Q(N-i-1.\Lambda)}{Q(N-i,\Lambda)}$ also tend to zero in the limit (4.5.17) uniformly with respect to $0\le i\le N-3$ according to the estimate

$$a\,(N-i,\Lambda)\leqslant\frac{1}{V\,(\Lambda)-(N-i-1)\,\frac{4}{3}\pi a^3}\leqslant\frac{1}{V\,(\Lambda)-(N-1)\,\frac{4}{3}\pi a^3}.\qquad(4.5.16')$$

4.5.4 The limit of $F^{(N)}$.

Theorem 4.5.1. *The sequence $F^{(N)}$ tends to zero in the norm of the space E_ξ if the thermodynamic limit and the Boltzmann–Grad limit (4.5.17) are performed simultaneously.*

Proof. To prove this, we consider relations (4.5.13). Using them many times, one gets

$$F^{(N)}=\sum_{i=1}^{N-3}a\,(N,\Lambda)\,K^{(N)}a\,(N-1,\Lambda)\,K^{(N-1)}\dots a\,(N-i+1,\Lambda)\,K^{(N-i+1)}$$
$$\times\,a\,(N-i,\Lambda)\,F_0+a\,(N,\Lambda)\,K^{(N)}$$
$$\times\,a\,(N-1,\Lambda)\,K^{(N-1)}\dots a\,(3,\Lambda)\,K^{(3)}F^{(2)}+a\,(N,\Lambda)\,F_0.\qquad(4.5.18)$$

For the operators $K^{(N-i)}$, $0\leqslant i\leqslant N-3$ estimates (4.5.12) hold. The value $a\,(N-i,\Lambda)$ satisfies estimate (4.5.16') which implies that this value tends to zero in limit (4.5.17). For the sake of simplicity, we take $V\,(\Lambda)$ so large and a so small that, according to (4.5.12) and (4.5.16), one can put

$$\left\|K^{(N-i+1)}\right\|a\,(N-i,\Lambda)\leqslant k<1,\quad 1\leqslant i\leqslant N-3.\qquad(4.5.19)$$

Let us estimate $F^{(2)}$. It is easy to see that

$$F_1^{(2)}\,(q_1)<\frac{V\,(\Lambda)}{-V\,(\Lambda)\,\frac{4}{3}\pi a^3+V\,(\Lambda)^2},\quad F_2^{(2)}\,(q_1,q_2)<\frac{1}{-V\,(\Lambda)\,\frac{4}{3}\pi a^3+V\,(\Lambda)^2}.\qquad(4.5.20)$$

It follows from (4.5.20) that $\left\|K^{(3)}\right\|\left\|F^{(2)}\right\|\leqslant k<1$ for sufficiently large $V\,(\Lambda)$. By using estimates (4.5.12), (4.5.19), (4.5.20), we get from (4.5.18)

$$\left\|F^{(N)}\right\|\leqslant a\,(N,\Lambda)\sum_{i=0}^{N-2}k^i\leqslant a\,(N,\Lambda)\,\frac{1}{1-k}.\qquad(4.5.21)$$

According to (4.5.16), $\| F^{(N)} \|$ tends to zero in limit (4.5.17). Note that, on formal level, this property of $F^{(N)}$ immediately follows from expression for $D^{(N)}(q_1, \dots, q_N)$, because $e^{-\beta \sum_{i<j=1}^{N} \Phi(q_i - q_j)} \leqslant 1$, and $Q(N, \Lambda) \sim V^N$, thus $D^{(N)}(q_1, \dots, q_N) \sim \frac{1}{V^N}$. For the same reason $F_s^{(N)}(q_1, \dots, q_s) \sim \frac{1}{V^s}$.

Thus, if one performs simultaneously, according to (4.5.17), the thermodynamic limit $N \to \infty$, $\Lambda \nearrow \mathbb{R}^3$, $(V(\Lambda) \to \infty)$, $\frac{N}{V} = \frac{1}{v} = \text{const}$, and the Boltzmann–Grad limit $N \to \infty$, $a \to 0$, $Na^2 = \alpha = \text{const}$, then the sequence of the equilibrium distribution functions $F^{(N)}$ defined according to (4.5.7) tends to zero in the norm of the space E_ξ.

Note that the functions $F_s^{(N)}(q_1, \dots, q_s)$ satisfy the compatibility conditions

$$\int_\Lambda F_{s+1}^{(N)}(q_1, \dots, q_s, q_{s+1})\, dq_{s+1} = F_s^{(N)}(q_1, \dots, q_s), \quad \int_\Lambda F_1^{(N)}(q_1)\, dq_1 = 1,$$

for all N and Λ.

To avoid this extremely undesirable phenomenon of tending to zero of the distribution functions, it is necessary to fix bounded domain Λ and only consider the Boltzmann–Grad limit.

4.5.5. The Boltzmann–Grad limit for fixed bounded domain Λ. Let us return to the relations (4.5.9) or, in a operator form, (4.5.13) and fix the domain Λ in them. In this section, it is useful to introduce the notations

$$\frac{Q(N - i - 1, \Lambda)}{Q(N - i, \Lambda)} = a_i(N), \quad N - 3 \geqslant i \geqslant 0, \quad a_0(N) \equiv a(N), \tag{4.5.22}$$

where the dependence on Λ is omitted because Λ is fixed.

The sequence $a_i(N)$ is bounded according to (4.5.16') (for fixed i). Therefore, one can select a convergent subsequence $a_i(N_j)$

$$\lim_{N_j \to \infty} a_i(N_j) = A_i, \quad A_0 \equiv A.$$

By using the diagonal procedure, one can do it for all i. We conclude that the following limits exist

$$\lim_{N_j \to \infty} a_i(N_j) = A_i \tag{4.5.23}$$

for all $i = 0, 1, 2, \ldots$. Of course, this does not mean, that there are no other convergent subsequences with limits $A_i^{(1)}, \ldots, A_i^{(k)}, \ldots, i = 0, 1, 2, \ldots$. Let restrict ourselves to the subsequence $a_i(N_j)$ (4.5.23). By using this notation, relations (4.5.13) for $F^{(N-i)}$ take the form

$$F^{(N-i)} = a_i(N)\left(K^{(N-i)}F^{(N-i-1)} + F_0\right), \quad i = 0, 1, 2, \ldots. \tag{4.5.24}$$

Let formally pass to the limit as $N_i \to \infty$ in (4.5.24), assuming that all required limits exist. Then we obtain

$$\begin{aligned} F &= A\left(KF^1 + F_0\right) \\ F^i &= A_i\left(KF^{i+1} + F_0\right) \quad i = 1, 2, \ldots \end{aligned} \tag{4.5.25}$$

where

$$F^i = \lim_{N_j \to \infty} F^{(N_j - i)}, \quad K = \lim_{N_j \to \infty} K^{(N_j - i)}, \quad F^0 = F. \tag{4.5.26}$$

Determine the operator K. To do this, we represent the operators $K^{(N)}$ in the following form

$$K^{(N)} = K_1^{(N)} + K_2^{(N)},$$

where

$$\begin{aligned} \left(K_1^{(N)}f\right)_s((q)_s) &= \mathcal{X}_\Lambda((q)_s)\exp\left\{-\beta\sum_{i=1}^{s}\Phi(q_1 - q_i)\right\}f_{s-1}\left((q)_s^1\right) \\ \left(K_2^{(N)}f\right)_s((q)_s) &= \mathcal{X}_\Lambda((q)_s)\exp\left\{-\beta\sum_{i=1}^{s}\Phi(q_1 - q_i)\right\}\sum_{k=1}^{N-s}\frac{1}{k!}\prod_{j=0}^{k-1}\frac{\alpha_{s+j}}{a^2} \\ &\quad\times\int_{\Lambda^k}\prod_{i=1}^{k}\varphi_{q_1}(y_i)\,f_{s-1+k}\left((q)_s^1, (y)_k\right)d(y)_k, \\ &\quad f \in E_\xi, f_0 = 0 \\ \left(K_1^{(N)}f\right)_s((q)_s) &= 0, \quad s > N, \\ \left(K_2^{(N)}f\right)_s((q)_s) &= 0, \quad s > N - 1. \end{aligned} \tag{4.5.27}$$

The following estimates are obvious

$$\left\| K_1^{(N)} \right\| \leqslant 1,$$

$$\left\| K_2^{(N)} \right\| \leqslant \xi^{-1} \sum_{k=1}^{N-s} \frac{1}{k!} \left(\frac{4}{3}\pi\, a\xi\alpha\right)^k \leqslant \xi^{-1} \left(\exp\left\{\frac{4}{3}\pi\, a\xi\, \alpha\right\} - 1\right),$$

which implies that the operator $K_2^{(N)}$ tends to zero in norm as $N \to \infty$, $a \to 0$, $Na^2 = \alpha = \text{const}$ (in the Boltzmann–Grad limit)

$$\lim_{\substack{N\to\infty, a\to 0\\ Na^2=\alpha}} \left\| K_2^{(N)} \right\| = 0.$$

It is obvious that operators $K_2^{(N-i)}$ tends to zero in norm as $N \to \infty$, $a \to 0$, $Na^2 = \alpha = \text{const}$ uniformly with respect to $i \geq 0$. Taking into account that for $|q_1 - q_i| > 0, \quad i = 2, \ldots, s$

$$\lim_{a\to 0} \exp\left\{-\beta \sum_{i=2}^{s} \Phi\left(q_1 - q_i\right)\right\} = 1,$$

one obtains that the operator K acts in E_ξ in the following way

$$(Kf)_s\left((q)_s\right) = \mathcal{X}_\Lambda\left((q)_s\right) f_{s-1}\left((q)_s^1\right). \tag{4.5.28}$$

For the sake of simplicity, we require that numbers A_i satisfy the conditions $A_i < 1$. It is easy to see from (4.5.16) that this condition is satisfied for sufficiently large $V\left(\Lambda\right)$ and sufficiently small $N\frac{4}{3}\pi a^3 = \alpha\frac{4}{3}a$.

Now we are able to formulate the problem, which should be solved in order to justify the existence and uniqueness of limit (4.5.26). For this, it is necessary

1) *To prove that the limits A_i, F^i (4.5.26) exist in certain sense.*
2) *To prove that they satisfy relations (4.5.25).*
3) *To show that the limits A_i, F^i do not depend on i*

$$A_i = A, \quad F^i = F, \quad i = 1, 2, \ldots.$$

The similar problem arose in the paper [BPK] devoted to justification of the existence of the thermodynamic limit for equilibrium distribution functions in framework of canonical ensemble. Therefore, we will solve these problems by modification of methods developed in [BPK] (see also monograph [PGM1]).

By using relations (4.5.24), (4.5.25) we get

$$F^{(N)} = \sum_{i=1}^{N-3} a(N)\, K^{(N)} a_1(N)\, K^{(N-1)} \dots a_{i-1}(N)\, K^{(N-i+1)} a_i(N)\, F_0 + a(N)\, F_0$$
$$+ a(N)\, K^{(N)} a_1(N)\, K^{(N-1)} \dots a_{N-3}(N)\, K^{(3)} F^{(2)},$$
$$F = \sum_{i=1}^{\infty} AKA_1K \dots A_{i-1}KA_iF_0 + AF_0. \qquad (4.5.29)$$

The following estimate holds

$$\left\| \mathcal{X}_{a_0} \left(F - F^{(N)} \right) \right\| \leqslant \Bigg\| \mathcal{X}_{a_0} \sum_{i=1}^{n_0} AKA_1K \dots A_{i-1}KA_iF_0 + \mathcal{X}_{a_0} AF_0$$
$$- \sum_{i=1}^{n_0} a(N)\, K^{(N)} a_1(N)\, K^{(N-1)} \dots a_{i-1}(N)\, K^{(N-i+1)}$$
$$\times\, a_i(N)\, F_0 + a(N)\, F^0 \Bigg\| + R(n_0), \qquad (4.5.30)$$

where a_0 is an arbitrarily fixed number and $\mathcal{X}_a$ is an operator in the function space E_ξ defined according to the formula

$$(\mathcal{X}_a f)_s\left((q)_s\right) = \mathcal{X}_a((q)_s) f_s\left((q)_s\right), \quad f \in E_\xi$$

where $\mathcal{X}_a\left((q)_s\right)$ is characteristic function of admissible configurations, *i.e.* $|q_i - q_j| \geqslant a$ for all $(i,j) \in (1,\dots,s)$. It is obvious that $\|\mathcal{X}_a\| = 1$

The series (4.5.29) representing $F^{(N)}$ and F converge in the norm of the space E_ξ if $\left| a_i^{(N)} \right| \| K^{(N-i)} \| < k < 1$, $| A_i | \| K \| < k < 1$ and, therefore, for sufficiently large n_0, $\| R(n_0) \| < \frac{\varepsilon}{2}$, where ε is as small as desired. These inequalities for $\| K^{(N-i)} \|$, $\| K \|$ hold if $\xi^{-1} e^{\frac{4}{3}\pi a\xi\alpha} < k$, i.e., if $\xi^{-1} < k$ and a is sufficiently small.

Substituting $K^{(N)} = K_1^{(N)} + K_2^{(N)}$ into (4.5.30), taking into account that $a_i(N) \to A_i$, $\left\| K_2^{(N)} \right\| \to 0$ as $N \to \infty$, $a \to 0$, $Na^2 = \alpha$ and, moreover,

$$\mathcal{X}_{a_0} AK_1^{(N)} A_1 K_1^{(N-1)} \dots A_{i-1} K_1^{(N-i+1)} A_i F_0 = \mathcal{X}_{a_0} AKA_1K \dots A_{i-1}KA_iF_0,$$

we conclude that the first term in (4.5.30) can also be made less than $\frac{\varepsilon}{2}$ for sufficiently large N. This means that

$$\lim_{\substack{N\to\infty, a\to 0\\ Na^2=\alpha}} \left\| \mathcal{X}_{a_0}\left(F - F^{(N)}\right)\right\| = 0.$$

Note that $F_s\big((q)_s\big) = \text{const} \neq 0$ if $q_i \in \Lambda$, $i = 1, \ldots, s$ for arbitrary $(q)_s$, but $F_s^{(N)}\big((q)_s\big) = 0$ on forbidden configurations $\mathbb{W}_s$. This is the reason why we consider the difference $\mathcal{X}_{a_0}\left(F - F^{(N)}\right)$ but not $F - F^{(N)}$.

Let us show that $A_i = A$, $F^i = F$. For this purpose, we note that (4.5.25), (4.5.28) yield

$$F_1(q_1) = A, \quad F_1^{(1)}(q_1) = A_1, \ldots F_1^{(i)}(q_1) = A_i, \ldots, \qquad q_1 \in \Lambda.$$

We also have the normalization conditions

$$\int\limits_{\Lambda} F_1(q_1)\, dq_1 = 1, \cdots \int\limits_{\Lambda} F_1^{(i)}(q_1)\, dq_1 = 1, \ldots,$$

which yield

$$A = A_1 = \cdots = A_i = \cdots = \frac{1}{V(\Lambda)}.$$

This means that relations (4.5.25) reduce to a single equation

$$F = A\big(KF + F_0\big) \tag{4.5.31}$$

and

$$F = F^1 = \cdots = F^i = \ldots.$$

The proof of equalities $A_i^{(1)} = \cdots = A_i^{(k)} = \cdots = A$ is analogous.

The above results can be summarized in the following theorem:

Theorem 4.5.2. *Sequences $F^{(N-i)}$, which satisfy relations (4.5.24), converge in Boltzmann-Grad limit $N \to \infty$, $a \to 0$, $Na^2 = \alpha = const$ and with fixed domain Λ to a sequence F, which satisfy equation (4.5.31), in the following sense*

$$\lim_{\substack{N\to\infty, a\to 0\\ Na^2=\alpha}} \left\| \mathcal{X}_{a_0}\left(F - F^{(N-i)}\right)\right\| = 0$$

where a_0 is arbitrary small fixed number.

Remark that sequence F can be represented, according to (4.5.29), by a series convergent in norm in E_ξ:

$$F = \sum_{i=0}^{\infty} (AK)^i AF_0$$

and from this representation it immediatly follows that sequence F satisfies equation (4.5.31).

4.5.6. The existence of the Boltzmann–Grad limit and thermodynamic limit for standardly normalized distribution functions on numbers of particles. Consider the equilibrium distribution functions with normalization (4.5.4). They are defined according to the formulae

$$F_s^{(N)}(q_1, \ldots, q_s) = N(N-1)\ldots(N-s+1) \times \int\limits_{\Lambda^{N-s}} D^{(N)}\big(q_1, \ldots, q_s, q_{s+1}, \ldots, q_N\big) dq_{s+1} \ldots dq_N \tag{4.5.32}$$

and satisfy the following Kirkwood–Salsburg relations

$$F_s^{(N)}(q_1, \ldots, q_s) = \frac{NQ(N-1,\Lambda)}{Q(N,\Lambda)} \exp\left\{-\beta \sum_{i=2}^{s} \Phi(q_1 - q_i)\right\} \Bigg[F_{s-1}^{(N-1)}(q_2, \ldots, q_s) + \sum_{k=1}^{N-s} \frac{1}{k!} \int\limits_{\Lambda^k} \prod_{i=1}^{k} \varphi_{q_1}(y_i)\, F_{s-1+k}^{(N-1)}\big(q_2, \ldots, q_s, y_1, \ldots, y_k\big) dy_1 \ldots dy_k\Bigg],$$

$$1 < s < N,$$

$$F_1^{(N)}(q_1) = \frac{NQ(N-1,\Lambda)}{Q(N,\Lambda)} \Bigg[1 + \sum_{k=1}^{N-1} \frac{1}{k!} \times \int\limits_{\Lambda^k} \prod_{i=1}^{k} \varphi_{q_1}(y_i)\, F_k^{(N-1)}(y_1, \ldots, y_k)\, dy_1 \ldots dy_k\Bigg],$$

$$F_N^{(N)}(q_1, \ldots, q_N) = \frac{NQ(N-1,\Lambda)}{Q(N,\Lambda)} \exp\left\{-\beta \sum_{i=2}^{N} \Phi(q_1 - q_i)\right\} \times F_{N-1}^{(N-1)}(q_2, \ldots, q_N). \tag{4.5.33}$$

The distribution functions are normalized on numbers of particles.

Define by $a(N,\Lambda)$ the ratio

$$a(N,\Lambda)=\frac{NQ(N-1,\Lambda)}{Q(N,\Lambda)}$$

and introduce the renormalized distribution functions

$$\widetilde{F}_s^{(N)}(q_1,\ldots,q_s)=a^{2s}F_s^{(N)}(q_1,\ldots,q_s). \tag{4.5.34}$$

(We stress that the value $a(N,\Lambda)$ (4.5.34) differs by the multiplier N from the value $a(N,\Lambda)$, defined according to (4.5.9)).

They satisfy the following Kirkwood–Salsburg relations (in what follows, for the sake of simplicity, we preserve the previous notation for the renormalized distribution functions)

$$F_s^{(N)}(q_1,\ldots,q_s)=a(N,\Lambda)\,a^2\exp\left\{-\beta\sum_{i=2}^{s}\Phi(q_1-q_i)\right\}\Bigg[F_{s-1}^{(N-1)}(q_2,\ldots,q_s)$$
$$+\sum_{k=1}^{N-s}\frac{1}{k!}\frac{1}{a^{2k}}\times\int_{\Lambda^k}\prod_{i=1}^{k}\varphi_{q_1}(y_i)\,F_{s-1+k}^{(N-1)}(q_2,\ldots,q_s,y_1,\ldots,y_k)dy_1\ldots dy_k\Bigg],$$
$$N>s>1,$$
$$F_1^{(N)}(q_1)=a(N,\Lambda)\,a^2\Bigg[1+\sum_{k=1}^{N-1}\frac{1}{k!}\frac{1}{a^{2k}}$$
$$\times\int_{\Lambda^k}\prod_{i=1}^{k}\varphi_{q_1}(y_i)\;F_k^{(N-1)}(y_1,\ldots,y_k)\,dy_1\ldots dy_k\Bigg],$$
$$F_N^{(N)}(q_1,\ldots,q_N)=a(N,\Lambda)\,a^2\exp\left\{-\beta\sum_{i=2}^{N}\Phi(q_1-q_i)\right\}F_{N-1}^{(N-1)}(q_2,\ldots,q_N). \tag{4.5.35}$$

Let us represent relations (4.5.35) in an operator form.

It is useful to extract the term with $k=0$ and, by analogy with (4.5.27), to define the operators $K_1^{(N)}$ and $K_2^{(N)}$ in E_ξ as follows

$$\begin{aligned}
&\left(K_1^{(N)} f\right)_s ((q)_s) = \mathcal{X}_\Lambda((q)_s) \exp\left\{-\beta \sum_{i=2}^{s} \Phi(q_1 - q_i)\right\} f_{s-1}((q)_s^1), \\
&\qquad N \geqslant s \geqslant 1, \\
&\left(K_1^{(N)} f\right)_s ((q)_s) = 0, \qquad s > N, \\
&\left(K_2^{(N)} f\right)_s ((q)_s) = \mathcal{X}_\Lambda((q)_s) \exp\left\{-\beta \sum_{i=2}^{s} \Phi(q_1 - q_i)\right\} \sum_{k=1}^{N-s} \frac{1}{k!} \frac{1}{a^{2k}} \\
&\qquad\qquad \times \int_{\Lambda^k} \prod_{i=1}^{k} \varphi_{q_1}(y_i)\, f_{s-1+k}\left((q)_s^1, (y)_k\right) d(y)_k, \\
&N-1 \geqslant s \geqslant 1, \\
&\left(K_2^{(N)} f\right)_s ((q)_s) = 0, \qquad s > N-1.
\end{aligned} \tag{4.5.36}$$

The following estimates for the norms of these operators are obvious

$$\left\| K_1^{(N)} \right\| = \xi^{-1}, \quad \left\| K_2^{(N)} \right\| \leqslant \xi^{-1} \sum_{k=1}^{N-s} \frac{1}{k!} \left(\frac{4}{3}\pi a \xi\right)^k. \tag{4.5.37}$$

Relations (4.5.35) can be written as the single operator equation

$$\begin{aligned}
F^{(N)} &= a(N,\Lambda)\, a^2 \left[\left(K_1^{(N)} + K_2^{(N)}\right) F^{(N-1)} + F_0\right], \\
&\vdots \\
F^{(N-i)} &= a(N-i,\Lambda)\, a^2 \left[\left(K_1^{(N-i)} + K_2^{(N-i)}\right) F^{(N-i-1)} + F_0\right], \\
&\vdots \\
F_0 &= (1, 0, \ldots).
\end{aligned} \tag{4.5.38}$$

We are faced with the following problem:

1) *To show that for* $N \to \infty$, $a \to 0$, $Na^2 = \alpha$, $V(\Lambda) \to \infty$, $\dfrac{N}{V(\Lambda)} = \dfrac{1}{v}$, *(i.e., in the Boltzmann–Grad limit and the thermodynamic limit (4.5.17)) there exist the limits of the sequences* $F^{(N-i)}$, *of the operators* $K_1^{(N-i)}$, $K_2^{(N-i)}$ *and of the numbers* $a(N-i)\,a^2$.

2) *To show that corresponding limits* F^i, K_1, K_2, A_i, $i = 0,1,2,\ldots$, $A_0 = A$, $F^0 = F$ *satisfy the equations*

$$F^i = A_i\left[(K_1 + K_2)F^{i+1} + F_0\right] = A_i\left(KF^{i+1} + F_0\right) \quad i = 0,1,2,\ldots. \tag{4.5.39}$$

4.5.7. The proof of the existence of the limits of $F^{(N)}$, $K_1^{(N)}$, $K_2^{(N)}$ and $a(N,\Lambda)a^2$. It follows from the estimate (4.5.37) that the operator $K_2^{(N)}$ tends to zero in norm in limit (4.5.17), $K_2 = 0$.

Define the operators K_1 and $\mathcal{X}_a$, $\mathcal{X}_\Lambda$ in E_ξ as follows

$$\begin{aligned}
(K_1 f)_s\left((q)_s\right) &= f_{s-1}\left((q)_s\right),\\
(\mathcal{X}_a f)_s\left((q)_s\right) &= \mathcal{X}_a\left((q)_s\right) f_s\left((q)_s\right),\\
(\mathcal{X}_\Lambda f)_s\left((q)_s\right) &= \mathcal{X}_\Lambda\left((q)_s\right) f_s\left((q)_s\right),
\end{aligned} \tag{4.5.40}$$

where $\mathcal{X}_a\left((q)_s\right)$ is the characteristic function of admissible configurations.

Since for $|q_i - q_j| > 0$, $(i,j) \in (1,\ldots,s)$

$$\lim_{a\to 0} \mathcal{X}_a\left((q)_s\right) = 1,$$

$$\lim_{a\to 0} e^{-\beta\sum_{i=2}^{s}\Phi(q_1 - q_i)} = 1,$$

the operator $K_1^{(N)}$ converges to the operator K_1 in the following sense (in the limit indicated by (4.5.17))

$$\left\| \mathcal{X}_a\mathcal{X}_\Lambda\left(K_1 - K_1^{(N)}\right) f \right\| \to 0 \tag{4.5.41}$$

for arbitrary finite sequence $f \in E_\xi$, *i.e.* $f_s\left((q)_s\right) = 0$ for $s > n$, where n is some number.

Consider the sequence $a(N-i,\Lambda)a^2$. The following estimate holds

$$\begin{aligned}
a(N-i,\Lambda)a^2 &< \frac{(N-i)a^2}{V(\Lambda) - (N-i-1)\frac{4}{3}\pi a^3} \leqslant \frac{Na^2}{V(\Lambda) - Na^2\frac{4}{3}\pi a}\\
&= \frac{\alpha}{V(\Lambda) - \alpha\frac{4}{3}\pi a} = \frac{1}{v}a^2\frac{1}{1 - \frac{1}{v}a^2\frac{4}{3}\pi a},
\end{aligned} \tag{4.5.42}$$

which implies that, for fixed bounded $Na^2 = \alpha$ and $V(\Lambda) \to \infty$, the value $a(N-i,\Lambda) \to 0$, i.e. $a(N-i,\Lambda) \to 0$ in the limit (4.5.17).

Repeating the proof of the first section, one obtains that $\left\| F^{(N-i)} \right\| \to 0$ in limit (4.5.17).

The above results can be summarized in the following theorem:

Theorem 4.5.3. *The renormalized distribution functions (4.5.39), (4.5.34) tend to zero, i.e.* $\left\| F^{(N-i)} \right\| \to 0$, *in the Boltzmann–Grad limit and the thermodynamic limit.*

It is obvious that one can consider the fixed bounded domain Λ. In this case, the values $a(N-i,\Lambda)$ are bounded according to (4.5.42) and one can repeat the reasoning of the previous section and the theorem II is true for sequence of distribution functions defined according to (4.5.32) and (4.5.34).

4.5.8. The existence of the limit distribution functions for $Na^2/V = a^2/V =$ const. One can see from estimate (4.5.42) that the unique possibility to avoid the convergence of the distribution functions to zero consists in the substitution of the condition $\frac{N}{V}a^2 = \frac{1}{v}a^2 = \text{const}$ for the conditions $Na^2 = \alpha = \text{const}$ in the limit transition (4.5.17).

Now the Boltzmann–Grad limit and the thermodynamic limit mean that

$$N \to \infty, \quad \Lambda \nearrow \mathbb{R}^3, \quad V(\Lambda) \to \infty, \quad \frac{N}{V} = \frac{1}{v} \to \infty, \quad a \to 0, \quad \frac{1}{v}a^2 = \text{const}. \tag{4.5.43}$$

We put $\frac{1}{v}a^2 = \frac{N}{V(\Lambda)}a^2 < 1$. Then $a(N-i,\Lambda)\,a^2 < 1$, $A_i < 1$, for sufficiently small d. The following theorem is true.

Theorem 4.5.4. *The sequences $F^{(N-i)}$ and F^i exist and belong to E_ξ. The sequences $F^{(N-i)}$ converge to the sequences F^i in the limit (4.5.43) in the following sense*

$$\left\| \mathcal{X}_a \mathcal{X}_\Lambda \left(F^i - F^{(N-i)} \right) \right\| \to 0, \quad i = 0,1,2,\ldots$$

The sequences F^i satisfy relations (4.5.39).

Proof. Using (4.5.38) and (4.5.39) we get

$$F^{(N)} = \sum_{i=1}^{N-3} a(N,\Lambda)\,a^2K^{(N)}a(N-1,\Lambda)\,a^2K^{(N-1)}\dots a(N-i+1,\Lambda)\,a^2$$
$$\times K^{(N-i+1)}a(N-i,\Lambda)\,a^2F_0 + a(N,\Lambda)\,a^2K^{(N)}$$
$$\times a(N-1,\Lambda)\,a^2K^{(N-1)}\dots a(N-i+1,\Lambda)\,a^2$$
$$\times K^{(N-i+1)}\dots a(3,\Lambda)\,a^2K^3F^{(2)} + a(N,\Lambda)\,a^2F_0, \tag{4.5.44}$$

$$F = \sum_{i=1}^{\infty} AKA_1K\dots A_{i-1}KA_iF_0 + AF^0. \tag{4.5.45}$$

Since $\left\| K^{(N-i)} \right\| \leqslant \xi^{-1} e^{\frac{4}{3}\pi a\xi}$, $\| K \| = \xi^{-1}$, for sufficiently large ξ and sufficiently small a, one has $\left\| K^{(N-i)} \right\| < 1$ uniformly with respect to N and i (it suffices to consider $\xi > 1$). Moreover $a(N-i,\Lambda)\,a^2 < 1$, $A_i < 1$. Therefore, the series (4.5.44), (4.5.45) are convergent in the norm of E_ξ and the sequences $F^{(N)}$ and F exist and belong to E_ξ. The following estimate holds

$$\left\| \mathcal{X}_a\mathcal{X}_\Lambda\big(F - F^{(N)}\big) \right\| \leqslant \Big\| \sum_{i=1}^{n_0} \mathcal{X}_a\mathcal{X}_\Lambda AKA_1K\dots A_{i-1}KA_iF_0$$
$$- \sum_{i=1}^{n_0} \mathcal{X}_a\mathcal{X}_\Lambda a(N,\Lambda)\,a^2K^{(N)}a(N-1,\Lambda)\,a^2K^{(N-1)}$$
$$\dots a(N-i+1,\Lambda)\,a^2K^{(N-i+1)}a(N-i,\Lambda)\,a^2F_0$$
$$+ \mathcal{X}_\Lambda\big(AF^0 - a(N,\Lambda)\,a^2F_0\big) \Big\| + R(n_0), \tag{4.5.46}$$

where n_0 is some bounded number and $R(n_0)$ can be made as small as desired for sufficiently large n_0. In sum (4.5.46) with bounded n_0, we took the limit (4.5.43) (with fixed a^2/V). For this purpose, we represent all the operators $K^{(N-i)}$ in (4.5.46) as the sum $K^{(N-i)} = K_1^{(N-i)} + K_2^{(N-i)}$ and take into account that, according to (4.5.37), $\left\| K_2^{(N-i)} \right\| \to 0$ and the numbers $a(N-i,\Lambda) \to A_i$ in limit (4.5.43). Since the sequences $K^{(i)}F_0$ and $K^{(N)}K^{(N-1)}\dots K^{(N-i+1)}F_0$ are finite, using (4.5.41), we get that the first term in (4.5.46) tends to zero in the limit (4.5.43).

Thus, it is proved that $\mathcal{X}_a\mathcal{X}_\Lambda F^{(N)} \to \mathcal{X}_a\mathcal{X}_\Lambda F$ in norm in limit (4.5.43). The proof of the theorem for $F^{(N-i)}$ and $F^{(i)}, i = 1,2,\dots$ is the same. Show that

$$A = A_1 = \dots = A_i = \dots, \qquad F = F^1 = \dots = F^i = \dots.$$

To do this, we use the relations

$$F_1(q_1) = A, \quad F_1^1(q_1) = A_1, \dots F_1^i(q_1) = A_i, \dots \tag{4.5.47}$$

and the normalization conditions

$$\frac{1}{V(\Lambda)} \int\limits_{\Lambda} F_1(q_1)\, dq_1 = \frac{1}{v} a^2, \dots \frac{1}{V} \int\limits_{\Lambda} F_1^i(q_1)\, dq_1 = \frac{1}{v} a^2, \dots . \tag{4.5.48}$$

It follows from relations (4.5.47), (4.5.48) that

$$A = A_1 = \dots = A_i = \frac{1}{v} a^2$$

and, consequently,

$$F = F^1 = \dots = F^i = \dots .$$

The sequence F satisfies equation

$$F = A\big(KF + F_0\big).$$

It follows from representation (4.5.45) with $A_i = A$, $i > 1$. It remains to consider the case where the distribution functions are not subjected to the scale transformation.

4.5.9. The proof of the existence of the limit for $F^{(N)}$ without the scale transformation. In this case, the proof is almost the same as in the previous subsection. Estimates (4.5.37) are valid and it follows from them that $\left\| K_2^{(N)} \right\| \to 0$ in the limit (4.5.17). The differences will arise only for numbers $a(N, \Lambda)$, which satisfy the inequalities

$$a(N - i, \Lambda) < \frac{N}{V(\Lambda) - N\frac{4}{3}\pi a^3} = \frac{1}{v} \frac{1}{1 - \frac{1}{v}\frac{4}{3}\pi a^3}$$

and, therefore, are bounded for bounded density. According to this, one can select a convergent subsequence from them. All further proof for $\frac{1}{v} < 1$ is analogous to that performed in the previous section. The proof of the estimate (4.5.46) and of the uniqueness of the limit distribution functions are the same.

CHAPTER V

On the Derivation of Kinetic Equations from the BBGKY Hierarchy

5.1. Introduction: Kinetic Equations

In Chapter IV we have considered one method of rigorous derivation of the Boltzmann equation from the BBGKY hierarchy, namely the Boltzmann-Grad limit method. The aim of the present chapter is to develop a general approach to the problem of the derivation of kinetic equations.

A description of the evolution of infinite particle systems in the framework of kinetic theory is based on the hypothesis that the states of these systems depend on time through the probability density of one particle (known as the one-particle distribution function), satisfying a certain evolution equation called the kinetic equation.

It is obvious, however, that such hypothesis can not be valid under all circumstances since all possible states are described by infinite sequences of s-particle distribution functions. Some formal features of the system and the initial data which allow to ensure the validity of a description of the evolution by means of kinetic equations can already be given from the qualitative analysis of the solutions of the BBGKY hierarchy described in Chapters II and III.

In both representations of the solutions by the functional (2.3.3) and the iteration expansions (2.5.2), the first term is

$$F_s(t, x_1, \ldots, x_s) = S^s(-t, x_1, \ldots, x_s)\, F_s(0, x_1, \ldots, x_s) + o\left(\frac{1}{v}\right), \qquad (5.1.1)$$

and as it has been established above for $F(0) \in L^\infty$, the condition for the validity of these expansions in dimensionless variables was the following:

$$t < t_0 \equiv \left(\frac{r_0^3}{v}\right)^{-1} \frac{r_0}{u_0},$$

where r_0 is a typical microscopic length scale (for example, the radius of the interaction potential of finite range), r_0^3 / v the dimensionless density, u_0 the thermal speed ($u_0 = \beta^{-\frac{1}{2}}$ and β the inverse temperature).

It should be pointed out that the coefficient $\frac{r_0}{u_0} \equiv \tau_c$ can be interpreted as the collision (interaction) time, and since $\frac{v}{r_0^2} = \lambda$ is the mean free path, then $\frac{r_0^3}{v} = \frac{r_0}{\lambda}$. By introducing the parameter $\tau_\lambda \equiv \frac{\lambda}{u_0}$ - time between collisions (mean free time), one obtains $\frac{r_0^3}{v} = \frac{r_0}{\lambda} = \frac{\tau_c}{\tau_\lambda}$ and as a result $t_0 \equiv \left(\frac{r_0^3}{v}\right)^{-1} \frac{r_0}{u_0} = \tau_\lambda$. Therefore, the approximation (5.1.1) is valid as long as the time is less than that between collisions.

To analyse the properties of functions (5.1.1), one should indicate some dynamical features of a system of s particles. We restrict ourselves to the consideration of systems with a short-range repulsive interaction potential ($r_0 < +\infty$) (for example, a system of particles interacting like hard spheres of radius r_0). In such systems the particles move as non-interacting ones if the mutual distances between them are greater than r_0. Therefore the momenta of particles $S^s(-t)\,p_i$, $i \in (1, \ldots, s)$ approach certain limiting momenta (the effective duration of this process is of the order $\tau_c = \frac{r_0}{u_0}$)

$$\lim_{t \to +\infty} S^s(-t)\,p_i = P_i^\infty(x_1, \ldots, x_s)\,, \quad i \in (1, \ldots, s) \tag{5.1.2}$$

and the following relations hold

$$\lim_{t \to +\infty} \int_0^t d\tau\,(P_i^\infty - S^s(-\tau)\,p_i) = \int_0^\infty d\tau\,(P_i^\infty - S^s(-\tau)\,p_i)\,,$$

$$\lim_{t \to +\infty} t\,(P_i^\infty - S^s(-t)\,p_i) = 0.$$

These relations have as a consequence the validity of the following ones

$$\lim_{t \to +\infty} (S^s(-t)q_i + tP_i^\infty) = \lim_{t \to +\infty} \Big(q_i + \int_0^t d\tau\,(P_i^\infty - S^s(-\tau)p_i)\Big) = Q_i^\infty,$$

$$\lim_{t \to +\infty} (S^s(-t)q_i + tS^s(-t)p_i) = Q_i^\infty, \tag{5.1.3}$$

where

$$Q_i^\infty \equiv q_i + \int_0^\infty d\tau\,(P_i^\infty - S^s(-\tau)p_i)\,. \tag{5.1.4}$$

Let us now consider the formal properties of the initial data $F_s(0)$ in Eq. (5.1.1). At time $t = 0$, let particles be found in the phase states $S^s(-\tau)x_i$, $i \in$

$(1,\ldots,s)$, where $\tau_c \ll \tau \ll \tau_\lambda$ (if $\frac{r_0^3}{v} \ll 1$ it is correct because $\tau_\lambda = \left(\frac{r_0^3}{v}\right)^{-1}\tau_c$). In this case at initial instant and during a long period, compared with τ_c, before this instant the particles are not at the distances less than r_0. And this means that the correlations among them can be neglected, thus

$$\lim_{\tau\to+\infty}\Big(F_s(0,S^s(-\tau)x_1,\ldots,S^s(-\tau)x_s) - \prod_{i=1}^{s} F_1(0,S^s(-\tau)x_i)\Big)$$

$$= \lim_{\tau\to+\infty} S^s(-\tau)\Big(F_s(0,x_1,\ldots,x_s) - \prod_{i=1}^{s} F_1(0,x_i)\Big) = 0, \quad s \geqslant 2. \quad (5.1.5)$$

Let us suppose that the initial data in Eq. (5.1.1) satisfy the condition of weakening of correlations (5.1.5) or that the initial data possess the factorization property: $F_s(0,x_1,\ldots,x_s) = \prod_{i=1}^{s} F_1(0,x_i)$.

Thus, according to (5.1.2-3), (5.1.5) for the first approximation (5.1.1) of the cluster expansions (2.3.3), (2.5.2) it holds

$$\lim_{t\to+\infty}\Big(F_s(t,x_1,\ldots,x_s) - \prod_{i=1}^{s} F_1(t,Q_i^\infty,P_i^\infty)\Big)$$

$$= \lim_{t\to+\infty}\Big(S^s(-t)F_s(0,x_1,\ldots,x_s) - \prod_{i=1}^{s} F_1(0,Q_i^\infty - tP_i^\infty,P_i^\infty)\Big)$$

$$= \lim_{t\to+\infty} S^s(-t)\Big(F_s(0,x_1,\ldots,x_s) - \prod_{i=1}^{s} F_1(0,x_i)\Big)$$

$$+ \prod_{i=1}^{s} \lim_{t\to+\infty}\Big(F_1(0,S^s(-t)q_i,S^s(-tp_i)) - F_1(0,Q_i^\infty - tP_i^\infty,P_i^\infty)\Big) = 0,$$

$$s \geqslant 2.$$

It is obvious from these considerations that, already in the first term (5.1.1) of the cluster expansions (2.3.3), (2.5.2), the s-particle function converges to the expression $\prod_{i=1}^{s} F_1(t,Q_i^\infty,P_i^\infty)$ when time tends to infinity (the effective duration of this process is the order τ_c).

Hence, under the conditions imposed on a system of particles the stated hypothesis can be really justified.

Concluding our analysis we shall make some observations with respect to expression (5.1.1) in the case $s = 1$. Because the operator $S^{-1}(-t, x_1)$ is the evolution operator of a free particle, the function $F_1(t, x_1) = S^1(-t, x_1)F_1(0, x_1)$ is not essentially changed during a time interval of order τ_c, (for example, the distribution of momentum coincides with initial one) and in order to describe its change in time one needs the higher terms of the expansion (5.1.1). This means that the changing process of $F_1(t)$ has the effective duration of order $\left(\frac{r_0^3}{v}\right)^{-1} \tau_c = \tau_\lambda$ (mean free time).

Thus, under the evolution of the considered system one is faced with a situation which can be interpreted in the following way. There exist two kinds of evolutionary processes: a "fast" one with a duration of order τ_c and a "slow" one with a duration of order τ_λ (the kinetic stage of the evolution, where all distribution functions depend on the one-particle distribution function). Such interpretation, as is well known, has been proposed by Bogolubov [Bo]. We note, that analogous problems are present in the theory of differential equations with small parameter in nonlinear mechanics [BM!!!??], namely one has to deal with a problem of constructing the integral manifold.

In connection with what has been said there arose one of the most interesting and challenging problem of deriving a kinetic equation starting from BBGKY hierarchy. In what way such program can be realized is the subject of the present chapter.

There are two successful approaches to the problem of the derivation of the kinetic equations from the BBGKY hierarchy. They are essentially based on the two different methods of constructing the hierarchy's solutions by either the iteration or the functional (a cluster expansion) series (2.3.3), (2.5.2). The first approach was developed in Bogolubov's paper [Bo], the second one was established in Cohen's and Green's papers [Co1][Co2][Co3][Co4][Co5][Co6][Co7][Gr1][Gr2][GPi]. We have to note that the rigorous justification of these methods was lacking till now.

The next two sections contain a formal (algebraic) discussion of the above-mentioned methods of deriving the kinetic equations. The relations between them are investigated. Then in the final section 5.4 we present a rigorous derivation of the most general kinetic equation. Namely, we prove that the Cauchy problem for the BBGKY hierarchy, when the initial data in L^1 possess a certain factorization property is identical to the corresponding initial value problem in L^1 for a certain generalized kinetic equation which has an unique solution. A form of the Boltzmann collision integral, which is valid in L^1 to all orders of the density, is described.

5.2 Bogolubov's Method of Constructing Kinetic Equations

5.2.1 The special solution of the BBGKY hierarchy. Bogolubov proposed the following approach to the derivation of the Boltzmann equation.

Consider the BBGKY hierarchy (1.4.7)

$$\frac{\partial}{\partial t}F_s(t,x_1,\ldots,x_s) = \{\, H^s, F_s(t,x_1,\ldots,x_s)\,\}$$
$$+\frac{1}{v}\int dx_{s+1}\Big\{\sum_{i=1}^{s}\Phi(q_i-q_{s+1}), F_{s+1}(t,x_1,\ldots,x_{s+1})\Big\},\quad s\geqslant 1 \tag{5.2.1}$$

for a system of particles interacting via a repulsive potential with a short - range $r_0 < \infty$ (like hard spheres). The hierarchy in (5.2.1) is obtained from (1.4.7) by using new functions $F'_s(t,x_1,\ldots,x_s) = v^s F_s(t,x_1,\ldots,x_s)$. It is clear that in order to reduce the hierarchy to a kinetic equation one must construct such particular solution of hierarchy (5.2.1) which depends on the time variable only through the one-particle distribution function $F_1(t)$

$$F_s(t,x_1,\ldots,x_s) = F_s\big(x_1,\ldots,x_s \mid F_1(t)\big),\quad s\geqslant 2.$$

For brevity, we shall use the notation $F_s\big(\cdots \mid F_1(t)\big)$ for such functional $F_s\big(x_1, \ldots, x_s \mid F_1(t)\big)$. Then, by definition

$$\frac{\partial}{\partial t}F_s(\ldots \mid F_1(t)) = \int dx \frac{\delta F_s(\ldots \mid F_1(t))}{\delta F_1(t,x)}\frac{\partial F_1(t,x)}{\partial t},$$

where the time derivative of $F_1(t,x)$ can be re-expressed by using the right - hand side of the equations (5.2.1) for $s=1$.

Thus, to be a solution of the hierarchy (5.2.1), the functionals $F_s(\ldots \mid F_1(t))$, $s\geqslant 2$ have to satisfy the following hierarchy of functional equations

$$\int dx\,\frac{\delta F_s(\ldots \mid F_1(t))}{\delta F_1(t,x)}\,\{H^1(x), F_1(t,x)\} = \{H^s, F_s(\cdots \mid F_1(t))\}$$
$$+\frac{1}{v}\int dx_{s+1}\bigg(\Big\{\sum_{i=1}^{s}\Phi(q_i-q_{s+1}), F_{s+1}(\cdots \mid F_1(t))\Big\}$$
$$-\int dx\frac{\delta F_s(\ldots \mid F_1(t))}{\delta F_1(t,x)}\,\{\Phi(q-q_{s+1}), F_2(x,x_{s+1} \mid F_1(t))\}\bigg),$$
$$s\geqslant 2. \tag{5.2.2}$$

For the expression on the left-hand side of equality (5.2.2) the following representation (Bogolubov's transformation) holds

$$\int dx \, \frac{\delta F_s(\cdots \mid S^1(-\tau)F_1(t))}{\delta S^1(-\tau)F_1(t,x)} \, \{H^1(x), S^1(-\tau,x)F_1(t,x)\}$$
$$= \frac{\partial}{\partial \tau} F_s(\cdots \mid S^1(-\tau)F_1(t)), \quad s \geqslant 2, \qquad (5.2.3)$$

where $S^1(-\tau)$ is the evolution operator of a free particle. In fact, the identity (5.2.3) is the result of a direct calculation of the derivative with respect to the variable τ.

Since hierarchy (5.2.2) is a sequence of functional equations, one can make a change of variable from $F_1(t)$ to $S^1(-\tau)F_1(t)$ and, then, using the representation (5.2.3), we conclude that the functionals $F_s(\cdots \mid S^1(-\tau)F_1(t))$, $s \geqslant 2$ satisfy the following equations

$$\begin{aligned}
&\frac{\partial}{\partial \tau} F_s(\cdots \mid S^1(-\tau)F_1(t)) \\
&\quad = \{H^s, F_s(\cdots \mid S^1(-\tau)F_1(t))\} \\
&\quad + \frac{1}{v} \int dx_{s+1} \Big(\Big\{ \sum_{i=1}^{s} \Phi(q_i - q_{s+1}), F_{s+1}(\cdots \mid S^1(-\tau)F_1(t)) \Big\} \\
&\qquad - \int dx \, \frac{\delta F_s(\cdots \mid S^1(-\tau)F_1(t))}{\delta S^1(-\tau)F_1(t,x)} \qquad (5.2.4) \\
&\qquad \times \{\Phi(q - q_{s+1}), F_2(x, x_{s+1} \mid S^1(-\tau)F_1(t))\} \Big), \quad s \geqslant 2.
\end{aligned}$$

Let us try to find a solution of the functional hierarchy (5.2.4) in the form of of the following expansion :

$$F_s(\cdots \mid F_1(t)) = \sum_{n=0}^{\infty} \frac{1}{v^n} F_s^{(n)}(\cdots \mid F_1(t)), \quad s \geqslant 2. \qquad (5.2.5)$$

By substituting this formal series into Eq. (5.2.4) and equating the coefficients of equal powers of $1/v$, we obtain the following sequence of recursive equations:

$$\frac{\partial}{\partial \tau} F_s^{(0)}(\cdots \mid S^1(-\tau)F_1(t)) = \{H^s, F_s^{(0)}(\cdots \mid S^1(-\tau)F_1(t))\}, \qquad (5.2.6.a)$$

$$\vdots$$

$$\frac{\partial}{\partial \tau} F_s^{(n)}(\cdots \mid S^1(-\tau)F_1(t)) = \{H^s, F_s^{(n)}(\cdots \mid S^1(-\tau)F_1(t))\}$$
$$+ \Psi_s^{(n)}(\cdots \mid S^1(-\tau)F_1(t)), \quad n \geqslant 1, \qquad (5.2.6.b)$$

where we denoted by $\Psi_s^{(n)}$ the expressions containing functionals of lower order:

$$\Psi_s^{(n)}(\cdots \mid S^1(-\tau)F_1(t))$$
$$= \int dx_{s+1}\Bigg(\Big\{\sum_{i=1}^{s}\Phi(q_i - q_{s+1}), F_{s+1}^{(n-1)}(\cdots \mid S^1(-\tau)F_1(t))\Big\}$$
$$-\sum_{k=1}^{n}\int dx\, \frac{\delta F_s^{(n-k)}(\cdots \mid S^1(-\tau)F_1(t))}{\delta S^1(-\tau)F_1(t,x)} \tag{5.2.6.c}$$
$$\times \big\{\Phi(q - q_{s+1}), F_2^{(k-1)}(x, x_{s+1} \mid S^1(-\tau)F_1(t))\big\}\Bigg),$$

$n \geqslant 1, s \geqslant 2$.

Now, treating the functional $\Psi_s^{(n)}(\cdots \mid S^1(-\tau)F_1(t))$ as an inhomogeneous term, Eq. (5.2.6) is nothing else than an inhomogeneous (homogeneous just for $n = 0$) Liouville's equation. Eqs. (5.2.6.a) have the solutions:

$$F_s^{(0)}(\cdots \mid S^1(-\tau)F_1(t)) = S^s(-\tau)F_s^{(0)}(\cdots \mid F_1(t))$$

and, thus,

$$F_s^{(0)}(\cdots \mid F_1(t)) = S^s(-\tau)F_s^{(0)}(\cdots \mid S^1(\tau)F_1(t)). \tag{5.2.7}$$

To obtain a unique solution of Eq. (5.2.6) we need a corresponding boundary (or initial) conditions. Following Bogolubov's idea we choose as such condition a principle of weakening of correlations in the form

$$\lim_{\tau \to +\infty} S^s(-\tau)\Big(F_s(x_1, \dots, x_s \mid S^1(\tau)F_1(t)) - \prod_{i=1}^{s}\big(S^1(\tau)F_1(t)\big)(x_i)\Big) = 0,$$

or in terms of coefficients of the expansion (5.2.5)

$$\lim_{\tau \to +\infty} S^s(-\tau)\Big(F_s^{(0)}(x_1, \dots, x_s \mid S^1(\tau)F_1(t)) - \prod_{i=1}^{s}\big(S^1(\tau)F_1(t)\big)(x_i)\Big) = 0 \tag{5.2.8.a}$$

$$\lim_{\tau \to +\infty} S^s(-\tau)F_s^{(n)}(x_1, \dots, x_s \mid S^1(\tau)F_1(t)) = 0, \qquad n \geqslant 1. \tag{5.2.8.b}$$

(We remark that the weakening of correlations is by no means obvious. In fact, it is only for initially factored (and hence uncorrelated) initial data that we can assume that the random correlations which are produced by collisions tend not to persist.)

If one observes that the left-hand side of equality (5.2.7) does not depend on the parameter τ and this relation is valid for any τ, then, taking into account the condition (5.2.8.a), we obtain

$$F_s^{(0)}(x_1,\ldots,x_s \mid F_1(t)) = \lim_{\tau\to+\infty} S^s(-\tau)F_s^{(0)}(x_1,\ldots,x_s \mid S^1(\tau)F_1(t))$$
$$= \lim_{\tau\to+\infty} S^s(-\tau)\prod_{i=1}^{s}(S^1(\tau)F_1(t))(x_i).$$

Finally, introducing the operator

$$\widehat{S}_\infty^s(x_1,\ldots,x_s) = \lim_{\tau\to+\infty} S^s(-\tau,x_1,\ldots,x_s)\prod_{i=1}^{s} S^1(\tau,x_i), \tag{5.2.9}$$

one finds the expression for the first term of the expansion (5.2.5)

$$F_s^{(0)}(x_1,\ldots,x_s \mid F_1(t)) = \widehat{S}_\infty^s(x_1,\ldots,x_s)\prod_{i=1}^{s} F_1(t,x_i). \tag{5.2.10}$$

It should be remarked that, using the preceding notations (5.1.2-4) from Sec. 5.1, formula (5.2.10) takes on the form

$$F_s^{(0)}(x_1,\ldots,x_s \mid F_1(t)) = \prod_{i=1}^{s} F_1(t, P_i^\infty(x_1,\ldots,x_s), Q_i^\infty(x_1,\ldots,x_s)). \tag{5.2.10.a}$$

Analogously one can find other coefficients $F_s^{(n)}(\cdots \mid F_1(t))$ for $n \geqslant 1$ of the iteration series (5.2.5). From the formal solution of Eq. (5.2.6.b) it follows that

$$F_s^{(n)}(\cdots \mid F_1(t)) = S^s(-\tau)F_s^{(n)}(\cdots \mid S^1(\tau)F_1(t))$$
$$+\int_0^\tau d\tau' S^s(\tau'-\tau)\Psi_s^{(n)}(\cdots \mid S^1(-\tau'+\tau)F_1(t))$$

for an arbitrary value of the parameter τ. The change of the variable of integration $\tau'' = \tau - \tau'$ yields

$$F_s^{(n)}(\cdots \mid F_1(t)) = S^s(-\tau)F_s^{(n)}(\cdots \mid S^1(\tau)F_1(t))$$
$$+\int_0^\tau d\tau'' S^s(-\tau'')\,\Psi_s^{(n)}(\cdots \mid S^1(\tau'')F_1(t)), \tag{5.2.11}$$

where the functional $\Psi_s^{(n)}\big(\cdots \mid S^1(\tau'')F_1(t)\big)$ is defined by equation (5.2.6.c).

Moreover, taking into account the principle of weakening of correlations given by (5.2.8.b), we pass to the limit $\tau \to +\infty$ in equation (5.2.11). As a result, we obtain

$$F_s^{(n)}\big(\cdots \mid F_1(t)\big) = \int_0^\infty d\tau\, S^s(-\tau)\, \Psi_s^{(n)}\big(\cdots \mid S^1(\tau)F_1(t)\big), \quad n \geqslant 1. \tag{5.2.12}$$

This recursive relation implicitly defines the functionals $F_s^{(n)}\big(\cdots \mid F_1(t)\big)$, $n \geqslant 1$, and from it one can deduce consequently for $n = 1, 2, \ldots$ the explicit expression of every term of the expansion (5.2.5). As an example we exhibit the expression for the two particle functional, i.e. $s = 2$, in the simplest case $n = 1$. To make it more visible we use an abbreviated notations for the Poisson brackets $\{.,.\}$ as we did before in Chapter 1, namely

$$\mathcal{H}^n \equiv \mathcal{H}_0^n + \mathcal{H}_{\text{int}}^n \equiv \sum_{i=1}^n \mathcal{H}_0(x_i) + \sum_{1=i<j}^n \mathcal{H}_{\text{int}}(x_i, x_j), \tag{5.2.13}$$

where

$$\mathcal{H}^n \equiv \{H^n, .\} \quad , \quad \mathcal{H}_{\text{int}}(x_i, x_j) = \{\Phi(q_i - q_j), .\}.$$

Then, according to Eq. (5.2.6.c) and taking into account the definition (5.2.9) and the expression (5.2.10) for $F_s^{(0)}\big(\cdots \mid F_1(t)\big)$, we have

$$\begin{aligned} &F_2^{(1)}\big(x_1, x_2 \mid F_1(t)\big) \\ &\quad = \int dx_3 \int_0^\infty d\tau S^2(-\tau, x_1, x_2) \\ &\qquad \times \bigg(\Big(\mathcal{H}_{\text{int}}(x_1, x_3) + \mathcal{H}_{\text{int}}(x_2, x_3)\Big)\widehat{S}^3_\infty(x_1, x_2, x_3) \\ &\qquad\quad - \widehat{S}^2_\infty(x_1, x_2)\Big(\mathcal{H}_{\text{int}}(x_1, x_3)\widehat{S}^2_\infty(x_1, x_3) + \mathcal{H}_{\text{int}}(x_2, x_3)\widehat{S}^2_\infty(x_2, x_3)\Big)\bigg) \\ &\qquad \times S^1(\tau, x_1)S^1(\tau, x_2)S^1(\tau, x_3)F_1(t, x_1)F_1(t, x_2)F_1(t, x_3). \end{aligned} \tag{5.2.14}$$

In what follows (Subsect. 5.3.4) we shall transform this expression to another representation.

It is obvious that to complete the construction of the special solution (5.2.5) of the BBGKY hierarchy (5.2.1) it remains to satisfy the first equation for $s = 1$ of the hierarchy.

5.2.2 The general form of a kinetic equation. From the construction of the previous subsection 5.2.1 it follows that the functionals $F_s(\cdots \mid F_1(t))$, (5.2.5) satisfy the BBGKY hierarchy (5.2.1) if the function $F_1(t)$ satisfies the following kinetic equation

$$\frac{\partial}{\partial t}F_1(t,x_1) = -\left\langle p_1, \frac{\partial}{\partial q_1}\right\rangle F_1(t,x_1) + \frac{1}{v}\int dx_2 \left\{\Phi(q_1-q_2), F_2\big(x_1,x_2 \mid F_1(t)\big)\right\}. \tag{5.2.15}$$

Using the explicit form of the leading terms of series (5.2.5) the kinetic equation (5.2.15) can be also rewritten as follows

$$\frac{\partial}{\partial t}F_1(t,x_1) = -\left\langle p_1, \frac{\partial}{\partial q_1}\right\rangle F_1(t,x_1)$$
$$+\frac{1}{v}\int dx_2 \left\{\Phi(q_1-q_2), \widehat{S}^2_\infty(x_1,x_2)F_1(t,x_1)F_1(t,x_2)\right\} + o\left(\frac{1}{v}\right) \tag{5.2.16}$$

This kinetic equation is called the generalized Boltzmann equation.

In the expression of the collision integral appearing in Eq. (5.2.16) we have omitted the terms of higher order in the density $1/v$. But we point out their physical interpretation. Each term in the formal density expansion of the collision integral from the right - hand side of Eq. (5.2.16) describes pair, triple, etc. effects of interactions of particles (in the statistical sense). For example, the second order term of the collision integral determined by the expression (5.2.14) describes the effect of a triple interaction of particles.

When discussing the physical meaning of the generalized Boltzmann equation it should be indicated that this equation is irreversible in time. As a matter of fact, in the process of deriving a kinetic equation the principle of weakening of correlations in a certain form, namely (5.2.8), was used . As a result the operators $\widehat{S}^s_\infty$ defined in (5.2.9) appear in Eq. (5.2.16) and thus this equation is only valid for $t \geqslant 0$. For $t \leqslant 0$ we must exploit the principle of weakening of correlations in another form and as a result we obtain the kinetic equation which differs from (5.2.16) by the sign before collision integral.

Concluding this subsection we finally write out the generalized Boltzmann equation for a system of hard spheres. According to the definition of operators $\widehat{S}^s_\infty$ for hard spheres we have

$$\frac{\partial}{\partial t}F_1(t,x_1) = -\left\langle p_1, \frac{\partial}{\partial q_1}\right\rangle F_1(t,x_1) + \frac{1}{v}a^2 \int\limits_{\mathbb{R}^3\times\mathbb{S}^2_+} d\eta\, dp_2 \,\langle \eta, (p_1 - p_2)\rangle$$
$$\times \Big(F_1(t,q_1,p_1^*)F_1(t,q_1 - a\eta, p_2^*) - F_1(t,q_1,p_1)F_1(t,q_1 + a\eta, p_2)\Big)$$
$$+ o\left(\frac{1}{v}\right), \quad (5.2.17)$$

where we used the same notations as in Chapter 1. It is clear, that the first term of the collision integral precisely defines the Boltzmann-Enskog equation and the whole equation (5.2.17) formally corresponds to the Enskog equation.

5.2.3 Homogeneous Boltzmann equation. We begin by establishing some properties of the derived kinetic equation (5.2.15) mainly its relation to the classical representative of kinetic equations, namely, the Boltzmann equation.

Let us consider a space homogeneous system of particles. In this case the one- particle distribution function does not depend on space variables and

$$\widehat{S}^2_\infty(x_1,x_2) = \lim_{t\to+\infty} S^2(-t; p_1, p_2, q_1 - q_2).$$

Thus Eq. (5.2.15) in a first approximation takes on the form

$$\frac{\partial}{\partial t}F_1(t,p_1) = \frac{1}{v}\int dx_2 \left\{\Phi(q_1 - q_2), F_1(t,P_1^\infty)F_2(t,P_2^\infty)\right\}, \quad (5.2.18)$$

where

$$P_i^\infty = \widehat{S}^2_\infty(x_1,x_2)p_i = \lim_{\tau\to+\infty} S^2(-\tau; p_1,p_2,q_1-q_2)p_i, \quad i = 1,2. \quad (5.2.19)$$

We show that this equation coincides with the classical homogeneous Boltzmann equation.

In fact, from the definition (5.2.19) of the values P_1^∞, P_2^∞ it follows that they are integrals of motion

$$\left\{H_2(P_1^\infty,P_2^\infty), P_i^\infty\right\} = \left\{\widehat{S}^2_\infty(x_1,x_2)H_2(x_1,x_2), \widehat{S}^2_\infty(x_1,x_2)p_i\right\}$$
$$= \widehat{S}^2_\infty(x_1,x_2)\left\{H_2(x_1,x_2), p_i\right\} = 0, \quad i = 1,2,$$

where it was used that

$$H_2(x_1, x_2) \equiv \frac{1}{2}(p_1^2 + p_2^2) + \Phi(q_1 - q_2)$$
$$= \frac{1}{2}(P_1^\infty)^2 + \frac{1}{2}(P_2^\infty)^2 \equiv H_2(P_1^\infty, P_2^\infty). \qquad (5.2.20)$$

Because any function of integrals of motion is an integral of motion too, the following identity holds

$$\{H_2(x_1, x_2), F_1(t, P_1^\infty)F_1(t, P_2^\infty)\} = 0$$

from which we obtain

$$\{\Phi(q_1 - q_2), F_1(t, P_1^\infty)F_1(t, P_2^\infty)\} = -\Big\{\frac{1}{2}(p_1^2 + p_2^2), F_1(t, P_1^\infty)F_1(t, P_2^\infty)\Big\}.$$

In virtue of the fact that the momenta P_1^∞, P_2^∞ depend on q_1, q_2 only through the difference $q_1 - q_2$ the collision integral in Eq. (5.2.18) takes on the form

$$\int\limits_{\mathbb{R}^3\times\mathbb{R}^3} dx_2 \{\Phi(q_1 - q_2), F_1(t, P_1^\infty)F_1(t, P_2^\infty)\}$$
$$= \int\limits_{\mathbb{R}^3\times\mathbb{R}^3} dx_2 \Big\langle (p_2 - p_1), \frac{\partial}{\partial q_2} F_1(t, P_1^\infty)F_1(t, P_2^\infty)\Big\rangle.$$

We transform the last expression by introducing the cylindrical coordinates with origin at q_1 instead of the Cartesian ones for point q_2. We direct the third axis ξ along the vector $p_2 - p_1$. The radius and the polar angle are denoted by r and φ respectively. Then, it is clear that

$$\Big\langle (p_2 - p_1), \frac{\partial}{\partial q_2} F_1(t, P_1^\infty)F_1(t, P_2^\infty)\Big\rangle = |p_2 - p_1| \frac{\partial}{\partial \xi} F_1(t, P_1^\infty)F_1(t, P_2^\infty),$$

and, therefore,

$$\int\limits_{\mathbb{R}^3\times\mathbb{R}^3} dx_2 \Big\langle (p_2 - p_1), \frac{\partial}{\partial q_2} F_1(t, P_1^\infty)F_1(t, P_2^\infty)\Big\rangle$$
$$= \int\limits_{\mathbb{R}^3} dp_2\, |p_2 - p_1| \int\limits_0^\infty dr\, r \int\limits_0^{2\pi} d\varphi \int\limits_{-\infty}^{\infty} d\xi\, \frac{\partial}{\partial \xi} F_1(t, P_1^\infty)F_1(t, P_2^\infty). \qquad (5.2.21.a)$$

To integrate over the variable ξ into this expression we remark that according to definition (5.2.19)

$$\lim_{\xi \to -\infty} P_i^\infty(p_1, p_2; r, \varphi, \xi) = p_i,$$
$$\lim_{\xi \to +\infty} P_i^\infty(p_1, p_2; r, \varphi, \xi) = p_i^*(p_1, p_2, r, \varphi), \quad i = 1, 2.$$

Here, p_1^*, p_2^* are solutions of the corresponding initial value problem for the Hamilton equations of the two particle system. In the generally accepted terminology of the theory of gases we can say that p_1^*, p_2^* are momenta of particles before the collision expressed as functions of momenta p_1, p_2 after collision and also parameters r, φ. As a result of the integration we have

$$\int_{\mathbb{R}^3} dp_2 \, |p_2 - p_1| \int_0^\infty dr \, r \int_0^{2\pi} d\varphi \int_{-\infty}^{+\infty} d\xi \, \frac{\partial}{\partial \xi} F_1(t, P_1^\infty) F_1(t, P_2^\infty)$$
$$= \int_{\mathbb{R}^3} dp_2 \, |p_2 - p_1| \int_0^\infty dr \, r \int_0^{2\pi} d\varphi \Big(F_1(t, p_1^*) F_1(t, p_2^*) - F_1(t, p_1) F_1(t, p_2) \Big). \tag{5.2.21.b}$$

Introducing the differential collision cross - section I through $r \, dr \, d\varphi = I(|p_1 - p_2|, \Theta) \times \sin\Theta \, d\Theta \, d\varphi$, where Θ is a collision angle (for hard spheres $I(|p_1 - p_2|, \Theta) = a^2 \cos\Theta$) we finally obtain from 5.2.21 the identity

$$\int_{\mathbb{R}^3 \times \mathbb{R}^3} dx_2 \left\langle (p_2 - p_1), \frac{\partial}{\partial q_2} F_1(t, P_1^\infty) F_1(t, P_2^\infty) \right\rangle$$
$$= \int_{\mathbb{R}^3} dp_2 \int\int d\Theta \, d\varphi \, |p_2 - p_1| \, I(|p_1 - p_2|, \Theta) \sin\Theta$$
$$\times \big(F_1(t, p_1^*) F_1(t, p_2^*) - F_1(t, p_1) F_1(t, p_2) \big).$$

Thus, the space homogeneous kinetic equation (5.2.18) can be indeed represented in the Boltzmann form.

In the nonhomogeneous case one can also transform the collision integral of the first approximation (5.2.15) as before [Bo] but its expression has a more general form than the usual Boltzmann collision integral. We note that the triple-collision integral has been discussed in [Co1][Co3][Co7] analogously to the second order term of the collision integral in equation (5.2.15).

5.3 The Nonequilibrium Cluster Expansion Method

5.3.1 A new representation of the BBGKY hierarchy solutions. Before considering the problem of deriving kinetic equations based on the representation of the solutions of the BBGKY hierarchy not through the iteration series (2.5.1), as above, but through the functional series (1.4.5), we must discuss the representation of such solutions in more detail.

As indicated in Chapters I and II, there are several ways to derive a solution formula of the BBGKY hierarchy. We shall now formulate another method, namely, the cluster expansions method for nonequilibrium distribution functions which is a generalization of the Ursell (or algebraic) expansion used to obtain the density expansions for equilibrium distribution functions [R1][Co4][Co5].

For the sake of simplicity we consider the definition (1.4.3) of the reduced distribution functions in the framework of a canonical nonequilibrium ensemble (a system with a fixed N number of particles)

$$F_s(t,x_1,\ldots,x_s)=\frac{N!}{(N-s)!}\int dx_{s+1}\ldots dx_N\, S^N(-t,x_1,\ldots,x_N)$$
$$\times D_N(0,x_1,\ldots,x_N),\quad s\in(1,\ldots,N). \quad (5.3.1)$$

To obtain the cluster expansion for these functions let us make a cluster expansion of the evolution operator

$$S^N(-t,x_1,\ldots,x_N)=U_t\,(x_1,\ldots,x_s)\,S^{N-s}(-t,x_{s+1},\ldots,x_N)$$
$$+\sum_{k=1}^{N-s}\ \sum_{i_1<\cdots<i_k=s+1}^{N} U_t(x_1,\ldots,x_s;x_{i_1},\ldots,x_{i_k}) \quad (5.3.2)$$
$$\times S^{N-(s+k)}(-t,x_{s+1},\overset{i_1}{\vee}\ldots\overset{i_k}{\vee},x_N),\quad s\in(1,\ldots,N),$$

where the functions $U_t(x_1,\ldots,x_s;x_{i_1},\ldots,x_{i_k})$ are defined from the recursive relation (5.3.2) for $N=s,s+1,\ldots$. The explicit expressions for these functions are given by the formula

$$U_t(X)=\sum_{P:X=\bigcup_\imath X_\imath}(-1)^{|P|+1}(|P|-1)!\prod_{X_\imath\subset P}S^{|X_\imath|}(-t,X_\imath), \quad (5.3.3)$$

where we use the following notations: $X \equiv Y \bigcup (x_{s+1}, \ldots, x_{s+n})$, $Y \equiv (x_1, \ldots, x_s)$, and the summation in equation (5.3.3) is carried out over all distinct partitions $X = \bigcup_i X_i$ of the set X into $|P|$ non-empty subsets $\{X_i\}$ which are mutually disjoint and the subset Y is completely included in one of the subsets X_i. For example, the solutions of the recursive relation (5.3.2) for $U_t(x_1, \ldots, x_s)$, $U_t(x_1, \ldots, x_{s+1})$ and $U_t(x_1, \ldots, x_{s+2})$ are given by the corresponding expression

$$\begin{aligned} &U_t(Y) = S^s(-t, Y), \\ &U_t(Y; x_{s+1}) = S^{s+1}(-t, Y, x_{s+1}) - S^s(-t, Y) S^1(-t, x_{s+1}) \\ &U_t(Y; x_{s+1}, x_{s+2}) = S^{s+2}(-t, Y, x_{s+1}, x_{s+2}) - S^{s+1}(-t, Y, x_{s+1}) S^1(-t, x_{s+2}) \\ &\qquad - S^{s+1}(-t, Y, x_{s+2}) S^1(-t, x_{s+1}) - S^s(-t, Y) S^2(-t, x_{s+1}, x_{s+2}) \\ &\qquad + 2 S^s(-t, Y) S^1(-t, x_{s+1}) S^1(-t, x_{s+2}). \end{aligned} \tag{5.3.3.a}$$

Inserting the cluster expansion (5.3.2) into the relation (5.3.1), and then using the Liouville theorem and the definition of the initial distribution functions

$$F_s(0, x_1, \ldots, x_s) = \frac{N!}{(N-s)!} \int dx_{s+1} \ldots dx_N \, D_N(0, x_1, \ldots, x_N), \tag{5.3.1.a}$$

we obtain the following expansion

$$\begin{aligned} F_s(t, x_1, \ldots, x_s) = \sum_{n=0}^{N-s} \frac{1}{n!} \int dx_{s+1} \ldots dx_{s+n} \, U_t(x_1, \ldots, x_s; x_{s+1}, \ldots, x_{s+n}) \\ \times F_{s+n}(0, x_1, \ldots, x_{s+n}). \end{aligned} \tag{5.3.4}$$

Thus, using the solution (5.3.3) of the recursive relation (5.3.2), we have in terms of notations of equation (5.3.3) the formula

$$\begin{aligned} F_s(t, Y) = \sum_{n=0}^{N-s} \frac{1}{n!} \int d(X \backslash Y) \sum_{P: X = \bigcup_i X_i} (-1)^{|P|+1} (|P|-1)! \\ \times \prod_{X_i \subset P} S^{|X_i|}(-t, X_i) F_{s+n}(0, X). \end{aligned} \tag{5.3.5}$$

It should be noted that at first sight the cluster expansion (5.3.5) differs from the previous expansion (2.3.3). Actually, as we can see, using the Liouville theorem (see equation (5.3.3.a)), the expansion (5.3.5) is the same as (2.3.3).

5.3.2. The initial distribution functions with factorization property

Let again the initial distribution functions possess the factorization property

$$F_n(0,x_1,\ldots,x_n)=\prod_{i=1}^{n}F_1(0,x_i).$$

Then it is possible to express the initial one-particle distribution functions $F_1(0,x_i)$ in formula (5.3.5) for the solution of the BBGKY hierarchy in the case $s\geqslant 2$ through the one-particle distribution function $F_1(t,x_1)$ defined by formula (5.3.5) for the case $s=1$

$$F_1(t,x_1)=\sum_{n=0}^{N-s}\frac{1}{v^n}\frac{1}{n!}\int dx_2\ldots dx_{n+1}\sum_{P:(x_1,\ldots,x_{n+1})=\bigcup_i X_i}(-1)^{|P|+1}(|P|-1)!$$
$$\times\prod_{X_i\subset P}S^{|X_i|}(-t,X_i)\prod_{i=1}^{n+1}F_1(0,x_i).$$

This procedure can be performed in a way analogous to the elimination of the activity from the equilibrium activity expansions of the density [R1][Co4][Co5]. As a result one obtains the functionals $F_s(t,x_1,\ldots,x_s\mid F_1(t))$, $s\geqslant 2$ which express the s-particle distribution function for $s\geqslant 2$ through the one–particle distribution function $F_1(t,x_1)$. By using the operators

$$\widehat{S}_t^s(x_1,\ldots,x_s)\stackrel{\text{def}}{=}S^s(-t,x_1,\ldots,x_s)\prod_{i=1}^{s}S^1(t,x_i)$$

and the formula

$$F_1(0,x_1)=S^1(t,x_1)\Big\{F_1(t,x_1)$$
$$-\frac{1}{v}\int\Big[S^2(-t,x_1,x_2)-S^1(-t,x_1)\Big]S^1(t,x_1)S^1(t,x_2)F_1(t,x_1)F_1(t,x_2)dx_2+o\big(\frac{1}{v}\big)\Big\}$$

the first few terms of the expansions in powers of the density $\frac{1}{v}$ of such functionals are given by the following expressions

$$F_s(t,x_1,\ldots,x_s\mid F_1(t))=\widehat{S}_t^s(x_1,\ldots,x_s)\prod_{i=1}^{s}F_1(t,x_i)+\frac{1}{v}\int_{\mathbb{R}^\nu\times\mathbb{R}^\nu}dx_{s+1}$$
$$\times\Big(\widehat{S}_t^{s+1}(x_1,\ldots,x_{s+1})-\widehat{S}_t^s(x_1,\ldots,x_s)\sum_{j=1}^{s}\widehat{S}_t^2(x_j,x_{s+1})$$
$$+(s-1)\widehat{S}_t^s(x_1,\ldots,x_s)\Big)\prod_{i=1}^{s+1}F_1(t,x_i)+o\big(\frac{1}{v}\big),\quad s\geqslant 2. \tag{5.3.6}$$

In the next section a rigorous derivation of the general formula for the functionals (5.3.6) will be given.

At this step of the derivation of the kinetic equation we must consider the evolution of the system after a time $t \gg t_c$, or on the kinetic scale of variation of the variables. According to papers [B][Col] let us make the physical assumption that for $t \gg t_c$ one can replace the operators $\widehat{S}_t^s$ in equation (5.3.6) by the limiting ones $\widehat{S}_\infty^s$ (scattering operators) defined by the formula (5.2.10). As a result we obtain the following kinetic cluster expansion in powers of the density $\frac{1}{v}$ for the limiting functionals $F_s(x_1, \ldots, x_s \mid F_1(t))$, $s \geqslant 2$

$$F_s(x_1, \ldots, x_s \mid F_1(t)) = \sum_{n=0}^{N-s} \frac{1}{v^n} \frac{1}{n!} \int dx_{s+1} \ldots dx_{s+n}$$
$$\times \widehat{U}_\infty^{(n)}(x_1, \ldots, x_s; x_{s+1}, \ldots, x_{s+n}) \prod_{i=1}^{s+n} F_1(t, x_i), \quad (5.3.7)$$

where

$$\widehat{U}_\infty^{(0)}(x_1, \ldots, x_s) = \widehat{S}_\infty^s(x_1, \ldots, x_s),$$
$$\widehat{U}_\infty^{(1)}(x_1, \ldots, x_s; x_{s+1}) = \widehat{S}_\infty^{s+1}(x_1, \ldots, x_{s+1}) - \widehat{S}_\infty^s(x_1, \ldots, x_s) \sum_{j=1}^{s} \widehat{S}_\infty^2(x_j, x_{s+1})$$
$$+ (s-1)\widehat{S}_\infty^s(x_1, \ldots, x_s), \text{ets.}$$

It should be indicated here that the expression (5.3.6) is an exact one in the sense that it was derived from corresponding solutions of the BBGKY hierarchy without any approximations. On the contrary, the expression of the kinetic expansion (5.3.7) is a suitable limit of the expression (5.3.6). Thus, as we will see, the assumption accepted above introduces irreversibility in the description of the evolution in a way similar to the condition of the weakening of correlationsin case of the Bogolubov's method.

By substituting the expansion (5.3.7) for $s = 2$ into the first equation of the BBGKY hierarchy (5.2.1) we deduce a closed equation for the one-particle distribution function:

$$\frac{\partial}{\partial t}F_1(t,x_1) = -\left\langle p_1, \frac{\partial}{\partial q_1}\right\rangle F_1(t,x_1)$$
$$+\frac{1}{v}\int\limits_{\mathbb{R}^\nu\times\mathbb{R}^\nu} dx_2\,\mathcal{H}_{\text{int}}(x_1,x_2)\widehat{U}^{(0)}_\infty(x_1,x_2)F_1(t,x_1)F_1(t,x_2)$$
$$+\frac{1}{v^2}\iint\limits_{(\mathbb{R}^\nu\times\mathbb{R}^\nu)^2} dx_2\,dx_3\,\mathcal{H}_{\text{int}}(x_1,x_2)\widehat{U}^{(1)}_\infty(x_1,x_2;x_3)F_1(t,x_1)F_1(t,x_2)$$
$$\times F_1(t,x_3) + o\left(\frac{1}{v^2}\right), \tag{5.3.8}$$

where the operators $\widehat{U}^{(i)}_\infty(x_1,x_2;\ldots x_i)$ are defined by equation (5.3.7) and one has used the notation (5.2.13). The kinetic equation (5.3.8) is called the generalized Boltzmann equation and is none other than equation (5.2.16). This fact will be established later (see Subsection 5.3.3).

As was mentioned in the introduction to the present Chapter it should be expected that, for continuous initial data $F_1(0)$ belonging to the space L^∞_ξ, the series (5.3.7) and, thus, the right-hand side of equation (5.3.8) contain secular terms. As is easily seen, such terms, which grow in time, have the same nature as the terms of the series representing the solutions of the BBGKY hierarchy (see Chapters II, III). A rigorous consideration of this problem will be given in Section 5.4.

Remark.

The definition of the cluster expansion terms in the explicit form has been used in Cohen's papers [Co1][Co2][Co3][Co4][Co5][Co6][Co7], where it was deduced the first few terms (5.3.6). In Green's papers [Gre1][Gre2][GPi] another approach was developed. It is based on the employment of recursion relations for the direct definition of the limiting (scattering) operators in every term of the expansion (5.3.7).

Concluding the formal introduction to the nonequilibrium cluster expansion method we shall make a digression in order to discuss the problem of the derivation of kinetic equations in the case of general initial data, which do not possess the factorization condition, namely

$$F_s(0,x_1,\ldots,x_s) = a_s(x_1,\ldots,x_s)\prod_{i=1}^{s}F_1(0,x_i).$$

where a_s is implicitly defined by this equation.

By introducing new functions $u_s(x_1, \dots, x_s)$ which characterize initial correlations

$$a_1(x_1) = u_1(x_1) = 1,$$
$$a_2(x_1, x_2) = u_2(x_1, x_2) + 1, \quad \text{ets.},$$

instead of the expansion (5.3.6) for the functional $F_s^{\text{cor}}(t, x_1, \dots, x_s \mid F_1(t))$ we obtain the sum of two expansions, one of which is the series (5.3.6) independent of initial correlations and another part that explicitly contains the initial correlations $u_s(x_1, \dots, x_s)$

$$F_s^{\text{cor}}(t, x_1, \dots, x_s \mid F_1(t)) = F_s(t, x_1, \dots, x_s \mid F_1(t)) + F_s^0(t, x_1, \dots, x_s \mid u),$$

where, for example,

$$F_s^0(t, x_1, \dots, x_s \mid u), = S^s(-t, x_1, \dots, x_s) u_s(x_1, \dots, x_s) \prod_{i=1}^{s} S^1(t, x_i) F_1(t, x_i)$$
$$+ O\left(\frac{1}{v}\right), \qquad s \geqslant 2.$$

On the kinetic scale, i.e. for times $t \gg t_c$, at least for a short range interaction potential, one would expect that the contribution of this expression can be neglected. Until now this problem has been discussed only from a physical point of view [Co1][Co2][Co3][Co4][Co5][Co6][Co7][DC][Pi][CU] and no rigorous results are available.

5.3.3 The comparison with Bogolubov's method. In the present subsection we shall establish that the special iteration expansion (5.2.5) and the functional kinetic expansion (5.3.7) are formally identical term by term, i.e. the kinetic equations (5.2.16) and (5.3.8) differ from each other only by form in which they are written. The idea of the proof is completely similar to the proof of the equivalence of the two representations of the BBGKY hierarchy solutions through iteration and functional series developed in Chapter II.

Evidently, the first terms from both expansions (5.2.5) and (5.3.7) coincide. Let us consider the term of the order $\frac{1}{v}$ from the special iteration series (5.2.5)

$$F_s^{(1)}(Y \mid F_1(t)) = \int_0^\infty d\tau\, S^s(-\tau, Y) \Psi_s^{(1)}\left(Y \mid S^1(\tau) F_1(t)\right), \quad s \geqslant 2, \tag{5.3.9}$$

where

$$\Psi_s^{(1)}\left(Y \mid S^1(\tau)F_1(t)\right) = \int\limits_{\mathbb{R}^\nu\times\mathbb{R}^\nu} dx_{s+1} \Big(\sum_{i=1}^{s} \mathcal{H}_{\rm int}(x_i, x_{s+1})\widehat{S}_\infty^{s+1}(Y, x_{s+1})$$

$$- \widehat{S}_\infty^{s}(Y)\sum_{i=1}^{s} \mathcal{H}_{\rm int}(x_i, x_{s+1})\widehat{S}_\infty^{2}(x_i, x_{s+1})\Big)$$

$$\times \prod_{j=1}^{s+1} S^1(\tau, x_j)F_1(t, x_j),$$

$Y \equiv (x_1, \ldots, x_s)$ and one has used the notations (5.2.13). Before establishing that this functional (5.3.9) is equivalent to the corresponding functional appearing in (5.3.7)

$$\widetilde{F}_s^{(1)}(Y, \mid F_1(t)) = \int\limits_{\mathbb{R}^\nu\times\mathbb{R}^\nu} dx_{s+1}\, \widehat{U}_\infty^{(1)}(Y, x_{s+1}) \prod_{i=1}^{s+1} F_1(t, x_i), \quad s \geqslant 2 \qquad (5.3.10)$$

where

$$\widehat{U}_\infty^{(1)}(Y, x_{s+1}) = \widehat{S}_\infty^{s+1}(Y, x_{s+1}) - \widehat{S}_\infty^{s}(Y)\sum_{j=1}^{s} S_\infty^2(x_j, x_{s+1}) + (s-1)\widehat{S}_\infty^{s}(Y),$$

we prove the validity of the following statement:

Lemma . *For the functional (5.3.9) the following representation holds*

$$F_s^{(1)}(Y \mid F_1(t)) = \int_0^\infty d\tau \int\limits_{\mathbb{R}^\nu\times\mathbb{R}^\nu} dx_{s+1}\, S^s(-\tau, Y)S^1(\tau, x_{s+1})$$

$$\times \Big(\big(\mathcal{H}_s(Y) + \mathcal{H}_0(x_{s+1})\big)\widehat{U}_\infty^{(1)}(Y, x_{s+1}) - \widehat{U}_\infty^{(1)}(Y, x_{s+1})\sum_{i=1}^{s+1} \mathcal{H}_0(x_i)\Big)$$

$$\times \prod_{j=1}^{s+1} S^1(\tau, x_j)F_1(t, x_j). \qquad (5.3.11)$$

Proof. According to the definitions of the operators appearing in equation (5.3.9-10) the following operator identities take place

$$\mathcal{H}_{s+1}(Y, x_{s+1})\widehat{S}_\infty^{s+1}(Y, x_{s+1}) = \widehat{S}_\infty^{s+1}(Y, x_{s+1})\mathcal{H}_{s+1}(Y, x_{s+1})$$

$$\widehat{S}_\infty^{s+1}(Y, x_{s+1}) \sum_{1=i<j}^{s+1} \mathcal{H}_{\rm int}(x_i, x_j) = 0.$$

For the first term of the functional (5.3.9) the following equalities are valid:

$$\sum_{i=1}^{s} \mathcal{H}_{\text{int}}(x_i, x_{s+1}) \widehat{S}_{\infty}^{s+1}(Y, x_{s+1})$$
$$= \sum_{i=1}^{s} \mathcal{H}_{\text{int}}(x_i, x_{s+1}) \widehat{S}_{\infty}^{s+1}(Y, x_{s+1}) + \mathcal{H}_{s+1}(Y, x_{s+1}) \widehat{S}_{\infty}^{s+1}(Y, x_{s+1})$$
$$- \widehat{S}_{\infty}^{s+1}(Y, x_{s+1}) \mathcal{H}_{s+1}(Y, x_{s+1}) - \widehat{S}_{\infty}^{s+1}(Y, x_{s+1}) \sum_{1=i<j}^{s+1} \mathcal{H}_{\text{int}}(x_i, x_j) \quad (5.3.12)$$
$$= \big(\mathcal{H}_s(Y) + \mathcal{H}_0(x_{s+1})\big) \widehat{S}_{\infty}^{s+1}(Y, x_{s+1}) - \widehat{S}_{\infty}^{s+1}(Y, x_{s+1}) \sum_{i=1}^{s+1} \mathcal{H}_0(x_i).$$

For the same reason it holds that

$$\widehat{S}_{\infty}^{2}(x_j, x_{s+1}) \mathcal{H}_{\text{int}}(x_j, x_{s+1}) = 0,$$

and we have the equalities

$$\mathcal{H}_{\text{int}}(x_j, x_{s+1}) \widehat{S}_{\infty}^{2}(x_j, x_{s+1})$$
$$= \big(\mathcal{H}_0(x_j) + \mathcal{H}_0(x_{s+1})\big) \widehat{S}_{\infty}^{2}(x_j, x_{s+1}) - \widehat{S}_{\infty}^{2}(x_j, x_{s+1}) \big(\mathcal{H}_0(x_j) + \mathcal{H}_0(x_{s+1})\big)$$
$$= \sum_{i=1}^{s+1} \mathcal{H}_0(x_j) \widehat{S}_{\infty}^{2}(x_j, x_{s+1}) - \widehat{S}_{\infty}^{2}(x_j, x_{s+1}) \sum_{i=1}^{s+1} \mathcal{H}_0(x_i),$$
$$\forall j = 1, \ldots, s. \quad (5.3.13)$$

Then, according to (5.3.13), for the remaining term in the functional (5.3.9) we obtain

$$\widehat{S}_{\infty}^{s}(Y) \sum_{i=1}^{s} \mathcal{H}_{\text{int}}(x_i, x_{s+1}) \widehat{S}_{\infty}^{2}(x_i, x_{s+1})$$
$$= \widehat{S}_{\infty}^{s}(Y) \Big(\sum_{i=1}^{s+1} \mathcal{H}_0(x_i) \sum_{j=1}^{s} \widehat{S}_{\infty}^{2}(x_i, x_{s+1}) - \sum_{j=1}^{s} \widehat{S}_{\infty}^{2}(x_i, x_{s+1})$$
$$\times \sum_{i=1}^{s+1} \mathcal{H}_0(x_i) \Big)$$
$$(5.3.14)$$
$$= \big(\mathcal{H}_s(Y) + \mathcal{H}_0(x_{s+1})\big) \widehat{S}_{\infty}^{2}(Y) \sum_{j=1}^{s} \widehat{S}_{\infty}^{2}(x_j, x_{s+1})$$
$$- \widehat{S}_{\infty}^{2}(Y) \sum_{j=1}^{s} \widehat{S}_{\infty}^{2}(x_i, x_{s+1}) \sum_{i=1}^{s+1} \mathcal{H}_0(x_i).$$

If we observe that

$$\big(\mathcal{H}_s(Y) + \mathcal{H}_0(x_{s+1})\big)\widehat{S}^s_\infty(Y) - \widehat{S}^s_\infty(Y)\sum_{i=1}^{s+1}\mathcal{H}_0(x_i) = 0, \tag{5.3.15}$$

then using the definition (5.3.10) of the operator $\widehat{U}^{(1)}_\infty(Y, x_{s+1})$, from the equalities (5.3.12), (5.3.14), (5.3.15) we establish the representation (5.3.11) of the functional (5.3.9).

By using the lemma just proved we can perform the time integration in the expression (5.3.11). In fact, since

$$\begin{aligned} S^s(-\tau, Y)S^1(\tau, x_{s+1})\Big(\big(\mathcal{H}_s(Y) + \mathcal{H}_0(x_{s+1})\big)\widehat{U}^{(1)}_\infty(Y, x_{s+1}) \\ - \widehat{U}^{(1)}_\infty(Y, x_{s+1})\sum_{i=1}^{s+1}\mathcal{H}_0(x_i)\Big)\prod_{j=1}^{s+1} S^1(\tau, x_j) \\ = -\frac{d}{d\tau}\Big(S^s(-\tau, Y)S^1(\tau, x_{s+1})\widehat{U}^{(1)}_\infty(Y, x_{s+1})\prod_{j=1}^{s+1} S^1(\tau, x_j)\Big), \end{aligned}$$

one formally gets that

$$\begin{aligned} F^{(1)}_s(Y \mid F_1(t)) = -\int dx_{s+1} \lim_{\tau\to+\infty} S^s(-\tau, Y)S^1(\tau, x_{s+1})\widehat{U}^{(1)}_\infty(Y, x_{s+1}) \\ \times \prod_{j=1}^{s+1} S^1(\tau, x_j)F_1(t, x_j) \\ + \int dx_{s+1}\, \widehat{U}^{(1)}_\infty(Y, x_{s+1})\prod_{j=1}^{s+1} F_1(t, x_j) \\ \equiv \lim_{\tau\to+\infty} S^s(-\tau, Y)\widetilde{F}^{(1)}_s(Y \mid S^1(\tau)F_1(t)) + \widetilde{F}^{(1)}_s(Y \mid F_1(t)). \end{aligned} \tag{5.3.16}$$

Thus, functionals (5.3.9) and (5.3.10) are identical if the functional (5.3.7) satisfies the principle of weakening of the correlations, condition (5.2.8), in the same way as the functional (5.2.5), i.e.

$$\lim_{\tau\to+\infty} S^s(-\tau, Y)\,\widetilde{F}^{(1)}_s\big(Y \mid S^1(\tau)F_1(t)\big) = 0.$$

5.4. Justification of the Generalized Kinetic Equation

5.4.1. The convergence of the kinetic cluster expansion. As we already mentioned, the method discussed in the previous section can only give a local in time result for general initial data. A global result will be proved if we can control globally the series representing solution for the BBGKY hierarchy. As is clear from Chapter II for suitable initial data one can prove existence of global solutions for initial data from the space L^1 of summable functions.

As before, we examine the Cauchy problem to the BBGKY hierarchy

$$\frac{\partial}{\partial t}F_s(t) = \{H_s, F_s(t)\} + \frac{1}{v}\int\limits_{\mathbb{R}^\nu\times\mathbb{R}^\nu} dx_{s+1}\Big\{\sum_{i=1}^{s}\Phi(q_i - q_{s+1}), F_{s+1}(t)\Big\}, \quad 1 \leqslant s \leqslant N, \tag{5.4.1}$$

with initial data, which possess the factorization property (the chaos property)

$$F_s(0, x_1, \dots, x_s) = \prod_{i=1}^{s} F_1(0, x_i), \quad s \geqslant 2, \tag{5.4.2}$$

where $F_1(0, x_i)$ is an arbitrary summable function $F_1(0, x_i) \in L^1(\mathbb{R}^\nu \times \mathbb{R}^\nu)$.

We shall assume that the function Φ (an interaction potential) satisfies certain conditions guaranteeing the existence (global in time) of solutions to the Hamilton equations for finite particle systems. For example, the function Φ is twice continuously differentiable with compact support, $\Phi(q) = 0$ if $|q| > r_0$, $r_0 < \infty$.

As was established in Chapter 2 there exists a global in time solution of the Cauchy problem (5.4.1-2) in L^1, given by the formula

$$\begin{aligned} F_s(t, x_1, \dots, x_s) &= S^s(-t, x_1, \dots, x_s)\prod_{i=1}^{s} F_1(0, x_i) \\ &+ \sum_{n=1}^{\infty}\frac{1}{v^n}\sum_{k=0}^{n}\frac{(-1)^k}{k!(n-k)!}\int\limits_{(\mathbb{R}^\nu\times\mathbb{R}^\nu)^n} dx_{s+1}\dots dx_{s+n} \\ &\times S^{s+n-k}(-t, x_1, \dots, x_s; x_{s+1}, \dots, x_{s+n-k})\prod_{i=1}^{s+n} F_1(0, x_i). \end{aligned} \tag{5.4.3}$$

Generally the sequence $F(0) = \big(F_1(0,x_1),\ldots,\prod_{i=1}^{s} F_1(0,x_i),\ldots\big)$ of functions (5.4.2) does not belong to the usual function space L^1 but it is clear that the sequence $F(t) = \big(F_1(t,x_1),\ldots,F_s(t,x_1,\ldots,x_s),\ldots\big)$ of functions (5.4.3) exists in the sense that $F_s(t,x_1,\ldots,x_s) \in L^1\big(\mathbb{R}^{\nu s}\times\mathbb{R}^{\nu s}\big)$ and they are solutions of the BBGKY hierarchy, if $F_1(0,x_i) \in L^1\big(\mathbb{R}^{\nu}\times\mathbb{R}^{\nu}\big)$ is a continuously differentiable function. In fact, the functional series (5.4.3) converge in norm in $L^1\big(\mathbb{R}^{\nu s}\times\mathbb{R}^{\nu s}\big)$ for arbitrary $t \in \mathbb{R}^1$ and in virtue of $\|S^{s+n-k}(-t)\| = 1$ they satisfy the inequalities

$$\|F_s(t)\| \leqslant \|F_1(0)\|^s \exp\Big(\frac{2}{v}\|F_1(0)\|\Big), \qquad s \geqslant 1. \tag{5.4.4}$$

The fact that the functions (5.4.3) are solutions of the BBGKY hierarchy (5.4.1) is verified directly.

Let us consider relation (5.4.3) for the case $s = 1$ as the equation in the space $L^1\big(\mathbb{R}^{\nu}\times\mathbb{R}^{\nu}\big)$ with respect to the function $F_1(0)$ with the given function $F_1(t,x_1) \in L^1\big(\mathbb{R}^{\nu}\times\mathbb{R}^{\nu}\big)$. To this aim we define the nonlinear operator A in the space $L^1\big(\mathbb{R}^{\nu}\times\mathbb{R}^{\nu}\big)$ which acts on the elements $F_1(0) \in L^1\big(\mathbb{R}^{\nu}\times\mathbb{R}^{\nu}\big)$ by the formula

$$\begin{aligned}\big(AF_1(0)\big)(x_1) = {} & S^1(t,x_1)F_1(t,x_1) \\ & - \sum_{n=1}^{\infty}\frac{1}{v^n}\sum_{k=0}^{n}\frac{(-1)^k}{k!(n-k)!}\int\limits_{(\mathbb{R}^{\nu}\times\mathbb{R}^{\nu})^n} dx_2\ldots dx_{n+1} S^1(t,x_1) \\ & \times S^{n+1-k}(-t,x_1;x_2,\ldots,x_{n+1-k})\prod_{i=1}^{n+1}F_1(0,x_i),\end{aligned} \tag{5.4.5}$$

for arbitrary $t \in \mathbb{R}^1$.

Thus, in the space $L^1\big(\mathbb{R}^{\nu}\times\mathbb{R}^{\nu}\big)$ we have the following equation

$$F_1(0) = AF_1(0). \tag{5.4.6}$$

We introduce the notation $S^1(t,x_1)F_1(t,x_1) \equiv F_1^0$ and let $\|F_1(t,x_1)\| \leqslant r$. Then $F_1^0 \in L^1(\mathbb{R}^{\nu}\times\mathbb{R}^{\nu})$,

$$\|F_1^0\| = \|S^1(t,x_1)F_1(t,x_1)\| \leqslant r < +\infty.$$

Let us consider in the Banach space $L^1(\mathbb{R}^{\nu}\times\mathbb{R}^{\nu})$ the domain $\mathbb{S}(F_1^0;R)$, which is a sphere of the certain radius $R < \infty$ with center in the point F_1^0

$$\mathbb{S}(F_1^0;R) \equiv \big\{F_1(0) \in L^1(\mathbb{R}^{\nu}\times\mathbb{R}^{\nu}) \,:\, \|F_1(0) - F_1^0\| \leqslant R\big\}.$$

We introduce the operators

$$\widehat{S}_t^n(x_1,\ldots,x_n) = S^n(-t,x_1,\ldots,x_n)\prod_{i=1}^{n} S^1(t,x_i).$$

For the nonlinear equation (5.4.6) the following statement holds.

Theorem 5.4.1. *If*

$$\frac{1}{v} < \frac{1}{2(R+r)}\min\{x_0\,;\ln\frac{2R+r}{R+r}\},$$

where x_0 is the solution of $e^{x_0}(1+x_0)=2$, then in the domain $\mathbb{S}(F_1^0\,;R)\subset L^1(\mathbb{R}^\nu\times\mathbb{R}^\nu)$ there exists a unique solution of equation (5.4.6) which is determined by the formula

$$F_1(0,x_1) = S^1(t,x_1)F_1(t,x_1) + \sum_{n=1}^{\infty}\frac{1}{v^n}\int\limits_{(\mathbb{R}^\nu\times\mathbb{R}^\nu)^n} dx_2\ldots dx_{n+1}\sum_{\mathcal{B}:X=\bigcup_i X_i}(-1)^{|\mathcal{B}|+1}$$

$$\times\prod_{X_i\subset\mathcal{B}}\sum_{k_j=0}^{|X_i|-1}\frac{(-1)^{k_i}}{k_i!\,(|X_i|-k_i)!}S^1(t,x_1)$$

$$\times\,\widehat{S}_t^{|X_i|+1-k_i}(x_1,x_2,\ldots,x_{|X_i|+1-k_i})\prod_{j=1}^{n+1}F_1(t,x_j), \tag{5.4.7}$$

where $\sum_{\mathcal{B}:X=\bigcup_i X_i}$ is the sum over all possible representations of the set $X=(x_1;x_2,\ldots,x_{n+1})$ through disjoint ordered subsets X_i that correspond to the connected graphs labeled by growing trees with the root in the point x_1; $|\mathcal{B}|$ is the number of branches of the tree.

Proof. For the proof of the existence of solution for the equation (5.4.6) we use the contraction mapping principle [RS]. We must then stablish that operator (5.4.5) maps the domain $\mathbb{S}(F_1^0\,;R)$ into itself and is a contraction operator.

In fact, in virtue of the fact that

$$\|F_1(0)\| \leqslant \|F_1(0)-F_1^0\| + \|F_1^0\| \leqslant R+r,$$

from the definition (5.4.5) we have

$$\|AF_1(0) - F_1^0\| \leqslant \sum_{n=1}^{\infty} \frac{1}{v^n} \frac{2^n}{n!} \|F_1(0)\|^{n+1} \leqslant (R+r)\left(e^{\frac{2}{v}(R+r)} - 1\right).$$

Thus, the operator A maps the domain $\mathbb{S}(F_1^0\,; R)$ into itself if

$$(R+r)\left(e^{\frac{2}{v}(R+r)} - 1\right) \leqslant R,$$

that is the condition on the density

$$\frac{1}{v} \leqslant \frac{1}{2(R+r)} \ln \frac{2R+r}{R+r}. \tag{5.4.8.a}$$

Now we shall find the condition under which the operator (5.4.5) is a contraction. Let Y_1, Y_2 are arbitrary elements from the $\mathbb{S}(F_1^0\,; R)$. Then, according to the definition (5.4.5), we obtain

$$\begin{aligned}\|AY_1 - AY_2\| &\leqslant \sum_{n=1}^{\infty} \frac{1}{v^n} \frac{2^n}{n!}(n+1)(R+r)^n \|Y_1 - Y_2\| \\ &= \left(e^{\frac{2}{v}(R+r)}\left(\frac{2}{v}(R+r)+1\right) - 1\right)\|Y_1 - Y_2\|.\end{aligned}$$

A is a contraction operator if the estimate holds

$$e^{\frac{2}{v}(R+r)}\left(\frac{2}{v}(R+r)+1\right) < 2,$$

which give the following condition on the density

$$\frac{1}{v} < \frac{x_0}{2(R+r)}. \tag{5.4.8.b}$$

where x_0 is the solution of $e^{x_0}(1+x_0) = 2$.

Thus, under the conditions (5.4.8) in the domain $\mathbb{S}(F_1^0\,; R) \subset L^1(\mathbb{R}^\nu \times \mathbb{R}^\nu)$ there exists a unique solution of the equation (5.4.6). This solution is determined as the limit of successive approximations $F_1^{(n)}(0) = AF_1^{(n-1)}(0)$ with the first approximation $F_1^{(0)}(0) = F_1^0 \equiv S^1(t)F_1(t)$. The explicit form of the limit of successive approximations $\lim_{n\to\infty} F_1^{(n)}(0) = F_1(0)$ is given by (5.4.7).

Remark.

If we use the cluster expansion (5.3.5) instead of equation (5.4.3), we obtain, according to Theorem 5.4.1, an expansion different from equation (5.4.7) (the first few terms were obtained in the previous section). But in order to show their equivalence it is sufficient to use the Liouville theorem.

The formula (5.4.7) makes it possible to express the solutions (5.4.3) for $s \geqslant 2$ of the BBGKY hierarchy (5.4.1) in the form of functionals of the one-particle distribution function $F_1(t)$. In fact, let us form the products of the functions $F_1(0, x_i)$, by using the convergent series (5.4.7) and substitute the result into the series (5.4.3). By collecting the terms with the same number of variables of integration, we construct the expansions in power of the density for such functionals $F_s\big(t, x_1, \ldots, x_s \mid F_1(t)\big)$, $s \geqslant 2$. Their explicit form is determined by the formula

$$F_s\big(t, x_1, \ldots, x_s \mid F_1(t)\big) = \widehat{S}_t^s\,(x_1, \ldots, x_s) \prod_{i=1}^{s} F_1(t, x_i)$$
$$+ \sum_{n=1}^{\infty} \frac{1}{v^n}\frac{1}{n!} \int\limits_{(\mathbb{R}^\nu \times \mathbb{R}^\nu)^n} dx_{s+1} \ldots dx_{s+n} \sum_{T: X = \bigcup_i X_i} (-1)^{|T|+s}$$
$$\times \prod_{X_i \subset T} \widehat{S}_t^{s+|X_i|}\big(x_1, \ldots, x_s\, ; x_{s+1}, \ldots, x_{s+|X_i|}\big) \prod_{j=1}^{n+s} F_1(t, x_j), \tag{5.4.9}$$

where the sum $\sum\limits_{T: X = \bigcup_i X_i}$ has the same meaning as in (5.4.7).

Under the hypothesis formulated in Theorem 5.4.1 the functionals (5.4.9) exist. In fact, according to the estimate (5.4.4) and the statement of the theorem we conclude that the functional series (5.4.9) converges in the norm of the space $L^1(\mathbb{R}^{\nu s} \times \mathbb{R}^{\nu s})$ for the arbitrary function $F_1(t) \in \mathbb{S}(F_1^0, R) \subset L^1(\mathbb{R}^\nu \times \mathbb{R}^\nu)$ because the corresponding $F_1(0) \in L^1$ and $\|F_1(0)\| < r + R$.

5.4.2. An existence theorem for the generalized kinetic equation

It is clear that functionals (5.4.9) satisfy the BBGKY hierarchy (5.4.1) if the one-particle distribution function $F_1(t)$ is a solution of the following kinetic equation

$$\frac{\partial}{\partial t} F_1(t, x_1) = -\left\langle p_1, \frac{\partial}{\partial q_1} \right\rangle F_1(t, x_1)$$
$$+ \frac{1}{v} \int\limits_{\mathbb{R}^\nu \times \mathbb{R}^\nu} dx_2 \left\{ \Phi(q_1 - q_2), F_2\big(t, x_1, x_2 \mid F_1(t)\big) \right\}, \tag{5.4.10}$$

where the functional $F_2(t, x_1, x_2 \mid F_1(t))$ is explicitly defined by the formula (5.4.9) for $s = 2$. This equation is called the generalized kinetic equation.

From the fact that in L^1 there exists the solution of the initial value problem for the BBGKY hierarchy and from the statement that if $F_1(0) \in \mathbb{S}(F_1^0; R)$, then the solution of equation (5.4.6) for low densities (5.4.8) belongs to the domain $\mathbb{S}(F_1^0; R) \subset L^1(\mathbb{R}^\nu \times \mathbb{R}^\nu)$ it follows that the following theorem holds:

Theorem 5.4.2. *If $F_1(0) \in \mathbb{S}(F_1^0; R) \subset L_0^1(\mathbb{R}^\nu \times \mathbb{R}^\nu)$ and inequality (5.4.8) holds, then there exists a unique, global in time, strong solution of the equation (5.4.10) which is represented by formula (5.4.3) for $s = 1$.*

Thus, for the initial data (5.4.2) the BBGKY hierarchy (5.4.1) is identical to the generalized kinetic equation (5.4.10) with initial data $F_1(t)\big|_{t=0} = F_1(0) \in \mathbb{S}(F_1^0; R) \subset L_0^1(\mathbb{R}^\nu \times \mathbb{R}^\nu)$. Equation (5.4.10) completely describes all possible types of kinetic equations with initial data (5.4.2). It should be noted that the equation (5.4.10) is reversible in time as the BBGKY hierarchy (5.4.1). To obtain the irreversible kinetic equation one needs to pass in the equation (5.4.10) to the kinetic scale of variation of the variables.

5.4.3. The connection between the Boltzmann equation and a stochastic dynamics. Random processes and stochastic equations were used to construct solutions of the smoothed Boltzmann equation [Ta][Sz][Ars1][Ars2]. However the Boltzmann hierarchy was not studied in these works.

Probably the first hierachy for the distribution functions that corresponds to certain stochastic equations was derived in Skorokhod's monograph [Sk]. Further progress in the approach formulated in the latter book is connected with the works [LP][CDPP][Lac1], where the iteration method in the mean field model was used.

Recently, a stochastic dynamics was proposed [PP], which can be considered as the Boltzmannn-Grad limit of the Hamiltonian dynamics of systems of hard spheres. The corresponding stochastic Boltzmann hierarchy was obtained and proved to have global in time solutions in L^1, local in time solutions in E_ξ. For some subsets of E_ξ, the solutions can be continued to arbitrarily long time intervals.

References

Aizenman, M., Goldstein, Sh., and Lebowitz, J. L.
[AGL] *Ergodic properties of infinite systems,* Lecture Notes in Physics, **38**, Springer-Verlag, Berlin (1975), 112-144.

Alexander, R. K.
[Al1] *Dynamics of ideal hard-core particles,* Comm. Math. Phys. (1976), **49**, 81-98.
[Al2] *Time evolution for infinitely many hard spheres,* Comm. Math. Phys. (1976), **49**, 217-232.

Arnold, V. I.
[Arn1] Mathematical Methods of Classical Mechanics [in Russian], Nauka, Moscow, 1974.
[Arn2] Ordinary Differential Equations [in Russian], Nauka, Moscow, 1975.
[Arn2] *Proof of Kolmogorov's theorem on the invariance of quasi-periodic motion under small perturbation of the Hamiltonian,* Usp. Mat. Nauk (1963), **18**, 9-36.

Arsen'ev, A. A.
[Ars1] *On the approximation of the solution of the Boltzmann equation by solution of Ito stochastic differential equations,* Z. Vychisl. Mat. i Mat. Fiz. (1987), **27**, 400-410.
[Ars2] *Approximation of the Boltzmann equation by stochastic equations,* Z. Vychisl. Mat. Fiz. (1988), **28**, 560-567.

Balescu, R.
[Ba] Equilibrium and Non-equilibrium Statistical Mechanics, John Wiley and Sons, London, 1975.

Belavkin, V. P., Maslov, V. P., and Tariverdiev, S. E.
[BMT] *Asymptotic dynamics of a system of a large number of particles described by Kolmogorov-Feller equations* Teor. Mat. Fiz. (1982), **49**, 298-306.

Bogolubov, N. N.
[Bo] Problems of a Dynamical Theory in Statistical Physics [in Russian], Gostekhisdat, Moscow, 1946; **see also:** Selected Papers [in Russian], Vol. 2, Naukova Dumka, Kiev, (1970), 99-196.

Bogolubov, N. N., Bogolubov, N. N. (jr.)
[BB] Introduction to Statistical Mechanics [in Russian], Nauka, Moscow, 1984.

Bogolubov, N. N., Petrina, D. Ya., and Khatset, B. I.
[BPK] *Mathematical description of the equilibrium state of classical systems based on the canonocal ensemble formalism,* Theor. Mat. Fiz. (1969), **1**, 251-274.

Boltzmann, L.
[Bol1] *Über die Beziehung eines allgemeinen mechanischen Satzes zum zweiten Hauptsatze der Wärmetheorie,* Sitzungsberichte Akad. Wiss. (1877), **75**, 67-73.
[Bol2] *Entgegnung auf die wärmetheoretischen Betrachtungen des Hrn. E. Zermelo,* Annalen der Physik (1896) **57**, 773-784.

Born, M., and Green, H. S.
[BG1] *A general kinetic theory of liquids,* Proc. R. Soc. Lond.(1947), **A188**, 168-201.
[BG2] A general kinetik theory of fluids, Cambridge University Press, 1949.

Caprino, S., De Masi, A., Presutti, E., and Pulvirenti, M.
[CDPP] *A stochastic particle system modeling the Carleman equation,* J. Stat. Phys. (1989), **55**, 625-638.

Cercignani, C.
[Ce1] Theory and Application of the Boltzmann Equation, Scottish Academic Press, Edinburgh and London, 1975.
[Ce2] The Boltzmann Equation and Its Application, Springer, Berlin-Heidelberg, 1988.
[Ce3] *On the Boltzmann equation for rigid spheres,* Transp. Theory Stat. Phys. (1972), **2**, 211-225.
[Ce4] *The Grad limit for a system of soft spheres,* Comm. Pure Appl. Math. (1983), **36**, 479-494.
[Ce5] *Le radici fisiche e matematiche dell'irreversibilità temporale,* (accompanied by a translation into English), Alma Mater Studiorum (1989) **2**, 37-52.

Cercignani, C., Illner, R., and Pulvirenti, M.
[CIP] The Mathematical Theory of Dilute Gases, Springer-Verlag, New York, 1994.

Cercignani, C. and Lampis, M.
[CL] *On the kinetic theory of a dense gas of rough spheres,* J. Stat. Phys. 1988, **53**, 655.

Choh, S. T. and Uhlenbeck, G. F.
[CU] The Kinetic Theory of Dense Gases, University of Michigan, 1958.

Cohen, E. G. D.
[Co1] *The Boltzmann equation and its generalization to higher density,* Fundamental Problems in Statistical Mechanics, North Holland (1961).

[Co2] *On the connection between various derivations of the Boltzmann equation,* Physica (1961), **27**, 163.
[Co3] *On the generalization of the Boltzmann equation to general order in the density,* Physica (1962), **28**, 1025.
[Co4] *Clustewr expansions and the hierarchy. I. Nonequilubrium distribution functions,* Physica (1962), **28**, 1045.
[Co5] *Cluster expansions and the hierarchy. II. Equilibrium distribution functions,* Physica (1962), **28**, 1060-1073.
[Co6] *On the kinetic theory of dense gases,* J. Math. Phys. (1962), **4**, 183.
[Co7] *The kinetic theory of dense gases,* Fundamental Problems in Statistical Mechanics, Vol. 11, Noth Holland (1968), 228-275.

Dobrushin, R. L.
[D1] *Description of a random field by conditional probabilities and conditions for its regularity,* Theor. Ver. i Prim. (1968), **13**, 201-229.
[D2] *Gibbs states. General case,* Funk. Anal. i Pril. (1969), **3**, 27-35.

Dobrushin, R. L. and Fritz, J.
[DF] *Nonequilibrium dynamics of one-dimensional infinite particle systems with hard core interactions,* Comm. Math. Phys. (1977), **55**, 275-292.

Dobrushin, R. L., Sinai, Ya. G., and Sukhov, Yu. M.
[DSS] *Dynamic systems of statistical mechanics,* VINITI AN USSR, **2**, Moscow (1985), 233-284.

Dobrushin, R.L., Maslova, N. B., Sinai, Ya. G. and Sukhov, Yu. M.
[DMS] *Dynamical systems in statistical mechanics and kinetic equations,* VINITI, Series on Contemporary Problems in Mathematics. Fundamental Directions, Vol. 2, Moscow (1985), 233-307.

Dorfman, J. R. and Cohen, E. G. D.
[DC] *Difficulties in the kinetic theory of dense gases,* J. Math. Phys. (1967), **8**, 282-297.

Dorfman, J. R. and Ernst, M. H.
[DE] *Hard-sphere binary-collision operators,* J. Stat. Phys. (1989), **57**, 581-193.

Ernst, M., Dorfman, J., Hoegy, W., and van Leeuwen, J. M. J.
[EDHL] *Hard-sphere dynamics and binary collision operators,* Physica (1969), **45**, 127-146.

Fritz, J. and Dobrushin, R. L.

[FD] *Nonequilibrium dynamics of two-dimensional infinite particle systems with singular interactions,* Comm. Math. Phys. (1977), **55**, 67-82.

Gallavotti, G., Lanford, O. and Lebowitz, J. L.
[GLL] *Thermodynamic limit of time-depdndent correlation functions for one-dimensional systems,* J. Math. Phys. (1970), **11**, 2898-2905.

Galperin, G.A.
[Ga] *The elastic scattering of particles,* Usp. Math. Nauk (1978), **33**, 199.

Georgii, H.- O.
[Geo1] *Canonical and grand canonical Gibbs states for continuum systems,* Comm. Math. Phys. (1976), **48**, 31-51.
[Geo2] *Canonical Gibbs measures,* Lecture Notes in Math., **760**, Springer (1979).

Gerasimenko, V. I.
[Ger1] *Evolution of the infinite particle system with nearest-neighbours interaction,* Dokl. Akad. Nauk USSR, Ser. A, (1982), **5**, 10-13.
[Ger2] *On the solution of BBGKY hierarchy for a one-dimensional hard sphere particle system,* Theor. Mat. Fiz. (1992), **91**, 120-128.

Gerasimenko, V. I. and Petrina, D. Ya.
[GP1] *Thermodynamic limit for non-equilibrium states of three-dimensional system of hard spheres,* Theor. Mat. Fiz. (1985), **64**, 130-149.
[GP2] *Existence of Boltzmann-Grad limit for an infinite hard sphere particle system,* Teor. Mat. Fiz. (1990), **83:1**, 92-114.

Grad, H.
[Gra1] *On the kinetic theory of rarified gases,* Comm. Pure Appl. Math. (1949), **2**, 331-407.
[Gra2] *Principles of the kinetic theory of gases,* Handbuch der Physik, Vol. 12, Springer, Berlin (1958), 205-294.

Green, M. S.
[Gre1] *The non-equilibrium pair distribution function at low densities,* Physica (1958), **24**, 393.
[Gre2] *Surface integral form for three-body collisions in the Boltzmann equation,* Phys. Rev. (1964), **136**, 905.

Green, M. S. and Piccirelli, R. A.
[GPi] *Basis of the functional assumption in the theory of the Boltzmann equation,* Phys. Rev. (1963), **132**, 1388-1410.

Gurevich, B. M. and Sukhov, Yu. M.
[GS] *Stationary solutions of the Bogolubov hierarchy equations in classical statistical mechanics,* Comm. Math. Phys. (1976), **49**, 69-96; (1977), **54**, 81-96; (1977), **56**, 225-236; (1982), **84**, 336-376.

Hurd, A.E.
[H1] *Existence of time evolutions of non-equilibrium states in statistical mechanics,* Arch. Ration. Mech. and Analysis (1984), **85**, 201-212.
[H2] *Global existence and validity for the BBGKY hierarchy,* Arch. Ration. Mech. and Analysis (1987), **98**, 191-209.

Illner, R.
[I] *On the number of collisions in a hard sphere particle system in all space,* Transp. Theor. Stat. Phys. (1989), **18**, 71-86.

Illner, R. and Neunzert, H.
[IN] *The Concept of Irreversibility in the Kinetic Theory of Gases,* Transport Th. and Stat. Phys. (1897), **16 (1)**, 89-112.

Illner, R. and Pulvirenti, M.
[IP] *A derivation of the BBGKY hierarchy for hard sphere particle system,* Transp. Theor. Stat. Phys. (1987), **16:7**, 997-1012.

Illner, R. and Shinbrot, M.
[IS] *The Boltzmann equation. Global existence for a rare gas,* Comm. Math. Phys. (1984), **95**, 217-226.

Kato, T.
[Ka] Perturbation Theory for Linear Operators, Springer-Verlag, New York, 1966.

Kirkwood, J. G.
[Ki] *The statistical mechanical theory of transport process,* J. Chem. Phys. (1946), **14**, 180-201; (1947), **15**, 72-76.

Kolmogorov, A. N.
[Kol] *On the conservation of conditionally periodic motions under small perturbations of the Hamiltonian,* Dokl. Akad. Nauk SSSR. (1954), **98**, 527-530.

Kotchubey, A. N.
[Kot] *The Liouville operator,* Ukrain. Math. Journ. (1991), **43**, 1664-1671.

Kornfeld, I.P., Sinai, Ya. G., and Fomin, S. V.
[KSF] Ergodic Theory [in Russian], Nauka, Moscow, 1981.

Krein, S. G.
[Kr] Linear Differential Equations in Banach Space [in Russian], Nauka, Moscow, 1967.

Lachowich, M.
[Lac1] *A system of stochastic differential equations modeling the Euler and the Navier-Stokes hydrodynamic equations,* Japan Journ. of Industr. and Appl. Math. (1993), **10**, 109-131.
[Lac2] *Stochastic differential equations for the kinetic and hydrodynamic equations,* Nonlinear Dynamics (1994), **5**, 393-399.

Lachowich, M. and Pulvirenti, M.
[LPu] *A stochastic particle system modeling the Euler equations,* Arch Rational Mech. Anal. (1990), **109**, 81-93.

Lampis, M. and Petrina, D.Ya.
[LPe] *The Boltzmann-Grad limit for equilibrium states of systems of hard spheres in framework of canonical ensemble,* Math. Phys. Analysis Geometry (1996), to be published.

Lanford, O. E.
[Lan1] *The classical mechanics of one-dimensional systems of infinitely many particles. 1. Existence theorem,* Commun. Math. Phys. (1968), **9**, 169-180;
[Lan2] *The classical mechanics of one-dimensional systems of infinitely many particles. 2. Kinetic theory,* Commun. Math. Phys. (1968), **11**, 257-292.
[Lan3] *Time evolution of large classical systems,* Lectures Notes in Physics, **38**, Springer-Verlag, Berlin (1975), 1-111.

Lanford, O. E. and Ruelle, D.
[LR] *Observables at infinity and states with short range correlations in statistical mechanics,* Comm. Math. Phys. (1969), **13**, 194-215.

Loschmidt, J.
[Lo] *Über den Zustand des Wärmegleichgewichtes eines Systems von Körpern mit Rucksicht auf die Schwerkraft Wien. Ber.* (1876) **73**, 128 .

Malyshev, P. V.
[Ma] *Mathematical description of the evolution of an infinite classical systems,* Teor. Mat. Fiz. (1980), **44**, 63-74.

Malyshev, P. V. and Petrina, D. Ya.
[MP] *Thermodynamic limit for non-equilibrium distribution functions for three- dimensional classical systems of interacting particles,* Dokl. Acad. Nauk SSSR (1988), **301**, 585-589.

Marchioro, G., Pellegrinotti, A., and Presutti, E.
[MPP] *Existence of time evolution for ν-dimensional statistical mechanics,* Comm. Math. Phys. (1975), **40**, 175-186.

Maslov, V. P. and Tariverdiev, S. E.
[MT] *Asymptotics of the Kolmogorov-Feller Equation for a system of a large number of particles,* VINITI Series on Probability Theoty, Mathematical Statistics and Theoretical Cybernetics, **19**, Moscow (1982), 85-126.

Minlos, R. A.
[Mi1] *Limit Gibbs distribution,* Funk. Anal. i Pril. (1967), **1**, 60-73.
[Mi2] *Regularity of limit Gibbs distribution,* Funk. Anal. i Pril (1967), **3**, 40-53.

Moser, J.
[Mo] *On invariant curves of area-preserving mapping of an annulus*, Nachr. Akad. Wiss. Göttingen, Math phys. (1962), **1**, 1-20.

Mynbaev, K. T. and Otebaev L. A.
[MO] Weighted Functional Spaces and Spectra of Differential Operators [in Russian], Nauka, Moscow, 1988.

Murphy, T. J. and Cohen, E. G. D.
[MC] *Maximum number of collisions among identical hard spheres,* J. Stat. Phys. (1993), **71**, 1063.

Petrina, D. Ya.
[Pe1] *Mathematical description of the evolution of infinite systems of classical statistical phisics: locally perturbed one-dimensional systems,* Theor. Mat. Fiz. (1979), **38**, 230-250.
[Pe2] *On solutions of the kinetic Bogolubov equations: quantum statistics,* Teor. Mat. Fiz. (1972), **13**, 391-405.
[Pe3] Mathematical Foundations of Quantum Statistical Mechanics. Continuous Systems, Kluwer, Amsterdam, 1995.

Petrina, D. Ya. and Gerasimenko, V. I.
[PG1] *Mathematical description of the evolution of states of finite systems of classical statistical mechanics,* Usp. Mat. Nauk. (1983), **38**, 3-58.
[PG2] *Evolution of states of infinite systems in classical statistical mechanics,* Sov. Sci. Rev., Ser. C, Math. Phys. (1984), **5**, 1-60.
[PG3] *Mathematical problems of statistical mechanics of a hard sphere particle system,* Usp. Mat. Nauk. (1990), **45:3**, 159-211.

Petrina, D. Ya., Gerasimenko, V. I., and Malyshev, P. V.
[PGM1] *Thermodynamic limit for solutions of Bogolubov's equations,* Sov. Sci. Rev., Ser. C, Math. Phys. (1987), **7**, 280-335.
[PGM2] Mathematical Foundations of Classical Statistical Mechanics. Continuous Systems, Gordon and Breach, New York, 1989.

Petrina, D. Ya. and Petrina, K. D.
[PP1] *On existence of equilibrium states of systems of hard spheres in the Boltzmann-Enskog limit,* Mathem. Phys.,Analysis and Geometry (1996), to be published.
[PP2] Stochastic Dynamics and Boltzmann Hierarchy, Preprint, Institute of Mathematics, National Academy of Sciences of Ukraine, Kiev (1996).

Petrina, D. Ya. and Vidybida, A. K.
[PV] *Cauchy problem for the Bogolubov chain of equations,* Trudy MIAN SSSR (1975), **136**, 370-378.

Piccirelli, R. A.
[Pi] *Some properties of the long-time value of the probability densities for moderately dense gases,* J. Math. Phys. (1966), **7**, 922.

Poincaré, H.
[Poi1] *Sur le problème des trois corps et les équation de la dynamique*, Acta Mathematica (1890) **13**, 1-270.
[Poi2] *Le mécanisme et l'expérience*, Revue de Metaphysique et de Morale (1893) **1**, 534-537.

Povzner, A. Ya.
[Po] *The Boltzmann equation in the kinetic theory of gases,* Amer. Math. Soc. Transl. (1962), **47**, 193.

Presutti, E., Pulvirenti, M., and Tirozzi, E.
[PPT] *Time evolution of infinite classical systems with singular long range two-body interactions,* Comm. Math. Phys. (1976), **47**, 81-95.

Reed, M. and Simon, B.
[RS] Methods of Modern Mathematical Physics, Vols. 1 -4, Academic Press, New York - London, 1972, 1975, 1978, 1979.

Resibois, P. and De Leener, H.
[RD] Classical Kinetic Theory of Fluids, Wiley, London, 1977.

Ruelle, D.
[R1] Statistical Mechanics. Rigorous Results, Benjamin, New York–Amsterdam, 1969
[R2] *Superstable interactions in classical statistical mechanics*, Comm. Math. Phys. (1970), **18**, 127-159.

Sengers, J. V. and Cohen, E. G. D.

[SC] *Statistical mechanical derivation of the generalized Boltzmann equation for a fluid consisting of rigid spherical molecules,* Physica (1961), **27**, 230.

Sinai, Ya. G.
[Si1] *Construction of dynamics in one-dimensional systems of statistical mechanics,* Theor. Mat. Fiz. (1972), **11**, 248-258.
[Si2] *Construction of cluster dynamics for dynamical systems of statistical mechanics,* Vestnic Mosk. Univ., Ser. 1, Matem., Mech. (1974), **1**, 152-158.
[Si3] *Dynamical systems with elastic reflections*, Usp. Mat. Nauk. (1970), **25**, 141-192.

Sinai, Ya. G. and Sukhov, Yu. M.
[SS] *On the theorem of existence of solutions of the Bogolubov chain of equations,* Teor. Mat. Fiz. (1974), **19**, 344-363.

Shinbrot, M.
[Sh] *Local validity of the Boltzmann equation,* Math. Meth. in the Appl. Sci. (1984), **6**, 539-549.

Skorochod, A. V.
[Sk] Stochastic Equation for Complex Systems [in Russian], Nauka, Moscow, 1983.

Spohn, H.
[Sp1] *Boltzmann Hierarchy and Boltzmann Equation* in Kinetic Theories and the Boltzmann Equation, C. Cercignani, Ed., **LNM 1048**, Springer, Berlin, 1984.
[Sp2] Large Scale Dynamics of Interacting Particles, Springer-Verlag, 1991.

Szasz, D.
[Sza] *Boltzmann's ergodic hypothesis, a conjecture for centuries?* Preprint ESI98, 1994.

Sznitman, A. S.
[Sz] *Equations de type de Boltzmann, spatialment homogenes,* Z. Wahrsch. Verw. Gebiete (1984), **66**, 559-592.

Tanaka, H.
[Ta] *Probabilistic treatment of the Boltzmann equation of Maxwellian molecules,* Z. Wahrsch. Verw. Gebiete (1978), **46**, 67-105.

Thirring, W.
[Thi] A course in Mathematical Physics, Vol 1, 2, 3, 4, Springer-Verlag, Wien, New York, 1978, 1979, 1980, 1981.

Thomson, W. (Lord Kelvin)
[Tho] *The Kinetic Theory of the Dissipation of Energy,* Proc. Royal Soc. Edinburgh (1874), Vol.8, 325-334.

Truesdell, C., and Muncaster, R.C.
[TM] Fundamentals of Maxwell's Kinetic Theory of a Simple Monatomic Gas, Academic Press, New York (1980).

Uchiyama, K.
[U] *Derivation of the Boltzmann equation from particle dynamics,* Hirosima Math. Journ. (1988), **18**, 245–297.

Uhlenbeck, G. E. and Ford, G. W.
[UF] *Lectures in statistical mechanics,* in: Lectures in Applied Mathematics, Vol. 1, AMS, Providence (1963).

Yosida, K.
[Yo] Functional Analysis, Academic Press, New York, 1966.

Yvon, J.
[Yv] La Theorie Statistique des Fluids et l'Equation d'Etat, Hermann, Paris, 1935.

Zermelo, E.
[Zer] *Über einen Satz der Dynamik und die mechanische Warmetheorie*, Annalen der Physik (1896),**54**, 485-494.

Zessin, H.
[Zes] *Equilibrium time-evolution and BBGKY hierarchy equations of large classical systems,* Z. Wahrscheinlichkeitstheoric (1985), **68**, 461.

Zubarev, D. N.
[Zu] Non Equilibrium Statistical Thermodynamics, Nauka, Moscow, 1972.

Subject Index

Other *Mathematics and Its Applications* titles of interest:

F. Langouche, D. Roekaerts and E. Tirapegui: *Functional Integration and Semiclassical Expansions*. 1982, 328 pp. ISBN 90-277-1472-X

N.E. Hurt: *Geometric Quantization in Action. Applications of Harmonic Analysis in Quantum Statistical Mechanics and Quantum Field Theory*. 1982, 352 pp. ISBN 90-277-1426-6

C.P. Bruter, A. Aragnal and A. Lichnerowicz (eds.): *Bifurcation Theory, Mechanics and Physics. Mathematical Developments and Applications*. 1983, 400 pp. *out of print*, ISBN 90-277-1631-5

N.H. Ibragimov: *Transformation Groups Applied to Mathematical Physics*. 1984, 414 pp. ISBN 90-277-1847-4

V. Komkov: *Variational Principles of Continuum Mechanics with Engineering Applications. Volume 1: Critical Points Theory*. 1986, 398 pp. ISBN 90-277-2157-2

R.L. Dobrushin (ed.): *Mathematical Problems of Statistical Mechanics and Dynamics. A collection of Surveys*. 1986, 276 pp. ISBN 90-277-2183-1

P. Kree and C. Soize: *Mathematics of Random Phenomena. Random Vibrations of Mechanical Structures*. 1986, 456 pp. ISBN 90-277-2355-9

H. Triebel: *Analysis and Mathematical Physics*. 1987, 484 pp. ISBN 90-277-2077-0

P. Libermann and Ch.M. Marle: *Sympletic Geometry and Analytical Mechanics*. 1987, 544 pp. ISBN 90-277-2438-5 (hb), ISBN 90-277-2439-3 (pb).

E. Tirapegui and D. Villarroel (eds.): *Instabilities and Nonequilibrium Structures*. 1987, 352 pp. ISBN 90-277-2420-2

W.I. Fushchich and A.G. Nikitin: *Symmetries of Maxwell's Equations*. 1987, 228 pp. ISBN 90-277-2320-6

V. Komkov: *Variational Principles of Continuum Mechanics with Engineering Applications. Volume 2: Introduction to Optimal Design Theory*. 1988, 288 pp. ISBN 90-277-2639-6

M.J. Vishik and A.V. Fursikov: *Mathematical Problems in Statistical Hydromechanics*. 1988, 588 pp. ISBN 90-277-2336-2

V.I. Fabrikant: *Applications of Potential Theory in Mechanics. A Selection of New Results*. 1989, 484 pp. ISBN 0-7923-0173-0

E. Tirapegui and D. Villarroel (eds.): *Instabilities and Nonequilibrium Structures (II). Dynamical Systems and Instabilities*. 1989, 328 pp. ISBN 0-7923-0144-7

V.G. Bagrov and D.M. Gitman: *Exact Solutions of Relativistic Wave Equations*. 1990, 344 pp. ISBN 0-7923-0215-X

Other *Mathematics and Its Applications* titles of interest:

O.I. Zavialov: *Renormalized Quantum Field Theory*. 1990, 560 pp.
ISBN 90-277-2758-9

S.S. Horuzhy: *Introduction to Algebraic Quantum Field Theory*. 1990, 360 pp.
ISBN 90-277-2722-8

V.G. Makhankov: *Soliton Phenomenology*. 1990, 461 pp. ISBN 90-277-2830-5

S. Albeverio, PH. Blanchard and L. Streit: *Stochastic Processes and their Applications in Mathematics and Physics*. 1990, 416 pp. ISBN 0-9023-0894-8

V.A. Malyshev and R.A. Minlos: *Gibbs Random Fields. Cluster Expansions*. 1991, 248 pp. ISBN 0-7923-0232-X

E. Tirapegui and W. Zeller (eds.): *Proceedings of the 3rd Workshop on Instabilities and Nonequilibrium Structures*. 1991, 370 pp. ISBN 0-7923-1153-1

C.A. Marinov and P. Neittaanmaki: *Mathematical Models in Electrical Circuits. Theory and Applications*. 1991, 160 pp. ISBN 0-7923-1155-8

C. Bartocci, U. Bruzzo and D. Hernandez-Ruiperez: *The Geometry of Supermanifolds*. 1991, 242 pp. ISBN 0-7923-1440-9

E. Goles and S. Martinez: *Statistical Physics, Automata Networks and Dynamical Systems*. 1992, 208 pp. ISBN 0-7923-1595-2

A. van der Burgh and J. Simonis (eds.): *Topics in Engineering Mathematics*. 1992, 266 pp. ISBN 0-7923-2005-3

L. Aizenberg: *Carleman's Formulas in Complex Analysis*. 1993, 294 pp.
ISBN 0-7923-2121-9

E. Tirapegui and W. Zeller (eds.): *Instabilities and Nonequilibrium Structures IV*. 1993, 371 pp. ISBN 0-7923-2503-6

G.M. Dixon: *Division Algebras: Octonions, Quaternions, Complex Numbers and the Algebraic Design of Physics*. 1994, 246 pp. ISBN 0-7923-2890-6

A. Khrennikov: p-*Adic Valued Distributions in Mathematical Physics*. 1994, 270 pp. ISBN 0-7923-3172-9

Z. Haba, W. Cegła and L. Jakóbczyk (eds.): *Stochasticity and Quantum Chaos*. (Proceedings of the 3rd Max Born Symposium, Sobótka Castle, September 15–17, 1993) 1995, 224 pp. ISBN 0-7923-3230-X

L. Stupelis: *Navier–Stokes Equations in Irregular Domains*. 1995, 582 pp.
ISBN 0-7923-3509-0

B. Jefferies: *Evolution Processes and the Feynman-Kac Formula*. 1996, 245 pp.
ISBN 0-7923-3843-X

Y.E. Gliklikh: *Ordinary and Stochastic Differential Geometry as a Tool for Mathematical Physics*. 1996, 206 pp. ISBN 0-7923-4154-6

Other *Mathematics and Its Applications* titles of interest:

E.S. Fradkin and M.Ya. Palchik: *Conformal Quantum Field Theory in D-dimensions*. 1996, 474 pp. ISBN 0-7923-4158-9

P. Bandyopadhyay: *Geometry, Topology and Quantization*. 1996, xxx pp. ISBN 0-7923-4305-0

N.E. Hurt: *Quantum Chaos and Mesoscopic Systems*. Mathematical Methods in the Quantum Signatures of Chaos. 1997, 347 pp. ISBN 0-7923-4459-6

C. Cercignani, V.I. Gerasimenko and D. Ya. Petrina: *Many-Particle Dynamics and Kinetic Equations*. 1997, 244 pp. ISBN 0-7923-4696-3

GPSR Compliance

The European Union's (EU) General Product Safety Regulation (GPSR) is a set of rules that requires consumer products to be safe and our obligations to ensure this.

If you have any concerns about our products, you can contact us on ProductSafety@springernature.com

In case Publisher is established outside the EU, the EU authorized representative is:

Springer Nature Customer Service Center GmbH
Europaplatz 3
69115 Heidelberg, Germany

Batch number: 10584448

Printed by Printforce, the Netherlands